7th Roma International Conference on AstroParticle Physics (RICAP18)

EPJ Web of Conferences Volume 209 (2019)

Roma, Italy
4-7 September 2018

Editors:

Mario De Vincenzi
Aldo Morselli
Antonio Capone

ISBN: 978-1-5108-8623-0

Printed from e-media with permission by:

Curran Associates, Inc.
57 Morehouse Lane
Red Hook, NY 12571

CURRAN ASSOCIATES INC. proceedings.com

Some format issues inherent in the e-media version may also appear in this print version.

Printed by Curran Associates, Inc. (2019)

For additional information, please contact EDP Sciences – Web of Conferences
at the address below.

EDP Sciences – Web of Conferences
17, Avenue du Hoggar
Parc d'Activité de Courtabœuf
BP 112
F-91944 Les Ulis Cedex A
France

Phone: +33 (0) 1 69 18 75 75
Fax: +33 (0) 1 69 28 84 91

contact-edps@webofconferences.org

Additional copies of this publication are available from:

Curran Associates, Inc.
57 Morehouse Lane
Red Hook, NY 12571 USA
Phone: 845-758-0400
Fax: 845-758-2633
Email: curran@proceedings.com
Web: www.proceedings.com

TABLE OF CONTENTS

RICAP 2018
©Maurizio PERCIBALLI

RICAP-18
Roma International
Conference on
Astroparticle Physics

Statement of Peer review

In submitting conference proceedings to EPJ Web of Conferences, I certify to the Publisher that I adhere to the **Policy on Publishing Integrity** of the journal in order to safeguard good scientific practice in publishing.

1. All articles have been subjected to peer review administered by the proceedings editors.
2. Reviews have been conducted by expert referees, who have been requested to provide unbiased and constructive comments aimed, whenever possible, at improving the work.
3. Proceedings editors have taken all reasonable steps to ensure the quality of the materials they publish and their decision to accept or reject a paper for publication has been based only on the merits of the work and the relevance to the journal.

Title, date and place of the conference

Ricap18 7th Roma International Conference on AstroParticle Physics
4 - 7 September 2018
Roma Tre University

Proceedings editor(s):

Mario De Vincenzi
Antonio Capone
Aldo Morselli

Date and editor's signature Roma 09 April 2019

The ASTRI Program

Salvatore Scuderi[1a], for the CTA ASTRI Project
[1]*INAF-IASF Milano, Via E. Bassini 15, Milan, Italy.*

Abstract. The ASTRI (*Astrofisica con Specchi a Tecnologia Replicante Italiana*) program was born as a collaborative international effort led by the Italian National Institute for Astrophysics (INAF) to design and realize, within the Cherenkov Telescope Array (CTA) framework, an end-to-end prototype of the Small-Sized Telescope (SST) in a dual-mirror configuration (2M). While the activities concerning the characterization of the prototype are under completion, the program entered a new phase. With the final aim of contributing at the production of the SST telescopes for the CTAO Southern site, we started the development of nine telescopes based on the evolution of the ASTRI prototype design to work as pathfinder for the CTAO. Furthermore, together with the CHEC (*Compact High Energy Camera*) collaboration, the ASTRI team presented a proposal, that will be evaluated with other proposals, to deliver to CTAO the complete set of SST telescopes.

1 Introduction

The Cherenkov Telescope Array Observatory [1] (CTAO) will be the largest ground-based very high-energy gamma-ray detection observatory in the world, with more than one hundred telescopes located in the Northern and Southern Hemispheres (La Palma, Canary Islands and currently proposed at the Atacama desert, Chile). The energy coverage, in the Southern CTA array, will extend up to 300 TeV thanks to a large number (up to 70) of small size telescopes (SST), with their primary mirrors of about 4 meters in diameter and large field of view of the order of 9 degrees.
In the CTA project framework Italy is planning an important contribution in the development of the telescope array through the ASTRI program. The program was born in 2011 with the final aim to contribute to the production and delivery of a number of SST telescopes for the CTAO Southern site. The effort is led by INAF in synergy with the Universidade de Sao Paulo (Brazil) and the North-West University (South Africa). The ASTRI program envisages three distinct but continuous phases. The first phase is the design, realization and deployment of a prototype of the SST telescopes named the ASTRI SST-2M. The prototype telescope has been built; it is now installed in Italy at the INAF Observing Station on Mt. Etna (Sicily) and is completing its science verification phase. The second phase of the project goes one-step further towards the CTAO construction through the implementation of an array of SST pathfinders. The third phase is the contribution to the construction of the final CTAO array. Dedicated funds from the Italian government and the international partners are already available for the implementation of the last two phases.

[a] Corresponding author: salvatore.scuderi@inaf.it

2 The ASTRI Prototype

The ASTRI SST-2M telescope [2] is characterized by an optical system based on a dual-mirror modified Schwarzschild-Couder design and by a curved focal plane covered by silicon photomultiplier sensors managed by a fast front-end electronics. The telescope has an alt-azimuthal mount with a segmented primary mirror of 4.3m diameter and a monolithic secondary mirror of 1.8m diameter. The focal plane of the prototype Cherenkov camera is populated with 21 Hamamatsu SiPM of the LCT5 series and the read-out electronics is based on CITIROC ASICs that implement a custom peak-detector operation mode to acquire the SiPM pulses. A comprehensive description of the technical characteristics of the telescope can be found in [3, 4] for the electromechanical structure, and in [5, 6] for the optics, while the camera and its electronics is thoroughly described in [7, 8]. Analogously, the control/acquisition hardware and software are described in [9], the data analysis software in [10], and the data archiving system in [11].

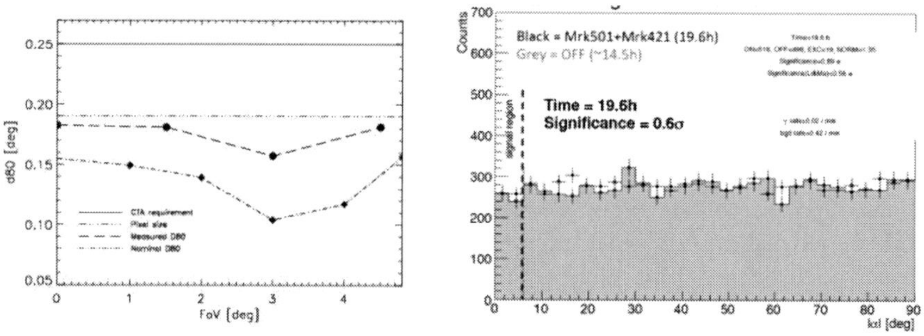

Figure 1. Left : Comparison of D80 (diameter of the circle corresponding to an encircled energy of 80%) as a function of field angle (off-axis) between its ideal (dot-dashed curve) and measured at ASTRI SST-2M (dashed curve) values. SiPM pixel size (dashed horizontal line) and CTA requirement (solid horizontal line) are also shown. The requirement of a flat PSF of ~10 arcmin along a large field of view (~10°) is clearly fulfilled. **Right** : |α| distribution of the stacked Mrk421 and Mrk501 signal and background estimation from data taken in June/July 2018, after an 80% efficiency-based gamma/hadron discrimination parameter cut (in the whole energy range). The region between zero and the vertical dashed line (at 6 deg) represents the fiducial signal region.

The telescope prototype is mainly a demonstrator used to validate the novel technology. However, it has been developed following an end-to-end approach that comprises all the work that should be done to obtain scientific products through Cherenkov observations [12].

The telescope prototype was installed in September 2014 at the INAF Catania Astrophysical Observatory on the slopes of the Etna volcano and, since then, has been operated on regular basis. A complete set of functional and verification tests was performed for the mechanical structure and for the optics. In particular, the pointing and tracking performance has been measured [1] and the results show that they are well within the CTA requirements of online and post calibration astrometric accuracy (1 arcmin and 7 arcsec respectively). The optical performance was also extensively studied. Figure 1 (left side) shows the most important result obtained, that is, the first validation ever of the SC optical design [6]. In addition, dedicated measurements show that the optical Point Spread Function is very stable against time and telescope elevation [13]. A somewhat unexpected result, coming from the mirrors' reflectivity characterization, was the fast ageing of their coating likely due to the aggressive volcanic atmosphere [2].

Since the first Cherenkov light obtained in May 2017, the camera has performed several observational campaigns at the telescope, most of them devoted to engineering tests and some to scientific observations of the brightest sources (Crab nebula, Mrk421, and Mrk501) observable from the prototype site. Figure 1 (right side) shows the results of the analysis of the stacked shower data on the two Markarians obtained in June/July 2018. We collected ~20 hours on the objects and ~15 hours of off data. No detection has been obtained. This is not unexpected due to the reduced efficiency of the telescope mirrors, and the low flux of the objects (<0.1 Crab units). The scientific observations obtained up to now allowed us to test the end-to-end concept of the prototype from the photon capture to the data analysis [10]. A complete report of the camera performance in the laboratory and on sky can be found in [8].

3 The ASTRI Pathfinder

The next step of the program is the production of a set of nine end-to-end ASTRI telescopes named ASTRI pathfinder based on the ASTRI design to be placed at the CTAO Southern site. The pathfinder is proposed as a precursor of the final array to prove technology and to test operational and maintenance procedures. In particular, the activities in which the pathfinder can give a fundamental contribution to the development of the CTAO are: the assessment of the array trigger; the test of the array control software and general operations; the test of the AIV procedures on site (time, personnel, infrastructure, and tools); the test of the array performance against Monte Carlo simulations and the test of the implementation of the intensity interferometry capability.
The implementation of the pathfinder will happen within the framework of CTAO development. In particular, it will have to pass, before deployment, all the CTAO technical reviews. If the currently proposed CTAO Southern site in Chile will be confirmed shortly, the current schedule will allow us to place the first complete pathfinder telescope at the end of 2020.

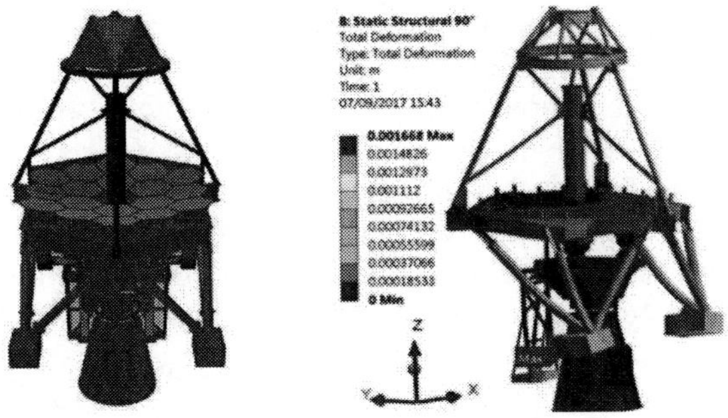

Figure 2. Left: the design of the ASTRI structure for the array of pathfinders (courtesy of GEC). Right: example of Finite Element Analysis for the new structure (courtesy of GEC).

Activities for the production of the array of pathfinders started already. In particular, the procurement of the telescope mirrors and the camera ASICs, items whose design is well consolidated and then considered at low risk, is well under way. Half of the primary mirror segments are ready and the secondary mirrors will be completed by the end of 2018. The production of the new version CITIROC 1A of the camera ASIC is also half way. Details of the procurement activities is given in [2]. On the other hand, the electromechanical structure and the Cherenkov camera are instead going through a phase of design consolidation and optimization. This is necessary to take into account the lessons learnt with the prototype, the technology improvements and the consolidation of CTA requirements due to the final choice of the South site. Figure 2 shows the consolidated design of the electromechanical structure. The main changes in the telescope design are the optimization of several parts of the mechanical structure to allow us lowering the total weight while keeping the stiffness unchanged. Furthermore, the stability of the measured optical PSF against gravity and time allowed a design without a permanent active mirror control. The optimized design is described in detail in [14]. Analogously, we will proceed in the design consolidation of the camera.
As it is conceived with an end-to-end approach, the pathfinder can be used not only for technological and operational aspects but also to make astrophysical observations. Preliminary Monte Carlo simulations (Figure 3) shows that the ASTRI pathfinder could surpass the H.E.S.S. sensitivity above 10 TeV, extending up to about 100 TeV with an energy resolution of about 10 – 15 % [10]. The ASTRI pathfinder will exploit its better sensitivity and extended spectral range to investigate the emission of prominent sources such as extreme blazars, nearby well-known BL Lac objects and radio galaxies, galactic pulsar wind nebulae, supernova remnants, the Galactic Centre and the search for cosmic ray (CR) PeVatrons.

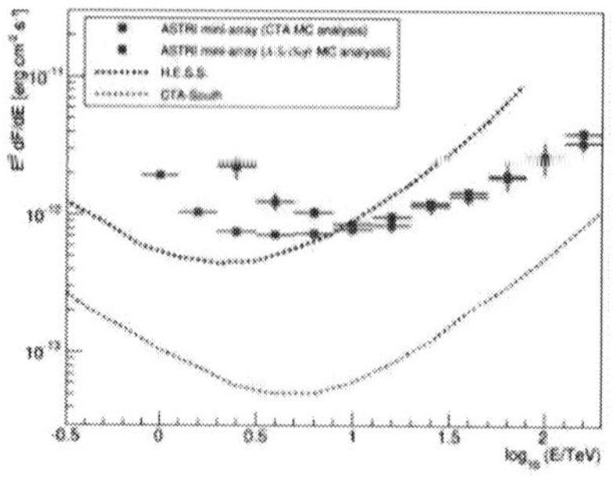

Figure 3. ASTRI pathfinder (9 telescopes, relative distance ~250 m, and square layout) differential sensitivity curves (5σ 50 hours) achieved with the A-SciSoft MC analysis (green points) and with the CTA MC analysis [10]. The differential sensitivities of CTA-South (blue line) and H.E.S.S (black line) are shown for comparison.

4 CTAO

Three different SST prototypes are currently being developed within the CTA Collaboration, namely the single mirror SST-1M [15] telescope and the dual-mirror GCT [16] and ASTRI telescopes. Recently, the CTAO project office started a harmonization process that includes the evaluation of these designs through an external panel of experts to be concluded by next Spring.
The ASTRI and the CHEC collaborations decided to answer jointly to this call proposing a design based on a combination of the developed work of the ASTRI project and the CHEC prototyping effort. The aim is to deliver to CTAO the complete set of SST telescopes including the pathfinder described previously as precursor of the final array. The telescope structure, including mirrors, will be the current ASTRI design while the final camera will be an improved version of the current CHEC camera [17]. The telescopes of the array of pathfinders will be part of the final delivery to CTAO while the CHEC-like cameras will replace the ASTRI-like cameras in the array of pathfinders.

References

1. R. Ong, RICAP 2018, these proceedings (2018).
2. S. Scuderi, for the CTA ASTRI project, Proc. SPIE, Volume 10700, id. 107005Z (2018).
3. G. Marchiori, et al., Proc. of the SPIE, Volume 9145, id. 91450L (2014).
4. R. Canestrari, et al, for the CTA ASTRI project, Proc. SPIE, Vol. 9603, id. 960303, (2015).
5. G. Rodeghiero, et al., PASP, Vol. 128, Issue 963, pp. 055001 (2016)
6. E. Giro E., et al., Astronomy & Astrophysics, Vol. 608, A86 (2017).
7. G. Sottile, et al., for the CTA ASTRI project, Proc. SPIE, Vol. 9906, id. 99063D, (2016).
8. O. Catalano, et al., for the CTA ASTRI project, Proc. SPIE, Vol. 10702, id. 1070237, (2018).
9. E. Antolini, et al., for the CTA ASTRI project, Proc. SPIE, Vol. 9913, id. 99131J, (2016).
10. S. Lombardi, et al., for the CTA ASTRI project, Proc. SPIE, Vol. 10707, id. 107070R, (2018).
11. A. Carosi, et al., for the CTA ASTRI project, Proc. SPIE, Vol. 9910, id. 99101V, (2016).
12. M.C. Maccarone, for the CTA ASTRI project, Proc. 35th International Cosmic Ray Conference, Busan, Korea, PoS(ICRC2017)855, (2017).
13. R. Canestrari, et al., for the CTA ASTRI project, Proc. SPIE, Vol. 9906, id. 990619, (2017).
14. G. Marchiori, et al., Proc. SPIE, Volume 10700, id. 107005W (2018).
15. M. Heller, et al., AIP Conference Proceedings, Vol. 1792, Issue 1, id.080003 (2017).
16. O. Le Blanc, et al., Proc. SPIE, Volume 10700, id. 1070010, (2018).
17. J.S., Lapington, et al. Nuclear Inst. and Methods in Physics Research, A, Vol. 876, p. 1, (2017).

Results from the Pierre Auger Observatory

Esteban Roulet[1,a], for the Pierre Auger Collaboration[2,b]

[1] *CONICET, Centro Atómico Bariloche, Av. Bustillo 9500, Bariloche, Argentina*
[2] *Full author list and acknowledgments at: http://www.auger.org/archive/authors_2018_09.html*

Abstract. I describe some of the results on ultrahigh-energy cosmic rays that have been obtained with the Pierre Auger Observatory. These include measurements of the spectrum, composition and anisotropies. Possible astrophysical scenarios that account for these results are discussed.

1 The Observatory

The Pierre Auger Observatory was built to understand the nature and origin of the highest energy cosmic rays, i.e. those with energies above 1 EeV $\equiv 10^{18}$ eV. It is the largest existing observatory, covering an area of 3000 km^2 near the town of Malargüe in Argentina, at a latitude of 35° South, and has been operating continuously since the year 2004. It consists of a surface array of water-Cherenkov detectors that can measure the lateral distribution of the air showers at ground level. This array is overlooked by 27 telescopes that detect, during clear moonless nights, the fluorescence emitted by the air nitrogen molecules that get excited by the passage of the charged particles in the showers. These last observations allow us to determine the longitudinal profile of the electromagnetic component of the showers and in particular to measure the depth of its maximum development, known as $X_{\rm max}$. It also allow us to determine the energy of the showers through a calorimetric determination of the deposited energy, providing a clean calibration of the energy estimator obtained with the surface array. The Collaboration involves more than 400 scientists from 17 countries [1].

Some of the main results obtained up to now have been, as will be discussed here, the measurement of the high-energy suppression of the cosmic ray (CR) spectrum [2], the observation of an increasingly heavier composition above the ankle [3] and the measurement of a dipolar anisotropy at energies above 8 EeV which indicates that the CRs above this threshold have an extragalactic origin [4]. Many other results have also been obtained, such as the determination of the proton-air cross section at energies beyond those achieved at colliders [5], the studies of the muon content of the showers that provide clues about models of hadronic interactions [6] or the bounds on neutrino and gamma fluxes that exclude exotic scenarios for the cosmic ray origin [7, 8]. Studies of large-scale anisotropies at different energies provided hints of a transition from an anisotropy of Galactic origin to one of extragalactic origin taking place at energies around the EeV [9]. Also hints of more localized anisotropies [10, 11], on scales of 10°–20°, appear at the energies for which the spectrum starts to become strongly suppressed and give the hope that with additional data, and with a better discrimination of the primary

[a] e-mail: roulet@cab.cnea.gov.ar
[b] e-mail: auger_spokespersons@fnal.gov

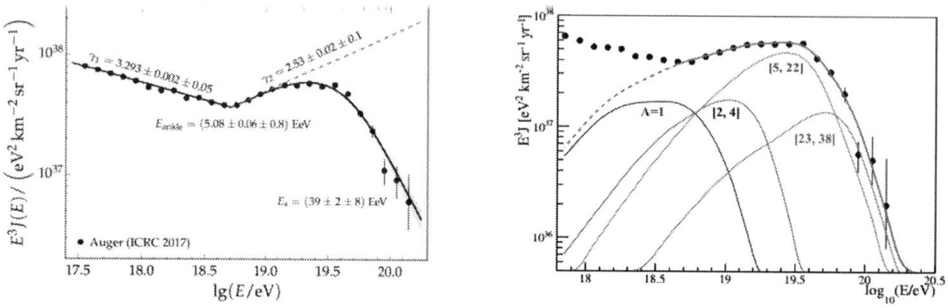

Figure 1. Spectrum measured by the Auger Observatory and fitted parameters obtained using a broken power-law with a high-energy supression (see text) [13]. The right panel shows the results obtained [14] by fitting the spectrum and $X_{\rm max}$ observations with a model of sources with a mixed composition of H, He, N and Si having a power-law spectra with a rigidity dependent cutoff. The individual curves depict the spectra at the Earth for different mass groups after the propagation and interactions with background radiation fields are taken into account. They correspond to the mass groups $A = 1$ (red), $2 \leq A \leq 4$ (grey), $5 \leq A \leq 22$ (green) and $23 \leq A \leq 38$ (cyan).

cosmic ray charges, the identification of the first individual ultrahigh energy CR sources may become feasible. There is also a strong participation in the new multi-messenger era that exploits the combination of results from cosmic-ray observatories with those from gravitational waves, high-energy neutrino and gamma-ray observatories [12].

2 The cosmic ray spectrum and composition

The left panel in figure 1 shows the measured CR flux, multiplied by E^3 in order to make it flatter, for energies above 3×10^{17} eV (the values below 1 EeV are obtained with a denser sub-array that covers an area of 23 km^2). The data points are fitted to a broken power-law with a high-energy suppression, according to [13]

$$
\begin{aligned}
J(E) &= J_0 \left(\frac{E}{E_{\rm ankle}}\right)^{-\gamma_1} && , E < E_{\rm ankle} \\
J(E) &= J_0 \left(\frac{E}{E_{\rm ankle}}\right)^{-\gamma_2} \left[1 + \left(\frac{E_{\rm ankle}}{E_s}\right)^{\Delta\gamma}\right] \left[1 + \left(\frac{E}{E_s}\right)^{\Delta\gamma}\right]^{-1} && , E \geq E_{\rm ankle}. \qquad (1)
\end{aligned}
$$

The so-called ankle break takes place at an energy $E_{\rm ankle} \simeq 5.1$ EeV, at which the spectrum hardens from a power law with index $\gamma_1 \simeq 3.29$ to one with $\gamma_2 \simeq 2.53$. At energies above $E_s \simeq 39$ EeV the spectrum becomes strongly suppressed, being reasonably well fitted with a power-law index $\gamma_3 \equiv \gamma_2 + \Delta\gamma \simeq 5.0$. It is important to keep in mind that the overall uncertainty on the energy scale is 14%.

From the spectrum results alone it is however not possible to understand the causes responsible for the features observed, for instance if the ankle reflects the transition between a Galactic component at low energies towards a harder extragalactic component at higher energies, or is instead reflecting the effects of a feature in the attenuation of an extragalactic component (such as a threshold for pair production for CR protons interacting with the CMB), or something different. It is also not clear whether the suppression is the result of the GZK effect due to photo-pion production of primary

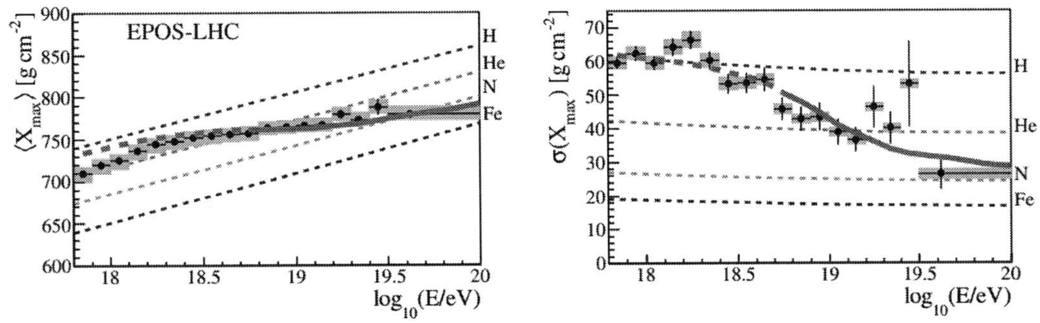

Figure 2. Average and standard deviation of the X_{max} distribution, compared to those predicted assuming EPOS-LHC interactions for pure H (red dashed lines), He (grey), N (green) and Fe (blue). Also shown with brown lines are the predictions [14] from the model with mixed composition illustrated in the right panel of figure 1, for which only the energy range where the brown lines are solid is included in the fit.

protons, the photodisintegration of heavy nuclei like Fe, or just a cutoff in the maximum energies achievable at the sources. To gain insight into these issues it is important to measure the composition of the CRs, i.e. what is the distribution of the CR masses as a function of the energy. The best tool to achieve this is the study of the distribution of the X_{max} values, since showers of energy E produced by nuclei of mass number A can in first approximation be considered as a superposition of A showers produced by nucleons of energy E/A. This means that for a given E the lighter primaries produce showers that are on average more penetrating than those produced by heavier primaries. Moreover, the fluctuations on the depth of shower maximum, $\sigma(X_{max})$, should decrease for increasing CR masses.

The results obtained for the average and dispersion of the X_{max} distributions at different energies are displayed in figure 2 (black dots with error bars) [3]. The main features are a break in the elongation rate (which is the change in X_{max} per decade of energy) which goes from about 80 g cm^{-2} below an energy of about 2 EeV to a value of about 26 g cm^{-2} above this energy. Since the elongation rate is expected to be about 56 g cm^{-2} if the composition were to remain constant (the predictions for a single mass component are illustrated as dashed lines for the cases of pure H, He, N and Fe primaries), the observed values give an indication that the average mass increases as the energy approaches the EeV. One should recall that at 10^{17} eV the CRs are believed to be predominantly of Galactic origin and are quite heavy. At a few EeV energies the X_{max} values suggest that the composition became light, with sizeable H and He components, while above these energies it should become increasingly heavier. The dispersion shown in the right panel is consistent with a light composition at EeV energies and becomes smaller above 2 EeV, in agreement with a composition that becomes heavier. Moreover, since a mixture of two different mass components would lead to values of X_{max} with a larger dispersion than any of the components alone, the fact that $\sigma(X_{max})$ becomes small above few EeV is an indication that there is no significant mixture of CRs with disparate masses.

Figures 1 (right panel) and 2 also display the results obtained by fitting the observations above 5 EeV with a scenario having a uniform distribution of identical sources with a mixture of four mass components (H, He, N and Si) [14]. Their spectra are assumed to be power-laws of index γ having a rigidity dependent cutoff R_{cut}, so that the spectra become exponentially suppressed above an energy $E_{cut} = Z_i R_{cut}$, with Z_i the corresponding atomic number of the nuclei of type i. The different nuclei are assumed to be produced with fractions f_i (defined at energies smaller than R_{cut}). Following the propagation of these nuclei from the sources up to the Earth and accounting for the interactions with the radiation backgrounds, the best fit to the spectra and X_{max} distributions lead to the results plotted in the figures, which correspond to values of $\gamma \simeq 1$ and $R_{cut} \simeq 5$ EeV. The fitted composition fractions correspond to about 67% He, 28% N and 5% Si, but note that the fit only includes data above 5 EeV and hence, given that the H cutoff obtained has a similar value, the H fraction is not really probed.

The value $\gamma \simeq 1$ obtained in the best fit corresponds to a spectrum much harder than what is expected in the Fermi diffusive shock acceleration scenarios, that predict values closer to $\gamma = 2$. One of the main reasons for the small values of γ inferred is the need to ensure that the heavier elements that dominate the spectrum at the highest energies become subdominant as the energy is decreased, so as to allow for a lighter composition and without leading to too large values of $\sigma(X_{max})$. One possible way to obtain a hardening of the observed spectrum for rigidities below a certain value, even if the source spectrum is softer, is if the CRs diffuse in the turbulent extragalactic magnetic fields [15]. If the density of sources is not too large so that the closest sources are at distances for which the CRs with rigidities below a certain value would take a time longer than the lifetime of the sources to reach the Earth, this 'magnetic horizon' effect would lead to a strong suppression of the observed spectrum at low energies, and will hence effectively harden it. Indeed, a fit to the spectrum and composition results in scenarios that include intergalactic magnetic fields led to a best fit source power-law index $\gamma \simeq 1.6$ [16], which is closer to the expectations.

3 Large scale anisotropies

Another crucial ingredient to understand the origin of the CRs, and to ultimately identify their sources, is the study of the distribution of their arrival directions to search for anisotropic signals. One of the main difficulties however is that, being charged, the CRs get deflected as they propagate through the Galactic and extragalactic magnetic fields and hence they do not point back to their sources when they are detected. However, since the deflections get smaller for increasing energies, anisotropies may show-up for sufficiently high energies. One has to keep in mind that the deflection of an extragalactic proton due to the effects of the Galactic magnetic field alone is about 10° to 30° at 10 EeV, and in general it scales approximately as Z/E for CRs with atomic number Z. The fact that above the ankle energy the average CR charges appear to increase makes the identification of individual sources more difficult, although anisotropies on large angular scales, such as a dipolar one, may still be observable. These kind of anisotropies are expected to result from the anisotropic distribution of the sources themselves, or also if some nearby sources contribute significantly and the CRs from them are sizeably deflected. Large angular scale anisotropies have already been searched for a long time with the Auger observatory. A technique which is often adopted to obtain the equatorial dipolar component is the harmonic analysis in right ascension, exploiting the almost uniform exposure achieved in this coordinate thanks to the Earth rotation. Different techniques exist to obtain the dipolar component along the Earth rotation axis, in particular one based on a harmonic analysis on the azimuth angle of the showers allows one to combine in a clean way different datasets, such as the so-called vertical one (including zenith angles $\theta \leq 60°$) with the inclined one ($60° < \theta \leq 80°$, that includes events with $E > 4$ EeV and relies on a different reconstruction than the one of the vertical sample). Many systematic effects have to be accounted for in order to obtain an accurate determination of the large scale anisotropies, such as the small variations of the exposure with time (due e.g. to dead times), the atmospheric effects on the reconstructed energies (which can induce spurious daily and seasonal variations on the rates above a given threshold), the geomagnetic effects on the air showers (which can induce a north-south asymmetry) or the slight tilt of the array towards the south-east.

The amplitude of the equatorial dipole is constrained to be not much larger than the percent level at EeV energies, and this excludes a dominant Galactic origin for the light component of the CRs that

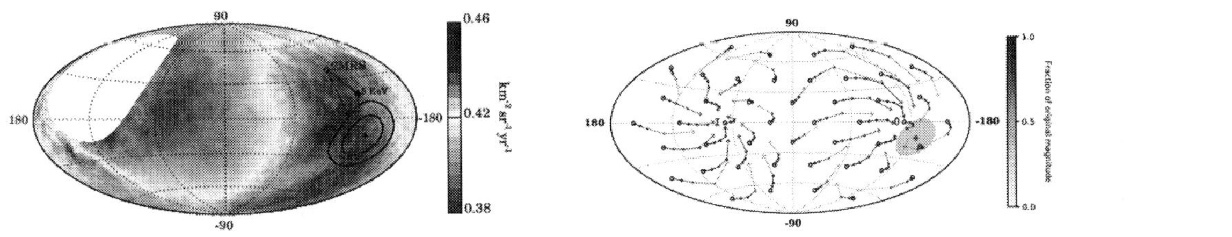

Figure 3. Measured CR flux for $E > 8$ EeV, averaged on top-hat windows of 45°, in Galactic coordinates (left panel) [4]. The direction of the reconstructed dipole direction is shown, together with the ellipses corresponding to the 68% and 95% CL regions. The right panel shows the changes in direction and amplitude of a dipolar extragalactic distribution after crossing the Galactic magnetic field, with circles being the original direction and the different points along the arrow corresponding to E/Z = 32 EeV, 16 EeV, 8 EeV and 4 EeV [19].

is observed at these energies [17]. Studying the energies above 4 EeV, for which the surface array is fully efficient up to zenith angles of 80° (allowing to cover 85% of the sky), and considering the two energy bins [4, 8] EeV and $E \geq 8$ EeV [18], a first harmonic modulation in right ascension with a significance larger than 5σ was found for $E \geq 8$ EeV [4]. The reconstructed CR flux in this energy range is shown in figure 3 (left panel). It shows a clear dipolar pattern and the reconstructed 3D dipole has an amplitude of 6.5%, pointing at Galactic coordinates $(\ell, b) = (233°, -13°)$. This direction is $\sim$ 125° away from the Galactic center, suggesting that this anisotropy has an extragalactic origin. Moreover, the dipole direction is not far from the direction of the outer spiral arm, which is actually an attractor for the directions of an extragalactic dipole after the effects of the Galactic magnetic field deflections are taken into account [19]. This is illustrated in the right plot in figure 3, where the dipole directions that would be observed at the Earth are shown for different original dipole directions outside the Galaxy, considering the effects of the deflections due to the Galactic field, modeled as in [20]. It is seen that the original directions that lie in the hemisphere of negative Galactic longitudes tend to be deflected towards the outer spiral arm (indicated with an **O** in the map), while the other hemisphere is instead deflected towards the inner spiral arm (indicated as **I**). The size of the deflections (and the resulting demagnifications of the dipole amplitude) depend on the CR rigidities. If the ultrahigh energy CR sources were to follow the distribution of the galaxies, one would expect that the anisotropy in the local distribution of galaxies should also reflect into an anisotropy in the CR arrival direction distribution. In particular, we show in the left panel of figure 3 the direction towards the flux weighted dipole of the galaxies in the 2MRS catalog [21], which is dominated by the contribution from galaxies closer than about 100 Mpc. If the resulting CR dipole were to point in a similar direction, after taking into account the deflections of the Galactic magnetic field (illustrated with the arrows corresponding to two representative values of E/Z) the resulting direction would be not far from the one observed above 8 EeV.

In ref. [19] a detailed study of the energy dependence of this dipolar anisotropy has been performed, considering the energy bins [4, 8] EeV, [8, 16] EeV, [16, 32] EeV and $E \geq 32$ EeV. The results are illustrated in figure 4. The left panel shows the 3D dipole amplitude as a function of the energy, as well as the expectations in scenarios in which extragalactic sources with a density of 10^{-4} Mpc^{-3} are distributed uniformly or following the observed galaxy distribution [22]. One can see that the overall trend of the amplitudes is quite consistent with the measured ones. The right panel shows the reconstructed dipole directions, which suggest an extragalactic origin for the anisotropies in all four bins.

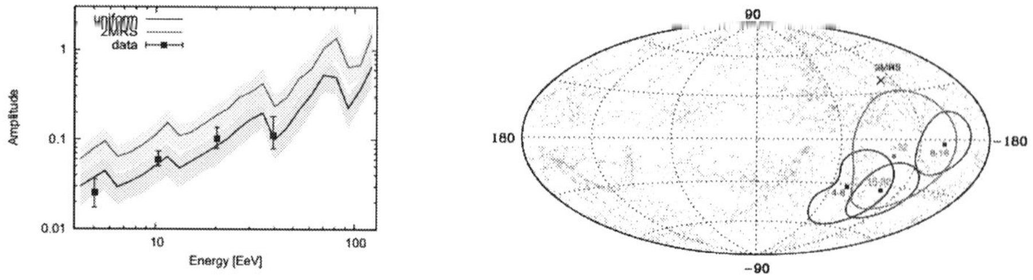

Figure 4. Left panel: Energy dependence of the dipole amplitude, together with the expectations from different extragalactic source scenarios [19]. Right panel: reconstructed dipole directions in the different energy bins.

An extensive program of upgrades to the Pierre Auger Observatory is taking place at the moment, including the deployment of scintillators on top of each surface detector, small PMTs to increase the dynamic range, radio antennas and new electronics. All this will help to measure more accurately the showers and should allow for mass discrimination of the primaries using the surface array, enhancing the capabilities to study the shower properties and to search for anisotropies restricting to the less deflected light component of the CR fluxes. Hence, the prospects for further improvements on our knowledge of the most energetic particles in nature are very promising.

References

[1] A. Aab et al. [Pierre Auger Coll.], NIM Phys. Res. A 798, 172 (2015)
[2] J. Abraham et al. [Pierre Auger Coll.], PLB 685, 239 (2010)
[3] A. Aab et al. [Pierre Auger Coll.], PRD 90, 122005 (2014); ibidem 122006.
[4] A. Aab et al. [Pierre Auger Coll.], Science 357, 1266 (2017)
[5] P. Abreu et al. [Pierre Auger Coll.], PRL 109, 062002 (2012)
[6] A. Aab et al. [Pierre Auger Coll.], PRD 91, 032003 (2015)
[7] A. Aab et al. [Pierre Auger Coll.], PRD 91, 092008 (2015)
[8] A. Aab et al. [Pierre Auger Coll.], JCAP 04, 009 (2017)
[9] P. Abreu et al. [Pierre Auger Coll.], Astropart. Phys. 34, 627 (2011)
[10] A. Aab et al. [Pierre Auger Coll.], ApJ 804, 15 (2015)
[11] A. Aab et al. [Pierre Auger Coll.], ApJ 853, L29 (2018)
[12] ANTARES, IceCube, Pierre Auger, LIGO and Virgo Coll., ApJ 850, L35 (2017)
[13] F. Fenu for the Pierre Auger Collaboration, PoS(ICRC2017) 486
[14] A. Aab et al. [Pierre Auger Coll.], JCAP 04, 038 (2017)
[15] S. Mollerach, E. Roulet, JCAP 10, 013 (2013)
[16] D. Wittkowski, for the Pierre Auger Collaboration, PoS ICRC2017, 563 (2017)
[17] P. Abreu et al. [Pierre Auger Coll.], ApJL 762, L13 (2013)
[18] A. Aab et al. [Pierre Auger Coll.], ApJ 802, 111 (2015)
[19] A. Aab et al. [Pierre Auger Coll.], ApJ (2018) in press
[20] R. Jansson, G.R. Farrar, ApJ 757, 14 (2012)
[21] P. Erdogdu et al., MNRAS 368, 1515 (2006)
[22] D. Harari, S. Mollerach, E. Roulet, PRD 92, 063014 (2015)

Towards an anagraphical picture of high-energy Galactic neutrinos

Antonio Marinelli[1,a], Dario Grasso[1], and Sofia Ventura[2]

[1] *INFN and Dipartimento di Fisica, Università di Pisa, Largo B. Pontecorvo 3, I-56127 Pisa, Italy*
[2] *INFN and Dipartimento di Fisica, Università di Siena, Pian de Mantellini 44, 53100 Siena, Italia*

Abstract. The TeV/PeV neutrino emission from our Galaxy is related to the distribution of cosmic-ray accelerators, their maximal energy of injection as well as the propagation of injected particles and their interaction with molecular gas. In the last years Interesting upper limits on the diffuse hadronic emission from the whole Galaxy, massive molecular clouds and Fermi Bubbles were set by the IceCube and ANTARES as well as HAWC and Fermi-LAT observations. On the other hand no evidence of Galactic point-like excess has been observed up to now by high-energy neutrino telescopes. This result can be related to the short duration of the PeV hadronic activity of the sources responsible for the acceleration of primary protons, possibly including supernova remnants. All these aspects will be discussed in this work.

1 Introduction

The astrophysical neutrino flux measured by the IceCube collaboration in the last 8 years gives rise to the need of estimating the Galactic and the extra-Galactic high-energy neutrino fluxes. While for the second component the debate about the source candidates is still open and more years of multi-messenger observations are needed to provide a solid answer, for the first component we already have a defined picture. Stringent constrains have been in fact obtained on the diffuse Galactic neutrino emission through IceCube and ANTARES analyses and important bounds come from the very-high-energy (VHE) gamma-ray observations of the Fermi Bubbles, the central molecular zone (CMZ) and the possible Galactic Pevatrons. For the diffuse Galactic emission new upper limits (ULs) were obtained with the joint ANTARES/IceCube analysis considering 2780 days of the Mediterranean-based experiment and 2431 days of the South Pole-based experiment [1]. These ULs set the maximal neutrino flux produced by the interaction of Galactic cosmic rays with the gas and consequently also the relative contribution to the full sky astrophysical neutrino flux measured by IceCube. This was possible thanks to a maximum likelihood analysis using the data of neutrino experiments and the map of the neutrino flux expected from the *Gamma Model* [2], as discussed in more detail in Sec. 2. Another component that deserve a special focus in our Galaxy is the possible neutrino flux associated with the Fermi Bubbles region [3, 4]. Fermi-LAT observations from the π sr solid angle of the Fermi Bubbles leave space for both leptonic and hadronic interpretations [5]. However the ULs posed recently by the HAWC observatory on the northern Bubble are compatible with a maximal

[a] e-mail: antonio.marinelli@pi.infn.it

 https://doi.org/10.1051/epjconf/201920901003

cutoff in the gamma-ray spectral energy distribution (SED) of few tens of TeV [6]. This is showed in more details in this proceeding using the gamma-ray observations collected up to now (see fig. 2). An interesting observation of the low latitude Fermi Bubbles SED showed up in 2017 with a new Fermi-LAT analysis [7]. The Fermi-Bubbles SED for $\theta \leq 10^{\circ}$[1] does not present any cutoff and may be suggesting a different emitting process respect to the outer Bubbles region. In this work we compare the SED obtained by the Fermi-LAT collaboration for the low latitudes region, after the subtraction of point-like and local diffuse contributions, with the SEDs of the expected diffuse Galactic emission. This comparison is reported to verify if a different diffuse Galactic contribution may change the interpretation of the Fermi-LAT data. Other considerations are reported in this proceeding about the possibility to observe Galactic point-like neutrino sources with IceCube and ANTARES using the high energy track-like events.

2 News on the Galactic diffuse emission

The work recently published by ANTARES and IceCube collaborations about the searches for a diffuse Galactic signal gives new interesting ULs about this astrophysical neutrino flux. Using 218 shower-like and 7850 track-like events of ANTARES and 730130 track-like events of IceCube with a joint analysis it was possible to reach an UL on the diffuse emission due to the Galactic cosmic-ray (CR) sea very close to the *Gamma model* prediction under the assumption of a cutoff of 5 PeV for CR primary proton energy. This limit, which is shown in Fig. 1, amounts to less than 8.5% of the astrophysical flux measured by IceCube on the whole sky, hence reducing drastically the possibility for the Galactic emission to create a considerable anisotropy in the "extraterrestrial neutrino" sky.

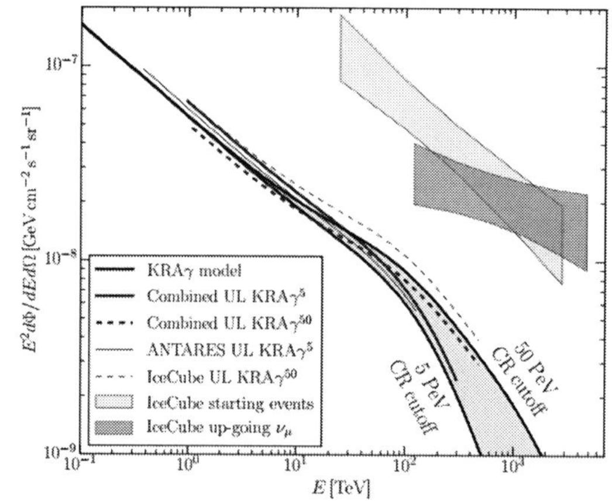

Figure 1: Combined ULs at 90% confidence level (blue lines) on the three-flavor neutrino flux of the KRAγ model (*Gamma model*) with the 5 and 50 PeV cutoffs (black lines) for the primary CRs. The boxes represent the diffuse astrophysical neutrino fluxes measured by IceCube using an isotropic flux template with high energy starting events (HESE) (yellow) and up-going tracks (green) [1, 2].

[1] With θ representing the Galactic declination.

3 Fermi Bubbles emission

The SED of gamma-ray emission observed by the Fermi-LAT collaboration from the Fermi Bubbles was fitted with a power law with slope α and a exponential cutoff at energy E_{cut}. While a hard spectrum ($\alpha \sim 2$) was observed from 1 to 100 GeV a exponential cutoff was set at E_{cut} of 113 GeV [4]. This SED was described through both hadronic and leptonic emission models, and several studies suggest the possibility to observe high energy neutrinos from this region of the sky assuming that the observed gamma rays are produced by the interaction of protons with photons along the Bubbles. An important constrain to the modeling of the Fermi Bubbles emission at high energies comes from the HAWC observations [6]. As showed in Fig. 2, the ULs posed by the HAWC observatory on the Northern Fermi Bubble leave space for a possible neutrino SED with a exponential cutoff at around 30 TeV. These ULs reduce considerably the possibility to detect TeV neutrinos from this Galactic source with a maximal corresponding neutrino SED showed in Fig.2. Moreover, the IceCube neutrino flux limits obtained considering the IceCube HESE events spatially correlated with this region do not allow a viable connection with the HAWC ULs, as visible in Fig 2. On the other hand new observations of the Fermi Bubbles were performed at low latitudes (previously excluded in the Fermi-LAT analysis), namely the Bubbles region with $\theta < 10°$. In this new analysis a power-law spectrum without any cutoff was used to describe the observed emission in the Fermi-LAT energy range. This observation suggests a different component in the inner part of the Bubbles and increases the possibility to observe a high energy neutrino signal in case this emission is related to a proton population. However, being this region of the Bubbles coincident with the inner Galactic plane, it is necessary to verify whether this hard SED is contaminated by the enhanced diffuse Galactic component of this part of the Galaxy. In Fig. 2 we show that the spectrum measured by Fermi-LAT for the low latitude Bubbles presents a spectral component that cannot be reproduced just considering the diffuse Galactic background. This test supports the presence of a different high energy emitting process in the freely-expanding wind zone [5].

4 Galactic Pevatrons

The data collected up to now by IceCube and ANTARES telescopes already set interesting ULs on the potential neutrino emission from Galactic point-like sources [10, 11], however the possibility to identify hadronic Pevatrons in our Galaxy is matter of debate. Recent studies on Supernova Remnants (SNRs) environment, including calculations on the protons and electrons energy losses and escaping times, show that hadronic Pevatrons can have, at maximum, a duration of few hundreds of years, while the leptonic Pevatrons can last much more [12]. In other words the VHE emission from a SNR can be dominated by the hadronic emission just during the early stages, whereas we should invoke a leptonic emission to describe the gamma-ray emission observed by the gamma-ray telescopes from the old SNRs. This description of the SNR activity would limit the possibility to observe Galactic point-like neutrino sources. In fact if we consider the angular resolution of 1° for $\nu_\mu + \bar{\nu_\mu}$, the time duration of the hadronic PeV emission [12] plus the diffusion time of protons in the space corresponding to 1° around the source we obtain at maximum of few thousands of years [2].

5 Conclusions

Among all the searches for VHE neutrino emitter candidates, our Galaxy deserves a special focus, with different components that can participate to a potential neutrino flux observable with the Ice-

[2] Using the local B/C for normalizing and a *Gamma Model* diffusion assumption for a Pevatron located in the CMZ we can roughly obtain $t_{diff}(E) = 1.500\ years\ (L/100pc)^2(E/1PeV)^{-1/3}$.

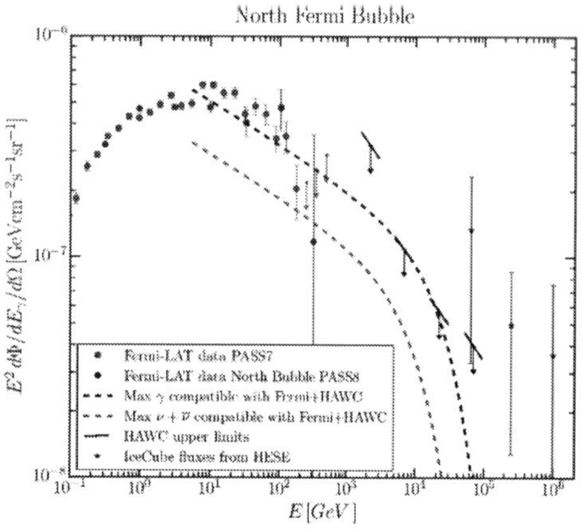

(a)

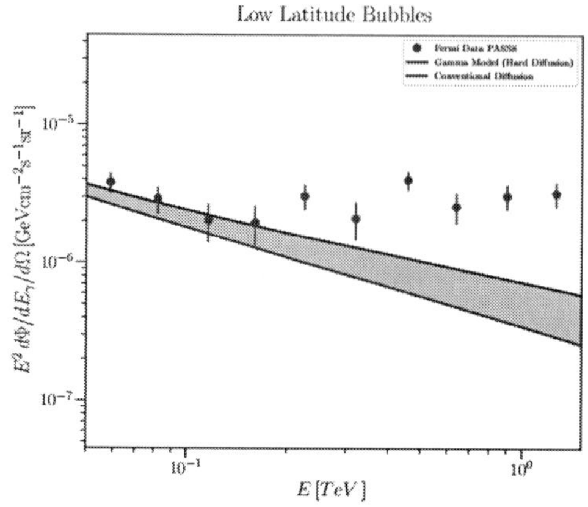

(b)

Figure 2: a) SED of the Fermi Bubbles obtained with PASS7 data sample [4] and North Fermi Bubble with PASS8 data sample [8]; the HAWC upper limits [6]; the IceCube fluxes limits obtained through the HESE events compatible with the Bubbles; the hadronic fitting for the gamma-ray data, compatible with Fermi-LAT and HAWC observations, with a exponential cutoff at 30 TeV (red dashed line) and the corresponding expected neutrino SED obtained considering the interaction of protons with the photons inside the bubbles (magenta dashed line). b) Fermi-LAT gamma-ray observations for the low latitude Bubbles ($\theta < 10°$) are shown as a data points while the green region represent the space of gamma-ray SEDs produced in the same region of the sky from the Galactic cosmic rays interaction with the gas, assuming different models. With *Gamma Model* (black line), we intend a harder diffusion due to a radial dependence of the scaling of the diffusion coefficient in addition with a spectral slope change around 300 GeV, as observed by PAMELA, AMS-02 and CREAM [2, 9]. Instead, with conventional diffusion (blue line), we intend the uniform CR-sea pervading all the Galaxy.

Cube and ANTARES telescopes. In this work we review the possible Galactic components of the astrophysical flux measured by the IceCube telescope. We report the recent results obtained thanks to the joint ANTARES/IceCube analysis of the diffuse Galactic emission and highlight the importance of these new constrains. The obtained UL of 8.5% of the total HESE flux (when a cutoff of 5 PeV is selected for the Galactic CR) tell us that the Galactic component can hardly produce a considerable anisotropy in the astrophysical neutrino sky. In this contribution we also compare the diffuse Galactic background with the Fermi-LAT measurements of the low latitude Fermi Bubbles, highlighting the possibility of a new component connected with the freely-expanding wind zone. We show that above 100 GeV the hard SED measured by Fermi-LAT for the low latitude Bubbles cannot be explained with a enhanced diffuse Galactic emission permeating the region. However a contribution of unresolved sources in the Fermi-LAT observed flux cannot be excluded. Finally we point out that the limited duration of the hadronic Pevatron phase of the SNR decreases the possibility to observe a Galactic point-like emission with IceCube and ANTARES telescopes.

References

[1] A. Albert, M. André, M. Anghinolfi, M. Ardid, J.J. Aubert, J. Aublin, T. Avgitas, B. Baret, J. Barrios-Martí, S. Basa et al., Astrophysical Journal Letter **868**, L20 (2018), 1808.03531

[2] D. Gaggero, D. Grasso, A. Marinelli, A. Urbano, M. Valli, Astrophysical journal Letter **815**, L25 (2015), 1504.00227

[3] C. Lunardini, S. Razzaque, K.T. Theodoseau, L. Yang, Physical Review D **90**, 023016 (2014), 1311.7188

[4] M. Ackermann, A. Albert, W.B. Atwood, L. Baldini, J. Ballet, G. Barbiellini, D. Bastieri, R. Bellazzini, E. Bissaldi, R.D. Blandford et al., Astrophysical Journal **793**, 64 (2014), 1407.7905

[5] R.M. Crocker, G.V. Bicknell, A.M. Taylor, E. Carretti, Astrophysical Journal **808**, 107 (2015), 1412.7510

[6] A.U. Abeysekara, A. Albert, R. Alfaro, C. Alvarez, J.D. Álvarez, R. Arceo, J.C. Arteaga-Velázquez, H.A. Ayala Solares, A.S. Barber, N. Bautista-Elivar et al., Astrophysical Journal **842**, 85 (2017), 1703.01344

[7] M. Ackermann, M. Ajello, A. Albert, W.B. Atwood, L. Baldini, J. Ballet, G. Barbiellini, D. Bastieri, R. Bellazzini, E. Bissaldi et al., Astrophysical Journal **840**, 43 (2017), 1704.03910

[8] S.A. Narayanan, T.R. Slatyer, Monthly Notices of the Royal Astronomical Society **468**, 3051 (2017)

[9] D. Gaggero, D. Grasso, A. Marinelli, M. Taoso, A. Urbano, Physical Review Letters **119**, 031101 (2017), 1702.01124

[10] A. Albert, M. André, M. Anghinolfi, G. Anton, M. Ardid, J.J. Aubert, T. Avgitas, B. Baret, J. Barrios-Martí, S. Basa et al., Physical Review D **96**, 082001 (2017), 1706.01857

[11] IceCube Collaboration, M.G. Aartsen, M. Ackermann, J. Adams, J.A. Aguilar, M. Ahlers, M. Ahrens, D. Altmann, K. Andeen, T. Anderson et al., arXiv e-prints (2018), 1811.07979

[12] D. Gaggero, F. Zandanel, P. Cristofari, S. Gabici, Montly Notices of the Royal Astronomical Society **475**, 5237 (2018), 1710.05038

Time dependence of the helium flux measured by PAMELA

N. Marcelli[1,20,a], O. Adriani[2,3], G. C. Barbarino[4,5], G. A. Bazilevskaya[6], R. Bellotti[7,8], M. Boezio[9], E. A. Bogomolov[10], M. Bongi[2,3], V. Bonvicini[9], S. Bottai[3], A. Bruno[8], F. Cafagna[7,8], D. Campana[5], P. Carlson[11], M. Casolino[1,12], G. Castellini[13], C. De Santis[1], V. Di Felice[1,14], A. M. Galper[15], A. V. Karelin[15], S. V. Koldashov[15], S. Koldobskiy[15], S. Y. Krutkov[10], A. N. Kvashnin[6], A. Leonov[15], V. Malakhov[15], L. Marcelli[1], M. Martucci[1,20], A. G. Mayorov[15], W. Menn[17], M. Mergè[1,20], V. V. Mikhailov[15], E. Mocchiutti[9], A. Monaco[7,8], N. Mori[3], R. Munini[9], G. Osteria[5], B. Panico[5], P. Papini[3], M. Pearce[11], P. Picozza[1,20], M. Ricci[16], S. B. Ricciarini[13], M. Simon[17], R. Sparvoli[1,20], P. Spillantini[2,3], Y. I. Stozhkov[6], A. Vacchi[9,19], E. Vannuccini[3], G. Vasilyev[10], S. A. Voronov[15], Y. T. Yurkin[15], G. Zampa[9], N. Zampa[9], M. S. Potgieter[21], O. P. M. Aslam[21], and D. Bisschoff[21]

[1] *INFN, Sezione di Rome "Tor Vergata", I-00133 Rome, Italy*
[2] *University of Florence, Department of Physics, I-50019 Sesto Fiorentino, Florence, Italy*
[3] *INFN, Sezione di Florence, I-50019 Sesto Fiorentino, Florence, Italy*
[4] *University of Naples "Federico II", Department of Physics, I-80126 Naples, Italy*
[5] *INFN,Sezione di Naples, I-80126 Naples, Italy*
[6] *Lebedev Physical Institute, RU-119991, Moscow, Russia*
[7] *University of Bari, Department of Physics, I-70126 Bari, Italy*
[8] *INFN, Sezione di Bari, I-70126 Bari, Italy*
[9] *INFN, Sezione di Trieste, I-34149 Trieste,Italy*
[10] *Ioffe Physical Technical Institute, RU-194021 St. Petersburg, Russia*
[11] *KTH, Department of Physics, and the Oskar Klein Centre for Cosmoparticle Physics, AlbaNova University Centre, SE-10691 Stockholm, Sweden*
[12] *RIKEN, EUSO team Global Research Cluster, Wako-shi, Saitama, Japan*
[13] *IFAC, I-50019 Sesto Fiorentino, Florence, Italy*
[14] *Space Science Data Center - Agenzia Spaziale Italiana, via del Politecnico, s.n.c., I-00133, Roma, Italy*
[15] *National Research Nuclear University MEPhI, RU-115409, Moscow, Russia*
[16] *INFN, Laboratori Nazionali di Frascati, Via Enrico Fermi 40, I-00044 Frascati, Italy*
[17] *Universitat Siegen, Department of Physics, D-57068 Siegen, Germany*
[18] *Istituto Nazionale di Astrofisica, Fosso del cavaliere 100, 00133 Roma, Italy*
[19] *University of Udine, Department of Mathematics, Computer Science and Physics Via delle Scienze, 206, Udine, Italy*
[20] *University of Rome "Tor Vergata", Department of Physics, I-00133 Rome, Italy*
[21] *North-West University, Centre for Space Research, 2520 Potchefstroom, South Africa*

Abstract. Precision measurements of the $Z = 2$ component in cosmic radiation provide crucial information about the origin and propagation of the second most abundant cosmic ray species in the Galaxy (9% of the total). These measurements, acquired with the PAMELA space experiment orbiting Earth, allow to study solar modulation in details. Helium modulation is compared to the modulation of protons to study possible dependencies on charge and mass. The time dependence of helium fluxes on a monthly basis

[a]e-mail: nadir.marcelli@roma2.infn.it

measured by PAMELA has been studied for the period between July 2006 to January 2016 in the energy range from 800 MeV/n to $\sim$ 20 GeV/n.

1 Introduction

Measurements of absolute fluxes and spectral shapes of primary cosmic rays are extremely important for better understanding both origin and propagation mechanisms of galactic cosmic rays (GCRs). During their propagation through the InterStellar Medium (ISM) from the source, cosmic rays lose energy, modifying their spectra and composition. Subsequently, by entering the heliosphere, GCRs are scattered by irregularities in the heliospheric magnetic field (HMF). They are transported by the expanding solar wind from the Sun to the outermost regions of the heliosphere and processes such as convection, diffusion, drifts and adiabatic deceleration become more important; the sum of these effects is known as solar modulation [1, 2].

At energies above $\sim$ 30 GeV for protons and $\sim$ 15 GeV/n for helium, solar modulation has a negligible influence on the observed spectra and intensity profiles which represent the ones measured beyond the heliopause [2, 3].

The solar activity possesses an approximately 11-year periodicity which adds to a 22-year periodicity associated to the solar magnetic field rearrangement and reversal that usually takes place around the time of solar maximum and reaches peak strength at the solar minimum. The record-setting minimum of activity of cycle 23rd (2006-late 2009) [4] caused extremely unusual minimum modulation conditions for GCRs and this translated into a perfect environment to study the various processes that affect particles inside the heliosphere. After 2009, solar activity started to rise slowly up to January 2014, however this was lower than the previous one and the 24th solar cycle has been classified as one of the weakest in recent years in terms of e.g. sunspot number, solar wind speed and number of solar events [5–7].

During its almost ten years of collecting data, PAMELA was well suited to measure the impact on solar modulation on many cosmic ray species, due to its high redundant detectors and high precision measurements which permitted to collect unprecedented volume of data. PAMELA collaboration published already the data on observed modulation for species like protons [8], [9], electrons [10] and the e^-/e^+ ratio [11].

2 The instrument

PAMELA (A Payload for Antimatter-Matter Exploration and Light-nuclei Astrophysics) is a satellite-borne experiment designed to make long duration observations of the cosmic radiation from Low Earth Orbit [12]. The instrument collected GCRs for almost 10 years from 2006 June 15 when it was launched from the Baikonur cosmodrome in Kazakhstan, up to January of 2016. PAMELA orbit was elliptical with altitudes ranging between 350 and 610 km with an inclination of 70°. After 2010 the orbit was changed to a circular one at a constant altitude of about 500 km. The payload comprises a number of highly redundant detectors capable of identifying particles by providing charge, rigidity and velocity measurements over a very wide energy range. Multiple sub-detectors were built around a magnetic spectrometer, composed of a silicon tracking system [13] placed inside a 0.43 T permanent magnet. A system of six layers of plastic scintillators, arranged in three double planes (S1, S2, and S3), provides a fast signal for triggering the acquired data, particle identification measuring the ionization energy loss and the time of flight (ToF) of particles passing through. An electromagnetic imaging W/Si calorimeter (16.3 radiation lengths and 0.6 interaction lengths deep) provides hadron-lepton discrimination [14]. A neutron counter [15] contributes to discrimination power by detecting the increased neutron production in the calorimeter associated with hadronic showers compared to electromagnetic ones, while a plastic scintillator, placed beneath the calorimeter, increases the identification of high-energy electrons. The whole apparatus is surrounded by an anti-coincidence system (AC) of three scintillators (CARD, CAS, and CAT) for the rejection of background events [16]. A comprehensive description of the instrument and main scientific results can be found in [17], [18]. The payload is schematically shown in Figure 1 left.

3 Event selection

Clean helium candidates are selected requiring a single track in the spectrometer and fully contained inside the instrumental acceptance. The tracker selection criteria require a lever-arm of at least 4 silicon planes and a minimum of 3 hit on both X and Y view. Down going particles are selected with time information measured by the Time of Flight (ToF) system. GCR helium nuclei was selected considering only particles with a rigidity greater by a factor 1.3 than the vertical cutoff, evaluated using the Störmer approximation, at a given orbital position of the satellite. Then the ionization energy loss over the silicon planes of the tracking system is used to identify the helium nuclei (see Figure 1 right). Finally the helium fluxes have been estimated taking in to account the livetime, acceptance and efficiencies of the instrument.

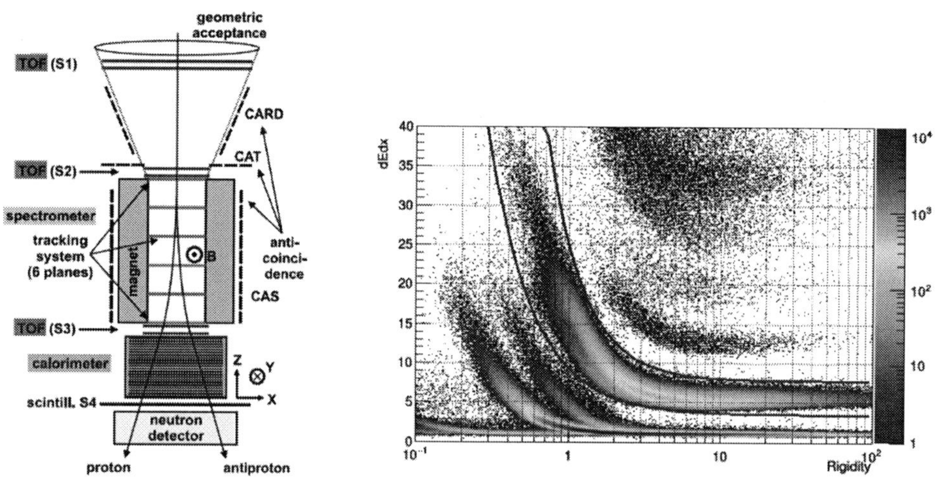

Figure 1. On the left: a schematic view of the PAMELA detector [17], [18]. On the right plot: the ionization energy loss distribution measured by the tracker as a function of rigidity. The region inside the two red curves represents the helium band considered for the selection criteria.

4 Conclusions

The time dependence of the helium fluxes measured by PAMELA offer the possibility to study the solar modulation of GCRs. The helium fluxes have been computed for a time period of one Carrington Rotation between July 2006 and January 2016, in the energy range from 800 MeV/n and $\sim 20\ GeV/n$; the results of this analysis will be published soon. Furthermore, comparison of protons and helium fluxes, through the time evolution of the p/He ratio in the overall period of PAMELA data-taking, offer the possibility to study possible differences in modulation between these two main CR species.

References

[1] E.N. Parker, Planetary and Space Science **13**, 9 (1965)
[2] M.S. Potgieter, Space Sci. Rev.**176**, 165 (2013)
[3] R.D. Strauss, M.S. Potgieter, Advances in Space Research **53**, 1015 (2014)
[4] R.A. Mewaldt, A.J. Davis, K.A. Lave, R.A. Leske, E.C. Stone, M.E. Wiedenbeck, W.R. Binns, E.R. Christian, A.C. Cummings, G.A. de Nolfo et al., Astrophys. J. Lett.**723**, L1 (2010)
[5] K.P. Schröder, M. Mittag, J.H.M.M. Schmitt, D. Jack, A. Hempelmann, J.N. González-Pérez, **470**, 276 (2017), 1705.03777
[6] O.P.M. Aslam, Badruddin, Solar Phys.**290**, 2333 (2015), 1507.07917
[7] J. Jiang, R.H. Cameron, M. Schüssler, Astrophys. J. Lett.**808**, L28 (2015), 1507.01764
[8] O. Adriani, G.C. Barbarino, G.A. Bazilevskaya, R. Bellotti, M. Boezio, E.A. Bogomolov, M. Bongi, V. Bonvicini, S. Borisov, S. Bottai et al., ApJ **765**, 91 (2013), 1301.4108
[9] M. Martucci, R. Munini, M. Boezio, V. Di Felice, O. Adriani, G.C. Barbarino, G.A. Bazilevskaya, R. Bellotti, M. Bongi, V. Bonvicini et al., Astrophys. J. Lett.**854**, L2 (2018), 1801.07112
[10] O. Adriani, G.C. Barbarino, G.A. Bazilevskaya, R. Bellotti, M. Boezio, E.A. Bogomolov, M. Bongi, V. Bonvicini, S. Bottai, A. Bruno et al., ApJ **810**, 142 (2015), 1512.01079
[11] O. Adriani, G.C. Barbarino, G.A. Bazilevskaya, R. Bellotti, M. Boezio, E.A. Bogomolov, M. Bongi, V. Bonvicini, S. Bottai, A. Bruno et al., Physical Review Letters **116**, 241105 (2016), 1606.08626
[12] P. Picozza, A.M. Galper, G. Castellini, O. Adriani, F. Altamura, M. Ambriola, G.C. Barbarino, A. Basili, G.A. Bazilevskaja, R. Bencardino et al., Astroparticle Physics **27**, 296 (2007), astro-ph/0608697
[13] O. Adriani, L. Bonechi, M. Bongi, G. Castellini, R. D'Alessandro, A. Gabbanini, M. Grandi, P. Papini, S.B. Ricciarini, P. Spillantini et al., Nuclear Instruments and Methods in Physics Research A **511**, 72 (2003)
[14] M. Boezio, V. Bonvicini, E. Mocchiutti, P. Schiavon, G. Scian, A. Vacchi, G. Zampa, N. Zampa, Nuclear Instruments and Methods in Physics Research A **487**, 407 (2002), physics/0109010
[15] Y.I. Stozhkov, A. Basili, R. Bencardino, M. Casolino, M.P. de Pascale, G. Furano, A. Menicucci, M. Minori, A. Morselli, P. Picozza et al., International Journal of Modern Physics A **20**, 6745 (2005)
[16] S. Orsi, P. Carlson, P. Hofverberg, J. Lund, M. Pearce, International Cosmic Ray Conference **3**, 369 (2005)
[17] O. Adriani, G.C. Barbarino, G.A. Bazilevskaya, R. Bellotti, M. Boezio, E.A. Bogomolov, M. Bongi, V. Bonvicini, S. Bottai, A. Bruno et al., Phys. Rep.**544**, 323 (2014)
[18] O. Adriani, G. Barbarino, G. Bazilevskaya, R. Bellotti, M. Boezio, E. Bogomolov, M. Bongi, V. Bonvicini, S. Bottai, A. Bruno et al., Rivista del Nuovo Cimento **40**, 473 (2017)

EPJ Web of Conferences **209**, 01005 (2019) https://doi.org/10.1051/epjconf/201920901005
RICAP-18

Search for tau neutrinos with the MAGIC telescopes: improving selection criteria*

Marina Manganaro[1,a], Leonora Kardum[1], Elisa Bernardini[2,3], Michele Doro[4], Dariusz Gora[5], Saša Mićanović[1], Tomislav Terzić[1], and Martin Will[6]on behalf of the MAGIC Collaboration[7]

[1]*University of Rijeka, Department of Physics, 51000 Rijeka*
[2]*Deutsches Elektronen-Synchrotron (DESY), D-15738 Zeuthen, Germany*
[3]*Humboldt University of Berlin, Institut für Physik D-12489 Berlin Germany*
[4]*Università di Padova and INFN, I-35131 Padova, Italy*
[5]*Institute of Nuclear Physics Polish Academy of Sciences, PL-31342 Krakow, Poland*
[6]*Max-Planck-Institut für Physik, D-80805 München, Germany*
[7]*the full list of MAGIC members is available at: wwwmagic.mppmu.mpg.de*

Abstract. MAGIC, a system of two Cherenkov telescopes located at the Roque de los Muchachos Observatory (2200 a.s.l.) in the Canary Island of La Palma, has lately been engaged in an unconventional task: the search for a signature of particle showers induced by earth-skimming cosmic tau neutrinos arising from the ocean, in the PeV to EeV energy range. When pointing at the sea, the MAGIC telescopes can collect data in a range of about 5 deg in zenith and 80 deg in azimuth: the analysis of the shower images from ~30 hours of data, together with the simulations of upward-going tau neutrino showers, shows that the air showers induced by tau neutrinos can be discriminated from the hadronic background coming from a similar direction. We have calculated the point source acceptance and the expected event rates, assuming an incoming ν_τ flux consistent with IceCube measurements, and for a sample of generic neutrino fluxes from photo-hadronic interactions in AGNs and GRBs. A 90% C.L. upper limit on the tau-neutrino point source flux of 2.0×10^{-4} GeV cm^{-2} s^{-1} has been obtained. The presented results can also be important for future Cherenkov experiments such as the Cherenkov Telescope Array. This next generation ground-based observatory can have a much better possibility to detect ν_τ, given its larger FOV and much larger effective area.

1 Introduction

The consolidated discovery by IceCube (of an astrophysical flux of high-energy neutrinos [1] has confirmed the theoretical expectations in[2–5]: indeed those very energetic neutrinos are produced in astrophysical objects by the decays of charged pions created in cosmic-rays interactions with radiation or gas. The pions produce then neutrinos which can travel long distances undisturbed and tracking back with an impressive precision the origin of the cosmic acceleration. In this exciting picture, recently made more interesting by the first multi-messenger detection including a neutrino

*original title was: Search for tau neutrinos with the MAGIC telescopes: the quest continues
[a]e-mail: marina.manganaro@phy.uniri.hr

[6], the MAGIC (Major Atmospheric gamma-ray Imaging Cherenkov) telescopes have played the role of Very-high-energy (VHE, E 100GeV) counterpart (together with H.E.S.S.–High Energy Stereoscopic System– and VERITAS–Very Energetic Radiation Imaging Telescope Array System–) in the collection of the broadband data. The observation by IceCube of the neutrino event 170922A in spatial coincidence with the γ-ray emitting blazar TXS 0506+056 during an active phase encourages the search for other possible blazars behaving as "neutrinos factories". After the IceCube detection of event 170922A, MAGIC went on observing the source for almost 40 hours, and the analysis of those data together with other quasi-simultaneous multi-wavelength data have been used to model the emission with a novel one-zone lepto-hadronic model, based on interactions of electrons and protons co-accelerated in the jet with external photons originating from a slow-moving plasma sheath surrounding the faster jet spine [7].

In general, the observed neutrino flux by IceCube and its composition is in agreement with equal fractions of all neutrino flavours [8, 9]. The possibility in IceCube to identify ν_τ is under investigation [10]: double cascade event topologies are only produced in tau-neutrino interactions above an energy threshold of ~100 TeV. The detection of ν_τ is very important from both the astrophysical and particle physics point of view. It would give new information about the astrophysical ν_τ flux and shed light on the emission mechanisms at the source, test the fundamental properties of neutrinos over extremely long baselines, and better constrain new physics models which predict significant deviations from equal fractions of all flavors.

2 Observations with MAGIC

MAGIC is a stereoscopic system of two Cherenkov telescopes, located on the Canarian island of La Palma at 2200m above the sea level. Devoted to the study of VHE γ-rays from extragalactic and galactic sources, they could also be pointed to a specific direction towards the surrounding Atlantic ocean and search for signatures of air showers induced by PeV-EeV ν_τs skimming from the water [11] in that special configuration. If ν_τs interact close to the Earth surface, the so-called Earth's skimming neutrinos [12–14] can produce tau leptons which can emerge from the Earth, decay and produce extended air showers. If the decay vertex of a τ lepton is close enough to the telescopes, it can be detected and distinguished from proton and nuclei induced showers through event parameter selection. Above PeV energies, the Earth becomes opaque to ν_e and ν_μ, while the ν_τ flux is regenerated through subsequent τ lepton decays to neutrinos. At high-energies, the ν_τ interacts in the Earth producing a τ lepton which in turn decays into a ν_τ with lower energy due to its short lifetime. The regeneration chain $\nu_\tau \rightarrow \tau \rightarrow \nu_\tau$, continues until the τ lepton reaches the detector. Observing on the "sea window" ([90-95]deg in Zenith angle and [-100 20] in Azimuth angle), MAGIC collected 30 hours of data in dark time and in good weather conditions at the horizon.

3 Monte Carlo simulations

To support this unconventional datataking, special Monte Carlo simulations of τ induced showers at different injection energies have been created. The software chain makes use of the ANIS (All Neutrino Interaction Simulation) code [15] to simulate signatures expected from neutrino-induced showers. CORSIKA [16] has been used to simulate the development of τ-induced showers and their Cherenkov light production. The decays of τ leptons have been simulated by PYTHIA [17]. The output of CORSIKA is then finally passed to a simulation of atmospheric extinction and MAGIC telescope response [18]. From the MonteCarlo studies it resulted that the main background is consisting in proton induced showers.

4 Signal-background discrimination

The details of the selection criteria and of the whole analysis performed on 30 hours of MAGIC data are reported in [20]. The main parameters characterizing the Cherenkov showers, called Hillas parameters [19] of the showers have been used to study the sample and to identify the selection criteria to be applied. In particular Fig. 2 of [20], left panel, shows the 2D distributions of measured events in the Hillas Length vs Size parameters plan and the corresponding cut. On this first sample of data, no events have survived the selection.

5 Improving selection criteria with machine learning techniques

In [20] a 90% C.L. upper limit on the tau-neutrino point source flux of 2.0×10^{-4}GeV cm^{-2} s^{-1} has been obtained: the result is not competitive with IceCube or Pierre Auger Observatory, due to the larger duty-cycle of those instruments (see Fig. 11 of [20]. However, MAGIC limits are almost

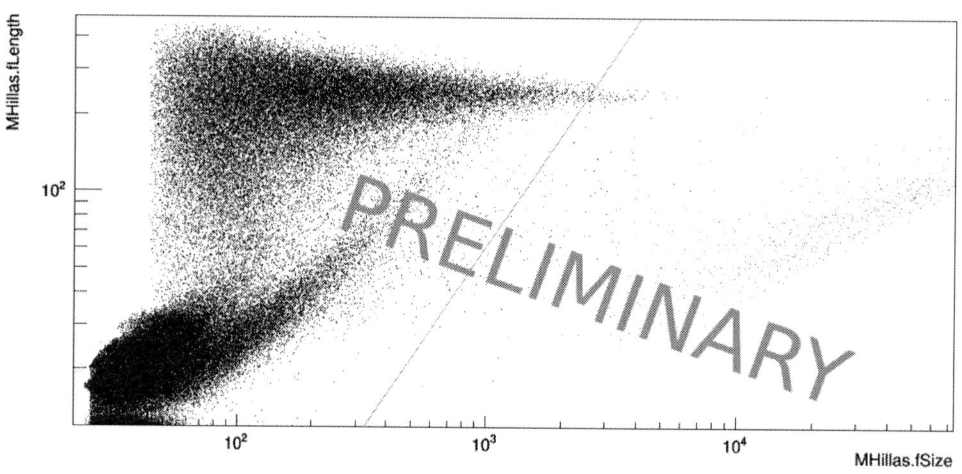

Figure 1. New selection cut (green line) on a larger sample of data (40 hours), obtained using a Fisher Discriminant analysis. Black points are the measured events (background + signal), orange the simulated signal. Survived events from the present cut will have to pass a further selection step which is shown in Fig.2.

background free, differently from IceCube detector, thus the tau neutrino sensitivity would increase linearly with the accumulated observation time. The next-generation Cherenkov telescopes, i.e. the Cherenkov Telescope Array, will have much larger FOV (in extended observation mode), and consequently a larger effective area. As recently shown in [21] the expected event rate will be at level of 1 event during 200 hours making CTA tau neutrino searches more promising. Thus, such additional data can be also useful when CTA prospects will be considered and quantitatively evaluated.

Further observations of the sea with MAGIC would help to improve and further develop the technique used in [20]. In particular a way of making the analysis more efficient and sensitive in prevision of more data to process is under development and includes the employment of machine learning techniques. Fisher Discriminant (see Fig. 1) and Genetic Algorithms are being tested to both observed and simulated data in order to create a more complex selection criteria based also on a further Hillas parameter, the Mean Time (Fig. 2).

The possibility to observe τ-induced showers tracing back to a neutrino from an astrophysical source are very low being the MAGIC sea window just 5 deg in Zenith angle, but still exciting enough to pursue the quest.

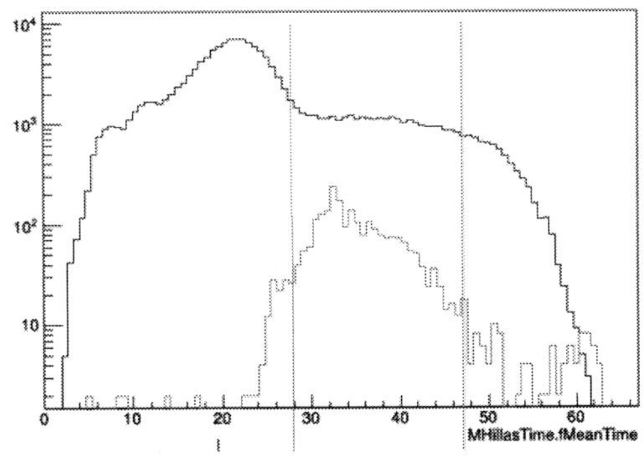

Figure 2. Distribution of the Hillas Mean time parameter (blue data, orange simulations). The green lines individuate identify the most efficient cut as found by the application of the Genetic algorithm.

References

[1] Aartsen M. G. et al., Phys. Rev. Lett. **113**, 101101 (2014)
[2] Margolis, S. H. Schramm, D. N. and Silberberg, R., Astrophys. J. **221**, 990 (1978)
[3] Stecker, F. W. Astrophys. J. **228**, 919 (1979)
[4] Biermann, P. L. Cosmic Gamma Rays, Neutrinos, and Related Astrophysics, in Proc. of the NATO Advanced Study Institute, Erice, Italy, pp. 21 (1989)
[5] Michalak, W., Wdowczyk, J. and Wolfendale, A. W. J. Phys. G **16**, 1917 (1990)
[6] The IceCube Coll., Science **361**, eaat1378 (2018)
[7] Ansoldi S., et al. ApJL, **863**, L10 (2018)
[8] Aartsen M. G. et al., Phys. Rev. Lett. **114**, 171102 (2015)
[9] Aartsen M. G. et al., ApJ, **809**, (1), 98 171102 (2015)
[10] Stachurska, J, et al., Neutrino 2018 - XXVIII International Conference on Neutrino Physics and Astrophysics, https://zenodo.org/record/1301122#.W-WBJhBRfCJ (2018)
[11] Gaug M. , et al., Proc. of I.C.R.C. 2007 (Merida) 1273 (2007)
[12] Fargion D., ApJ, **570**, 909 (2002)
[13] Feng, J.J., et al. Phys. Rev. Lett., **88**, 161102 (2002)
[14] Bertou, X., et al., Atropart. Phys. 17,183 (2002)
[15] Gazizow A. Z., and Kowalski M., Comput. Phys. Commun.,**172** 203 (2005)
[16] Heck D., et al., Report FZKA 6019 (1998)
[17] Sjöstrand, T., JHEP **0605**, 026 (2006)
[18] Zanin R. Proc. of 33rd I.C.R.C. (Rio de Janeiro) (2013)
[19] Hillas, M. A. Nucl. Phys. Proc. Suppl., **52B**, 29 (1997)
[20] Ahnen M. L., et al. , Astropart. Phys. **102**, 77 (2018)
[21] Góra D. , and Bernardini E., Astropart. Phys. **82**, 77 (2016)
[22] Góra D. et al., Proc of EPS-HEP (Venice) (2017), PoS(EPS-HEP2017)017, arXiv:1710.04165

EPJ Web of Conferences **209**, 01006 (2019) https://doi.org/10.1051/epjconf/201920901006
RICAP-18

Status and first results from the ARCA and ORCA lines of the KM3NeT experiment

Paolo Fermani[1,2,a] and Irene Di Palma[1,2,]

[1] *Sapienza Università di Roma, P.le Aldo Moro 5, 00185, Roma, Italy*
[2] *INFN - Sezione di Roma, P.le Aldo Moro 2, 00185, Roma, Italy*

Abstract. KM3NeT is a network of submarine Cherenkov neutrino telescopes under construction in two different sites in the Mediterranean Sea [1]. The detector at the Italian site, close to the Sicilian coast and named ARCA, will be devoted to the detection of high-energy astrophysical neutrinos coming from sources in the Universe, while the detector at the French site, in the Toulon bay and named ORCA, will exploit atmospheric neutrinos to determine the neutrino mass hierarchy. The telescopes are an array of flexible strings anchored to the sea floor and held close to vertical by submerged buoys. The strings are instrumented with digital optical modules hosted within pressure-resistant glass spheres, each housing 31 3" photomultipliers tubes and the readout electronics. The geometry of the detectors has been adapted to their physics goals. The first calibrations and results of ARCA and ORCA are presented.

Introduction

The KM3NeT experiment infrastructure is composed by two detectors: ARCA and ORCA. In ARCA, every string is 700 m high and has 18 multi-photomultiplier tubes (PMTs) digital optical modules (DOMs), one every 36 m, with the inter-string distance of about 90 m; while in ORCA the strings are 200 m high, with an inter-DOM spacing of only 9 m and an inter-line spacing of roughly 23 m. Each 3" PMT, thanks to its construction design, permits to take into account and optimise the directional information sensibility of the KM3NeT DOMs. The first two lines of ARCA have been deployed in December 2015 and May 2016 respectively, the first string of ORCA has been deployed in September 2017.

1 ARCA - Astroparticle Research with Cosmics in the Abyss

The design of the KM3NeT DOMs allows to select atmospheric muons over the environmental background looking for local coincidences between the PMTs [2]. The decay of Potassium 40 (^{40}K), dissolved in the sea water, in the vicinity of the DOM, causes the vast majority of coincident hits on 2 PMTs on the same DOM in a time window of 25 ns. For the ^{40}K individual events the distribution of the time difference (Δt) is expected to follow a Gaussian distribution, whose the mean is the difference between the transit times of the involved PMTs, the width is related to the PMTs transit time spreads

[a] e-mail: paolo.fermani@roma1.infn.it

 https://doi.org/10.1051/epjconf/201920901006

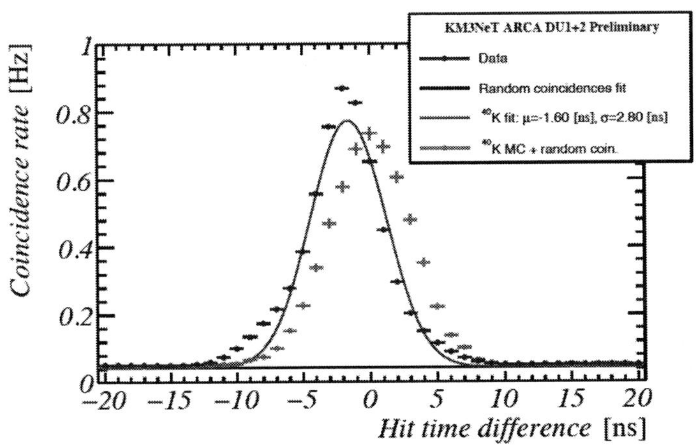

Figure 1: *Coincidence rate vs Δt for a pair of adjacent PMTs. Blue dots represent uncalibrated data. The red line is a Gaussian fit with mean equal to −1.6 ns and standard deviation equal to 2.8 ns. The green points show the expected hit time difference corrected with the addition of the background rate estimated from the tail of the data distribution (black solid line) [3].*

and the integral under the curve proportional to the product of the PMT efficiencies [3]. The values to use for calibration of these quantities are derived from the comparison between the data and the Monte Carlo (MC) simulations of the ^{40}K event distributions, as shown in figure 1.

The rates of the multifold coincidences (m) for the lowest and highest DOMs of the two strings of ARCA are shown in figure 2a together with the rates predicted by the MC simulations. No normalisation factor is applied to the MC simulations, showing an excellent absolute agreement between the data and the MC. At low m the signal from ^{40}K dominates the rates while the muon signature becomes dominant for $m \geq 8$.

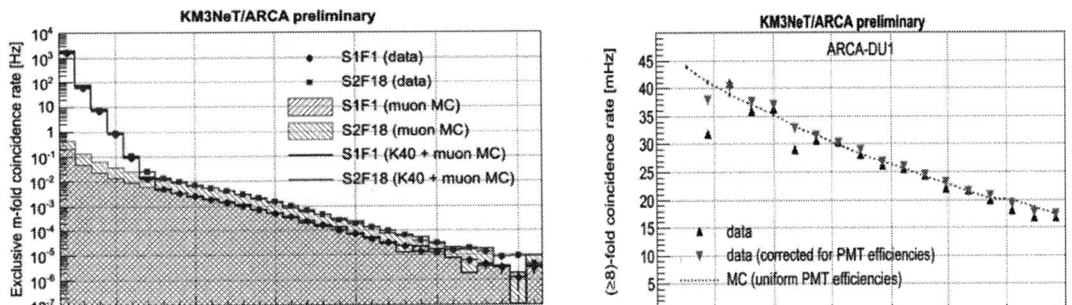

(a) Coincidence rate vs multiplicity.

(b) ≥8-fold coincidence rate vs sea depth.

Figure 2: ***(a)** Coincidence rate vs multiplicity. Two DOMs are shown: floor 1 (the lowest floor) of ARCA-DU1 and floor 18 (the highest floor) of ARCA-DU2. Data are compared to with the expectations from MC simulations of ^{40}K and atmospheric muons [4].*

***(b)** ≥8-fold coincidence rate vs depth below the sea surface for ARCA string 1. The statistical error bars are too small to be visible [4].*

The stability of the PMTs and the homogeneity of the detector performances have been verified through the study of the dependence of the multifold ($m \geq 8$) coincidence rate on the sea depth [5]. In figure 2b the red dashed line shows the coincidence rates as expected by MC simulations when uniform PMT efficiencies are considered neglecting the differences between individual PMTs. The black points show the uncorrected data and the green points show the data corrected according to the PMT efficiencies measured with in-situ calibration.

2 ORCA - Oscillation Research with Cosmics in the Abyss

The ORCA detector has a denser setup than the ARCA detector. This allows the detection of atmospheric neutrinos with an energy threshold $E_\nu \approx$ 5-10 GeV. Dedicated analysis cuts are applied to the reconstructed tracks to select a clean neutrino candidate sample. With the available livetime of 82 days, 13 upgoing neutrino candidate events are observed, in good agreement with the MC prediction of 8.3 ± 2.9 (stat.) atmospheric neutrinos plus 1 ± 1 (stat.) events most likely due to bad reconstructed tracks. The zenith angle distribution for the selected tracks in the ORCA detector is in good agreement with MC predictions as shown in figure 3. Here the MC atmospheric muons are represented with blue crosses, the MC atmospheric neutrinos with red crosses and the data with black crosses [6].

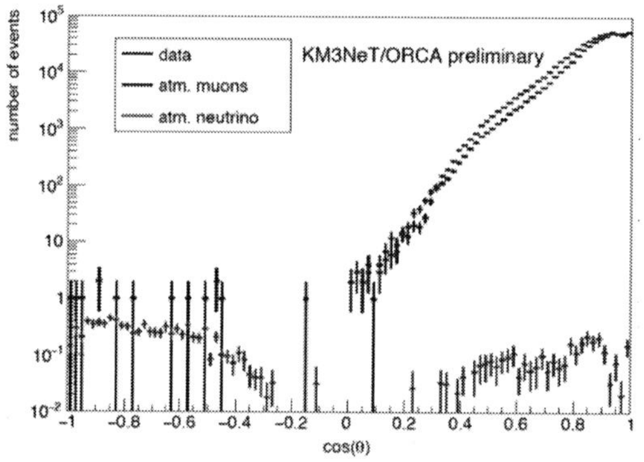

Figure 3: *Reconstructed zenith angle distribution for the selected tracks in the ORCA detector [6].*

Conclusions

These preliminary tests, both for ARCA and ORCA detectors, show the capabilities of the Cherenkov technique in sea water, verifiyng the KM3NeT technology and confirming the stability of the detector performances.

References

[1] S. Adrian-Martinez et al. (KM3NeT collaboration), J. Phys G **43**, 084001 (2016)

[2] S. Adrian-Martinez et al. (KM3NeT collaboration), Eur. Phys. J. C **76**, 54 (2016)

[3] K. Melis (on behalf of the KM3NeT collaboration), 35th ICRC, Busan, South Korea, PoS **301**, 1059 (2017)

[4] M. Jongen (on behalf of the KM3NeT collaboration), 35th ICRC, Busan, South Korea, PoS **301**, 1018 (2017)

[5] G. Ferrara, S. Biagi (on behalf of the KM3NeT collaboration), JPCS **1056**, 012021 (2018)

[6] I. Di Palma, J. Hofestädt, D. Zaborov (on behalf of the KM3NeT collaboration), Results from the first ARCA and ORCA detector units, Poster presented at 28th Neutrino conference, Heidelberg, Germany (2018)

EPJ Web of Conferences **209**, 01007 (2019) https://doi.org/10.1051/epjconf/201920901007
RICAP-18

Properties of Elementary Particle Fluxes in Primary Cosmic Rays Measured with the Alpha Magnetic Spectrometer on the International Space Station

Francesco Nozzoli on behalf of the AMS Collaboration[1,a]

[1] *INFN TIFPA, I-38123 Trento, Italy*

Abstract. Precision measurements by AMS of the fluxes of cosmic ray positrons, electrons, antiprotons, protons as well as their rations reveal several unexpected and intriguing features. The presented measurements extend the energy range of the previous observations with much increased precision. The new results show that the behavior of positron flux at around 300 GeV is consistent with a new source that produce equal amount of high energy electrons and positrons. In addition, in the absolute rigidity range 60–500 GV, the antiproton, proton, and positron fluxes are found to have nearly identical rigidity dependence and the electron flux exhibits different rigidity dependence.

1 Introduction

AMS-02 [1–14] is a state-of-the-art particle physics detector designed to operate as an external module on the International Space Station (ISS). It is studying the universe and its origin by searching for antimatter and dark matter, while performing precision measurements of Cosmic Ray (CR) composition and flux.

The high statistics of the measurements, along with the high precision of the experiment, allow to study the detailed variations with rigidity of the flux spectral indices and flux ratios, important in understanding the origin, acceleration and propagation of cosmic rays in our galaxy. In particular, experimental data on the cosmic ray antiprotons, $\bar{p}$, and positrons, e^+, are crucial for understanding the origin of antimatter in cosmos, providing an insight into new physics phenomena. While some of cosmic ray $\bar{p}$ and e^+ are produced by interactions of the cosmic ray nuclei with the interstellar gas, there could be a substantial contribution from annihilation of dark matter particles, acceleration/production by astrophysical objects, or evaporation of primordial black holes. The sensitivity of cosmic ray positrons to these new phenomena is complementary to the sensitivity of the measurements of antiprotons, whose production by nearby pulsars is expected to be negligible. Therefore AMS has now precisely measured the fluxes of all the elementary particles in the cosmos, this allows for new observations of their properties.

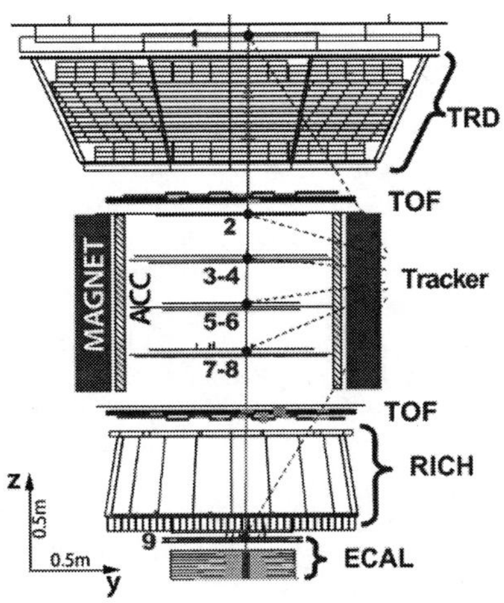

Figure 1. A 868 GeV positron event as measured by the AMS detector on the ISS in the (y-z) plane. Tracker planes 1-9 measure the particle charge, sign and momentum. The TRD identifies the particle as an electron/positron. The TOF measures the charge and ensures that the particle is downward-going. The RICH measures the charge and velocity. The ECAL independently identifies the particle as an electron/positron and measures its energy.

2 The AMS Detector

The layout of the AMS detector is shown in Fig.1. It consists of 9 planes of precision silicon Tracker; a Transition Radiation Detector, TRD; four planes of Time of Flight counters, TOF; a Magnet; an array of anti-coincidence counters, ACC, surrounding the inner Tracker; a Ring Imaging Čerenkov detector, RICH; and an Electromagnetic Calorimeter, ECAL. The figure also shows a high energy positron of 868 GeV recorded by AMS. Together, the tracker and the magnet measure charged cosmic rays with momentum p, charge Z and rigidity R = pc/Ze. The tracker has nine layers, the first at the top of the detector,

[a]e-mail: Francesco.Nozzoli@tifpa.infn.it

the second just above the magnet, six within the bore of the magnet, and the last just above the ECAL. Each layer contains double-sided silicon microstrip detectors that independently measure the x and y coordinates. The tracker accurately determines the particle trajectory by multiple measurements of the coordinates with a resolution in each layer of 10 μm in the bending (y) direction. The maximum detectable rigidity, MDR, is 2 TV for |Z| = 1, 3.2 TV for helium and 3.7 TV for carbon nuclei over the 3 m lever arm. Signal amplitude from each layer of the tracker also provides an independent measurement of charge |Z|. The overall charge resolution of the tracker, $\Delta Z/Z$, is 5% for |Z| = 1, 3.5% for helium and 2% for carbon nuclei. The TOF counters measure the velocity β = v/c and |Z| of cosmic rays. Two TOF planes are located above and two planes are located below the magnet. For |Z| = 1 particles, the average time resolution of each counter has been measured to be 160 ps and the overall velocity resolution to be $\Delta\beta/\beta^2$ = 4%. The time resolution improves for higher charges to reach 50 ps for oxygen nuclei with the corresponding improvement in the velocity resolution of $\Delta\beta/\beta^2$ = 1%. This discriminates between downward- and upward-going particles. The coincidence of signals from the four TOF planes provides a basis for charged particle trigger, whereas the coincidence of 3 out of the 4 TOF layers provides an unbiased trigger. The unbiased trigger, prescaled to 1%, is used to measure the trigger efficiency from the data. To distinguish antiprotons and positrons from protons and electrons which are reconstructed in the tracker with wrong rigidity sign due to the finite tracker resolution or due to interactions with the detector materials, a charge confusion estimator, Λ_{CC}, is defined using the boosted decision tree technique. The estimator combines information from the tracker such as the track χ^2/d.o.f., rigidities reconstructed with different combination of tracker layers, the number of hits in the vicinity of the track, and the charge measurements in the TOF and the tracker. With this method, antiprotons/positrons have Λ_{CC} ~+1 whereas charge confusion protons/electrons have Λ_{CC} ~-1. This ensures efficient separation of the signal from the charge confusion events. The measured charge confusion probability is less than 1% up to 300 GV and less than 10% up to TV. The TRD uses transition radiation to distinguish between $\bar{p}$(p) and $e^-(e^+)$ and dE/dx to independently identify nuclei. It consists of 5248 proportional tubes of 6 mm diameter with a maximum length of 2 m assembled in 16-tube modules. The 328 modules are mounted in 20 layers. There are 12 layers of proportional tubes along the y axis located in the middle of the TRD and, along the x axis, four layers located on top and four on bottom. To differentiate between $\bar{p}$(p) and $e^-(e^+)$ in the TRD, signals from the 20 layers of proportional tubes are combined into a TRD estimator, Λ_{TRD}, formed from the ratio of the log–likelihood probability of the $e^\pm$ hypothesis to that of the $\bar{p}$(p) hypothesis in each layer. Antiprotons and protons have Λ_{TRD} ~ 1 whereas electrons and positrons have Λ_{TRD} ~ 0.5. This provides efficient separation of $\bar{p}$(p) and $e^-(e^+)$. The TRD $\bar{p}$(p) rejection at 90% selection efficiency for $e^-(e^+)$ is larger than 700 up to 300 GV and larger than 10 up to TV. The ring imaging Čerenkov detector measures velocity and |Z|. It consists of two radiators, an expansion volume and a photodetection plane. The dielectric radiators induce the emission of a cone of Čerenkov photons when traversed by charged particles with a velocity greater than the velocity of light in the material. The central radiator is formed by 16 sodium fluoride, NaF, tiles, each 85×85×5 mm^3, with a refractive index n = 1.33. These are surrounded by 92 tiles, each 115×115×25 mm^3, of silica aerogel with a refractive index n = 1.05. This allows the detection of particles with velocities $\beta > 0.75$ with the NaF radiator and $\beta > 0.953$ with the aerogel radiator. The expansion volume has a distance along z of 470 mm and is surrounded by a high reflectivity mirror to increase detection efficiency. The photodetection plane is an array of 10880 photosensors with an effective spatial granularity of 8.5×8.5 mm^2. The sum of the signal amplitudes is proportional to Z^2. The opening angle of the Čerenkov radiation cone is a measure of the velocity of the incoming charged particle. Typical velocity resolution is $\Delta\beta/\beta^2$ = 0.1% for |Z|=1. As an example, Fig.2 shows clear separation of |Z|=1 events measured by AMS in the (β_{RICH} - sign(R)Λ_{TRD}) plane for the absolute rigidity range 5.4-6.5 GV. The contributions of primary $\bar{p}$, p, e^+, e^- and secondary π^+, π^- from interactions within the detector, are clearly seen.

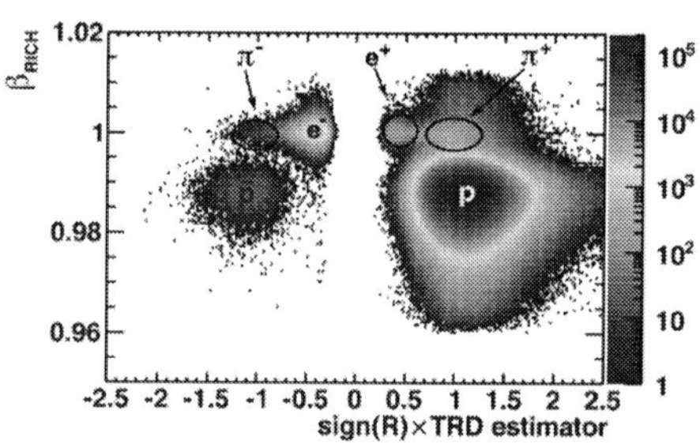

Figure 2. Example of negative rigidity and positive rigidity data samples in the (β_{RICH} - sign(R)Λ_{TRD}) plane for the absolute rigidity range 5.4–6.5 GV. The contributions of $\bar{p}$, p, e^+, e^-, π^+, π^- are clearly seen.

Finally, the three-dimensional imaging capability of the 17 radiation length ECAL allows for an accurate measurement of the $e^\pm$ energy E and shower shape. It consists of a multilayer sandwich of 98 lead foils and ~50000 scintillating fibers with an active area of 648×648 mm^2 and a thickness of 166.5 mm. The calorimeter is composed of 9 superlayers, with the fibers running in one direction only in each superlayer. The 3-D imaging capability of the detector is obtained by stacking alternate superlayers with fibers parallel to the x- and y-axes (5 and 4 superlayers, respectively). The energy resolution has been measured to be $\sigma(E)/E = \sqrt{(0.104)^2/E + (0.014)^2}$ (E in GeV). The uncertainty of the absolute $e^\pm$ energy scale has been verified to be 2% in the range 10-290 GeV. Below 10 GeV it increases to 5% at 0.5 GeV and above 290 GeV to 4% at 700 GeV. To cleanly separate protons from electrons and positrons, an ECAL estimator, is constructed using the 3-D shower shape in the ECAL. The proton rejection power of the ECAL estimator when combined with the energy-momentum matching requirement E/p > 0.75 reaches ~10000. The entire detector has been extensively calibrated in a test beam at CERN with e^+ and e^- from 10 to 290 GeV/c, with p at 180 and 400 GeV/c, and with π from 10 to 180 GeV/c, which produce transition radiation equivalent to p up to 1.2 TeV/c. Measurements with 18 different beam energies/particles scanning 2000 detector positions were performed. A Monte Carlo program based on the GEANT 4.10.1 package is used to simulate physics processes and signals in the detector.

3 Elementary Particles in Cosmic Rays

AMS has collected 1.07×10^{11} cosmic rays in 6.5 years of operation. A total of 1.9 million positrons are identified in the energy range from 0.5 GeV to 1 TeV and this allows for a detailed study of the properties of the positron flux.

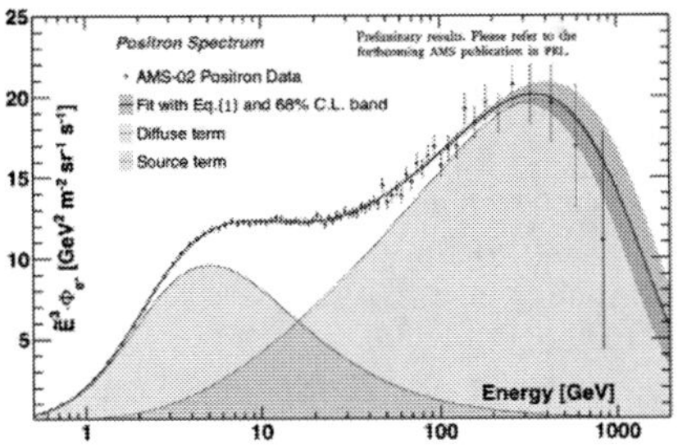

Figure 3. The fit of Eq.1 (green line) to the positron flux in the energy range [1 - 1000] GeV together with the 68% CL interval (green band). The red dots represent the measured average positron flux values over the measurement period scaled by E^3. The source term contribution is represented by the magenta area, and the diffuse term contribution by the grey area

As seen in Fig. 3, the positron spectrum, scaled by E^3, exhibits a complex behavior. It decreases rapidly with energy below ~7 GeV. In the energy range [7.10 - 27.25] GeV it is energy independent, with spectral index, γ = -2.986 ± 0.005, then it increases, with γ = -2.716±0.037, in the energy range [55.58 - 148.81] GeV. Above 148.81 GeV the spectral index experiences significant decrease reaching γ = -3.35 ± 0.32 in the highest energy range [290 - 1000] GeV. This behavior is consistent with a new source of high energy positrons with a characteristic cutoff energy, whether of dark matter or astrophysical origin; it is not consistent with exclusive secondary production of positrons in collisions of cosmic rays.

Moreover it is possible to describe the positron spectrum with a minimal model. In this model the positron flux Φ_{e^+} is parameterized as the sum of a diffuse term and a new source term:

$$\Phi_{e^+}(E) = \frac{E^2}{\hat{E}^2}\left[C_{e^+}\left(\hat{E}/E_1\right)^{\gamma_{e^+}} + C_s\left(\hat{E}/E_2\right)^{\gamma_s} e^{(-\hat{E}/E_s)}\right] \quad (1)$$

The diffuse term is a power law function, which is characterized by a normalization factor C_{e^+} and a spectral index γ_{e^+}. The source term is a power law function with an exponential cutoff, which is characterized by a normalization factor C_s, a spectral index γ_s and a cutoff energy E_s. In order to account for solar modulation effects the force-field approximation is used, with the energy of particles in the interstellar space $\hat{E} = E+\psi_{e^+}$ and the effective solar potential ψ_{e^+} accounting for the effects of the solar magnetic field. The constant E_1 is chosen to be 7.0 GeV to minimize the correlation between parameters C_{e^+} and γ_{e^+}. The constant E_2 is chosen to be 60.0 GeV to minimize correlation between the parameters C_s and γ_s. A fit to the data in the energy range [1 - 1000] GeV accounts for a correlated systematic error due to the acceptance uncertainty. This correlated error is estimated to be 1% in the fit energy range, it affects only normalization parameters C_{e^+} and C_s and is quoted as a systematic error on these two parameters. The fit yields χ^2/d.o.f. = 50/65 and ψ_{e^+} = 1.10 ± 0.04 GeV for the solar modulation potential; C_{e^+} = (6.52 ± 0.13 ± $0.07(syst)) \times 10^{-2}[m^2 sr\ s\ GeV]^{-1}$, $\gamma_{e^+} = -4.08 \pm 0.07$ for the diffuse term; and C_s = (6.80 ± 0.14 ± $0.07(syst)) \times 10^{-5}[m^2 sr\ s\ GeV]^{-1}$, $\gamma_s = -2.581 \pm 0.055$, and the inverse cutoff energy $1/E_s = 1.22\pm0.35$ TeV^{-1} for the source term. The last parameter corresponds to $E_s = 820^{+330}_{-180}$ GeV. The result of the fit is superimposed in Fig. 3. As seen, the source term dominates the positron spectrum at energies above 20 GeV, the region where solar modulation effects are negligible. It is the contribution of the source term that leads to the observed increase of the positron spectrum in the energy range from 27.25 to 290 GeV. The decrease of the positron spectrum above 290 GeV is very well described by the sharp exponential cutoff of the source term. Note that the choice of the E_1 and E_2 constants does not affect values and errors of the fit parameters ψ_{e^+}, γ_{e^+}, γ_s and $1/E_s$, i.e. neither the shapes nor the magnitudes of the diffuse and the source terms shown in Fig. 3 depend on the choice of E_1 and E_2. The significance of the finite energy cutoff is established at a level of 3.5σ from the errors on the $1/E_s$ fit parameter and this result is based on the positron data alone.

Another AMS measurement that is very interesting for dark matter indirect search is the antiproton ratio $\bar{p}/p$.

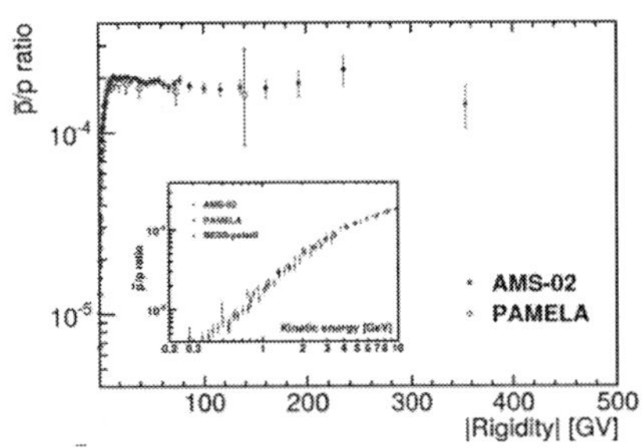

Figure 4. The $\bar{p}/p$ flux ratio as measured by AMS along with previous measurments.

Overall, in the absolute rigidity range 1-450 GV 3.49×10^5 antiproton events are selected. It is important to note that in the high rigidity range above 100 GV AMS has selected 2200 $\bar{p}$. This can be compared with 3 $\bar{p}$ detected before AMS in the same rigidity range.

The $\bar{p}/p$ flux ratio as measured by AMS along with previous measurments is presented in Fig. 4. Above ~60 GV the ratio is found to be independent on rigidity. This observation caused a major revision of modeling cosmic ray antiprotons produced in ordinary collisions of protons with interstellar media (see e.g. [15–17]). The revised conservative models still predicts a 30-40% drop of the $\bar{p}/p$ ratio from the maximum, that is not supported by the AMS data, providing room for the presence of exotic/new antiproton source within the large existing theoretical uncertainties of secondary $\bar{p}$ production/propagation models.

It is interesting to compare behavior of the spectra of all elementary particles among themselves. Traditionally, electrons and protons are assumed to be primary cosmic rays, i.e. particles produced directly at sources of cosmic rays like exploding supernovae. On the contrary, positrons and antiprotons are assumed to be secondary cosmic rays, i.e. coming from the interaction of primary cosmic rays with the interstellar media. In addition, electrons and positrons have much smaller mass than protons and antiprotons so they lose much more energy in the galactic magnetic fields. This is illustrated in Fig. 5, where the behavior of the electron spectrum is compared with that of the proton, prositron and antiproton spectrum. As expected, in the rigidity region free of solar modulation effects, above 20 GeV, the proton spectrum is much harder that the electron spectrum. However, the behavior of antiprotons (assumed to be secondaries) and protons (assumed to be primaries) at high rigidities are very similar, while the behavior of electrons (i.e. primaries) and positrons (secondaries) are very different.

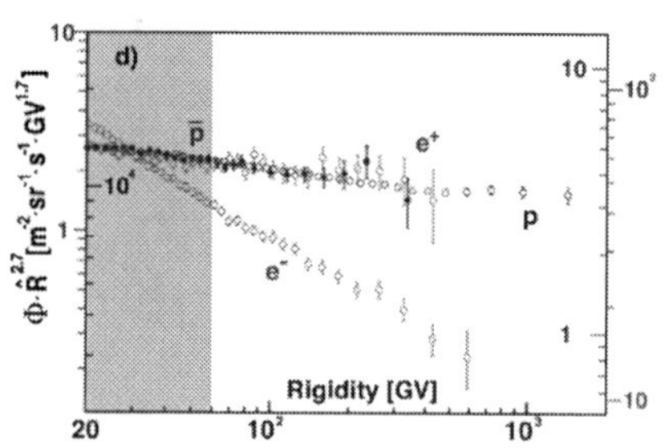

Figure 5. AMS measurements of the fluxes of elementary particles. The positron, antiproton and proton fluxes show the very same functional behavior in the absolute rigidity range 60–500 GV, whereas electrons exhibit distinctly different dependence.

Most surprising is that above 60 GeV, positrons, protons and antiprotons display an identical energy dependence whereas electrons exhibit a totally different energy dependence. A compelling explanation for this AMS observation within traditional models of galactic cosmic rays is yet to be found.

4 Conclusions

In six years on the ISS, AMS has recorded more than 100 billion cosmic ray events. The latest AMS measurements of the positron spectrum and positron fraction, the antiproton/proton ratio, the behavior of the fluxes of electrons, positrons and protons provide precise and unexpected information on the production, acceleration and propagation of cosmic rays. The accuracy and characteristics of the data, simultaneously from many different types of cosmic rays require the development of a comprehensive model of cosmic rays. As a magnetic spectrometer studying cosmic rays, AMS is unique in its precision and energy reach. For the foreseeable future this is the only magnetic spectrometer in space to perform precision measurements and to explore the unknown with high expectations for exciting discoveries. Published data of AMS fluxes and flux ratios can be downloaded from the ASI/SSDC cosmic ray database [18].

References

[1] M. Aguilar et al. (AMS), Phys. Rev. Lett. **110**, 141102 (2013)
[2] L. Accardo et al. (AMS), Phys. Rev. Lett. **113**, 121101 (2014)
[3] M. Aguilar et al. (AMS), Phys. Rev. Lett. **113**, 121102 (2014)
[4] M. Aguilar et al. (AMS), Phys. Rev. Lett. **113**, 221102 (2014)
[5] M. Aguilar et al. (AMS), Phys. Rev. Lett. **114**, 171103 (2015)
[6] M. Aguilar et al. (AMS), Phys. Rev. Lett. **115**, 211101 (2015)
[7] M. Aguilar et al. (AMS), Phys. Rev. Lett. **117**, 091103 (2016)
[8] M. Aguilar et al. (AMS), Phys. Rev. Lett. **117**, 231102 (2016)
[9] M. Aguilar et al. (AMS), Phys. Rev. Lett. **119**, 251101 (2017)
[10] M. Aguilar et al. (AMS), Phys. Rev. Lett. **120**, 021101 (2018)
[11] M. Aguilar et al. (AMS), Phys. Rev. Lett. **121**, 051103 (2018)
[12] M. Aguilar et al. (AMS), Phys. Rev. Lett. **121**, 051101 (2018)
[13] M. Aguilar et al. (AMS), Phys. Rev. Lett. **121**, 051102 (2018)
[14] M. Aguilar et al. (AMS), Phys. Rev. Lett. **submitted** (2018)
[15] N. Tomassetti, Phys. Rev. D **92**, 081301 (2015)
[16] J. Feng, N. Tomassetti, A. Oliva, Phys. Rev. D **94**, 123007 (2016)
[17] N. Tomassetti, J. Feng, A. Oliva, Research Notes of the AAS **1**, 35 (2017)
[18] (ASI/SSDC cosmic ray database https://tools.ssdc.asi.it/CosmicRays/)

Status of the AugerPrime upgrade of the Pierre Auger Observatory

Niraj Dhital[1,a] for the Pierre Auger Collaboration[b,c]

[1] *Institute of Nuclear Physics PAN, Krakow, Poland*

Abstract. The Pierre Auger Observatory has been very successful in determining many aspects of the highest-energy cosmic rays including, among others, the flux suppression at energies above 4×10^{19} eV, stringent upper limits on photon and neutrino fluxes at ultra-high energies and an unexpected evolution of the mass composition with energy. We expect an extension of the frontiers of our knowledge on these aspects from a major upgrade of the Observatory. The upgrade, known as AugerPrime, will include an addition of a 4 m^2 Surface Scintillator Detector atop each water-Cherenkov station of the Surface array. The new detectors will provide us with an unprecedented opportunity to perform a complementary measurement of the shower particles and thus determine the primary mass composition with good accuracy on an event-by-event basis. AugerPrime will also include an upgrade of electronics, installation of the AMIGA Underground Muon Detector and a change of observation mode of the Fluorescence Detector, which will increase its current duty cycle by about 50%. Current status of the upgrade with the main focus on the Surface Scintillator Detectors will be presented, following a brief description of the physics motivation for the upgrade.

1 Introduction

Measurements of the highest energy cosmic rays made by the Pierre Auger Observatory[1] over the past decade have helped us understand many aspects of ultra-high energy cosmic rays (UHECRs). Nonetheless, a number of important questions on UHECRs remain to be addressed, which require additional measurements from the Observatory after a major upgrade. The upgrade called AugerPrime[2] will allow the operation of the Observatory until the end of 2024 with new types of detectors as well as an enhancement of already existing ones. The new measurements from the upgraded Observatory will provide additional information on extensive air showers (EASs) which are of crucial importance to resolve these important concerns. The main objective of the upgrade is to elucidate the mass composition and the origin of the flux suppression at the highest energies. The upgrade will also allow us to search for a flux contribution of protons up to the highest energies, which in turn will open up the prospects for more accurate particle astronomy. Also, fundamental particle physics at energies beyond those accessible at man-made accelerators, especially EASs and hadronic interactions will be better understood. This improved physics understanding will help to resolve the discrepancy between the number of muons observed in data and that obtained from simulations.

[a] e-mail: niraj.dhital@ifj.edu.pl

[b] Av. San Martín Norte 304 (5613) Malargüe, Prov. de Mendoza, Argentina

[c] Full author list and acknowledgments at: http://auger.org/archive/authors_2018_09.html

AugerPrime comprises four main improvements: 1) addition of the Surface Scintillator Detector (SSD) atop each water-Cherenkov station of the Surface Detector (SD) array, 2) Surface Detector Electronics Upgrade, 3) extension of the operation of the Fluorescence Detector (FD) and 4) installation of Underground Muon Detector.

2 The Surface Scintillator Detector

An SSD unit consists of 48 bars of 160 cm $\times$ 5 cm $\times$ 1 cm plastic scintillators grouped into two modules, enclosed in an aluminium casing [3]. Each bar has two extruded holes for insertion of wavelength shifting (WLS) fibre. The WLS fibres, which collect the scintillation light produced by the shower particles passing through the bars, are grouped into a bundle before being glued to a photomultiplier tube (PMT). The assembled SSD is installed on the top of a WCD. A layout of an

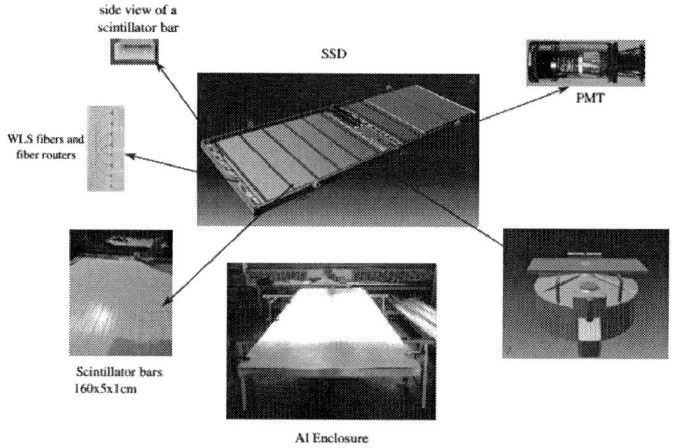

Figure 1: Layout of an SSD unit.

SSD unit and its placement above a WCD detector is shown in Fig. 1.

The installation of SSDs atop each water-Cherenkov detector (WCD) in the SD array will enable an additional measurement of particles in the EASs independent of the measurement from WCDs. An additional detector type (SSD) is a sensible approach for a better investigation of EASs since using scintillators is not only a robust and well-understood technique of particle detection but also provides measurements of shower particles which are complementary to the measurements performed using the water-Cherenkov technique and allows a good measurement of muons. Thanks to the different

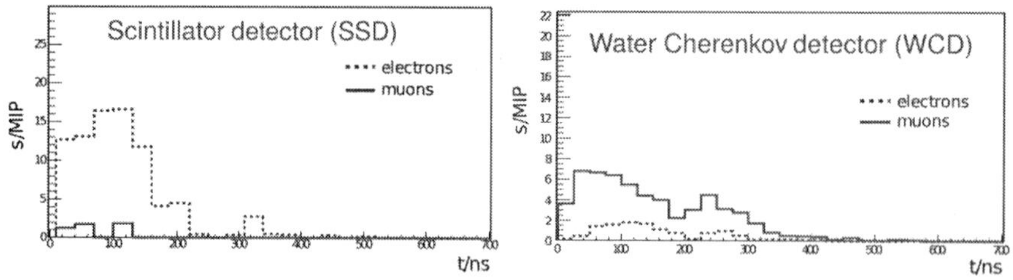

Figure 2: Detector response of electrons and muons for SSD (left panel) and WCD (right panel).

responses of SSDs and WCDs to the electrons and the muons in EASs, as can be seen in Fig. 2, the upgraded observatory will have much better sensitivity to the UHECR primary mass composition, even on an event-by-event basis.

3 SD Electronics Upgrade and additional small PMT

The upgraded electronics will process both the WCD and the SSD signals. It will increase the data quality by providing a faster sampling of ADC traces at a rate of 120 MHz compared to the current rate of 40 MHz. The faster sampling rate is better suited for counting muons. Also, a better absolute timing accuracy will be obtained from new GPS receivers. In addition, more powerful local station processor and FPGA will enhance the local trigger and processing capabilities. The new electronics will provide a facility for interfacing not only the SSD but also any other additional detector in future. Backwards compatibility with the current data-set will be maintained and the deployment have minimal impact on the continuous data taking of the SD.

Each WCD will have an additional 4th small PMT with an active area of 1% of the currently used standard PMT. This will greatly improve the dynamic range of WCDs. Consequently, there will be less than 2% saturated events at the highest energies and the determination of particle densities down to ~ 300 m from the shower core will be possible.

4 The Underground Muon Detector

The Underground Muon Detector (UMD) will consist of 61 AMIGA muon detectors[4], each with an effective area of 30 m^2, on a 750 m grid in the infill array over a total area of 23.5 km^2. The UMD will provide direct muon measurement of a sub-sample of showers detected by the SSDs and WCDs and serves for fine-tuning of different hadronic models and verification of the methods used to extract shower muon content using the SSDs and WCDs.

5 Extension of Duty Cycle of the FD Observations

The FD provides excellent information about EASs (model-independent energy reconstruction and direct measurement of the longitudinal development profiles). However, it is limited mainly by its duty cycle which is around 15%. The main goal of the extension is to increase the exposure for cosmic ray events above 10^{19}eV by extending the FD measurement into hours with a larger illuminated fraction of the moon when the night sky background is high. This extension will increase the duty cycle to ~ 23%. The extension is achievable by reducing the PMT gain lowering the supplied high-voltage to avoid an excessively high anode current. This procedure has been successfully tested in one telescope for a few nights.

6 Status of the Upgrade

The first twelve AugerPrime detectors were deployed at the Engineering Array (EA) site and are operational since October 2016 [5]. Nine of the EA stations are located in the standard SD array (1500 m spacing between the adjacent SD stations), while the remaining three are located in the infill region. The upgraded stations with WCDs and SSDs operate with good stability and signals from EA stations are as expected. SSDs in the EA are currently triggered by the corresponding WCDs in each station. The co-located SSD and WCD pairs sample the EASs essentially at the same positions. The correlation between the signals measured by the SSD and the corresponding WCD for same showers is shown in Fig. 3. Also, in Fig. 4, an example event triggering both the upgraded stations in the EA and WCDs in the regular 1500 m SD array is shown. The lateral distribution function (LDF) for the event is obtained by performing a fit using the regular array, while the signals measured by the upgraded stations are shown alongside. Signals measured by the WCDs of the upgraded stations are in a good agreement with the fitted LDF.

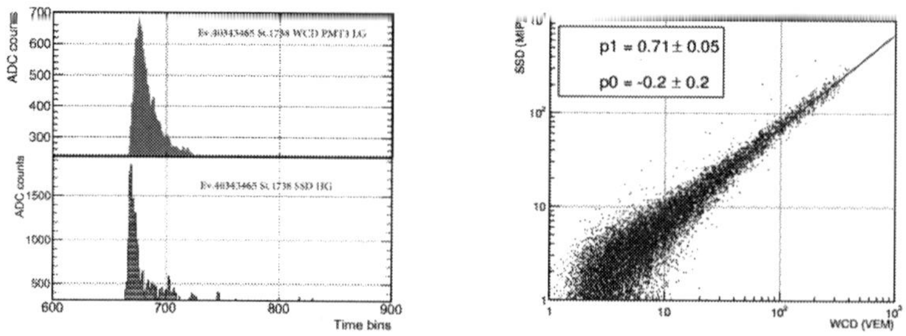

Figure 3: Left panel: WCD and SSD signals in the same station for an event. Right panel: Correlation between sizes of the SSD and WCD signals.

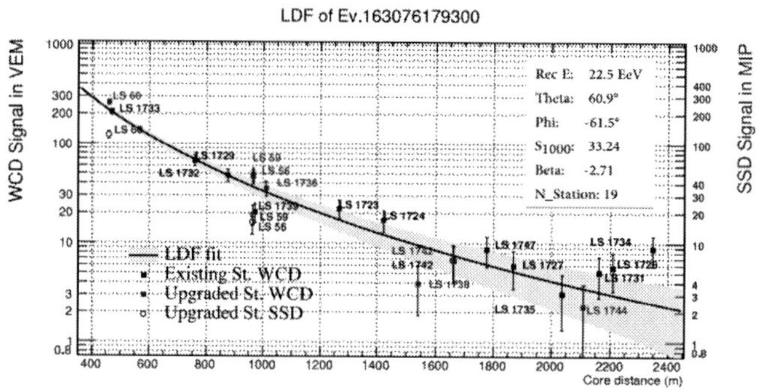

Figure 4: Signals in upgraded stations from an event compared with its LDF reconstructed with regular 1500 m array.

7 Summary

Definitive answers to some of the pressing questions on UHECRs are expected with the data from the upgraded Observatory. Its operation until 2024 will provide us not only better event statistics but also additional information on air showers. Measurement of air showers with the upgraded Observatory will provide high-quality data helpful to extend the current frontiers of our knowledge on UHECRs. The improved knowledge on UHECRs will allow reanalysis of the current data for more accurate energy evaluation and mass-composition studies. The AugerPrime EA has been operational since October 2016. The upgraded stations with WCDs and SSDs operate with good stability and signals from EA stations are as expected. SSDs are being assembled and tested at several sites, and the deployment has just begun.

References

[1] A. Aab et al. (Pierre Auger Collaboration), Nucl. Instrum. Meth. **A798**, 172 (2015), 1502.01323
[2] A. Aab et al. (Pierre Auger Collaboration) (2016), 1604.03637
[3] R. Šmída for the Pierre Auger Collaboration, PoS **ICRC2017**, 390 (2018)
[4] A. Aab et al. (Pierre Auger Collaboration), Journal of Instrumentation **11**, P02012 (2016)
[5] Z. Zong for the Pierre Auger Collaboration, PoS **ICRC2017**, 449 (2018)

Core-Collapse Supernova neutrino detection prospects with the KM3NeT neutrino telescopes.

Marta Colomer Molla[1,2,a] and Massimiliano Lincetto[3,b]

[1] *APC, Univ Paris Diderot, CNRS/IN2P3, CEA/Irfu, Obs de Paris, Sorbonne Paris Cité, France*
[2] *IFIC—Instituto de Física Corpuscular (CSIC—Universitat de València), Paterna, Valencia, Spain*
[3] *Aix Marseille Univ, CNRS/IN2P3, CPPM, Marseille, France*

Abstract. Core Collapse Supernovae (CCSN) are explosive phenomena that may occur at the end of the life of massive stars, releasing over 99% of the energy through neutrino emission with energies on the 10 MeV scale. While the explosion mechanism is not fully understood, neutrinos are believed to play an important role. The only detection as of today are the 24 neutrinos from supernova SN1987A. The observation of the next Galactic CCSN will lead to important breakthroughs across the fields of astrophysics, nuclear and particle physics.
For a Galactic CCSN, the KM3NeT ORCA and ARCA detectors in the Mediterranean Sea will observe a significant number of neutrinos via the detection of Cherenkov light, mostly induced by Inverse Beta Decay (IBD) interactions in sea water. The detection of coincident photons by the 31 photomultipliers of the KM3NeT digital optical modules (DOMs) allows to separate the signal from the optical background sources.
The KM3NeT detection sensitivity to a Galactic CCSN and the potential to resolve the neutrino light-curve have been estimated exploiting detailed Monte-Carlo simulations. Specific criteria are proposed for the online triggering and the participation in the SNEWS network.

1 Introduction

The KM3NeT experiment consists of two Cherenkov detectors under construction in the Mediterranean Sea, ORCA and ARCA, composed of Digital Optical Modules (DOMs) each housing 31 3-inch photomultipliers (PMTs) [1]. Groups of 18 DOMs are arranged into flexible vertical lines and an array of 115 lines constitutes a building block. ORCA, off the French coast, will consist of one block with 9 m vertical separation between DOMs and ARCA of two blocks with 36 m vertical spacing between DOMs, offshore Sicily. Together, they constitute an instrumented volume at the km^3 scale [2].

Although KM3NeT detectors are mainly designed for high-energy neutrino detection, the 10 MeV-scale neutrino signal from a CCSN can be identified as a simultaneous increase of the counting rate in the photomultipliers of the detector. The main interaction modes of these MeV neutrinos in water are Inverse Beta Decay, Elastic Scattering (ES) and interactions of neutrinos with oxygen atoms.

[a] e-mail: mcolomer@apc.in2p3.fr
[b] e-mail: lincetto@cppm.in2p3.fr

The outgoing electron/positron produced in these interactions induces Cherenkov light that can be detected by the PMTs.

2 Flux models and simulations

A state-of-the-art 3D simulation of the accretion phase of a 27 M⊙ and an 11 M⊙ CCSN provided by the Garching group have been considered this work [3, 4].

The time dependent CCSN neutrino spectrum is a quasi-thermal distribution, which depends on the average neutrino energy $\tilde{E}_\nu$, the neutrino luminosity $L(t)^\nu_{SN}$ and the spectral pinching shape parameter, α:

$$\frac{d\Phi^\nu}{dE_\nu}(E_\nu, t) = \frac{L(t)^\nu_{SN}}{4\pi d^2} \times f(E_\nu, \tilde{E}_\nu(t), \alpha(t)), \tag{1}$$

$$f(E_\nu, \tilde{E}_\nu(t), \alpha(t)) = \frac{E_\nu^\alpha}{\Gamma(\alpha+1)}(\frac{\alpha+1}{\tilde{E}_\nu})^{(\alpha+1)} \exp[-\frac{E_\nu(\alpha+1)}{\tilde{E}_\nu}], \tag{2}$$

The simulation output has been used to compute the CCSN neutrino interaction rate by multiplying the flux with the latest cross section estimates in [5, 6] as follows:

$$\frac{dR_{int,\kappa}}{dE} = N_\kappa \cdot \sigma_\kappa(E) \cdot \frac{d\Phi}{dE} \quad [\mathrm{s^{-1}MeV^{-1}}] \qquad \kappa \in \{\mathrm{p, e^-, ^{16}O}\} \tag{3}$$

A full MC simulation of CCSN neutrinos in KM3NeT has been produced and used to estimate the expected detection rate in the detector for a CCSN at an arbitrary distance, considering background rates as observed on the first ORCA and ARCA detection lines in the sea.

^{40}K decays in sea water are one of the main background sources and they can be distinguished from a CCSN signal using coincidences between the different PMTs of a given DOM. The number of PMTs in a DOM detecting a photon within a time window of 10 ns is called *multiplicity* (M) and is used to reject the ^{40}K background, dominating at low multiplicities. Dedicated simulations have shown that selecting events in the multiplicity range $6 \leq \mathrm{M} \leq 10$ optimizes the signal-to-noise ratio.

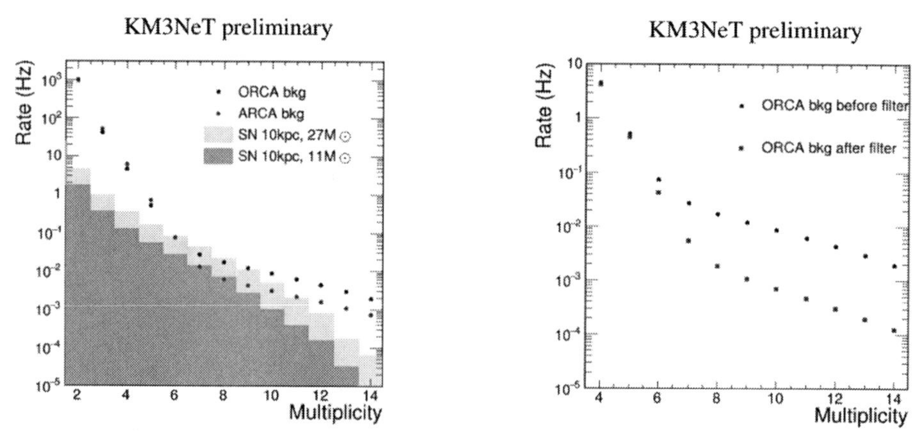

Figure 1. On the left, DOM background rates as a function of the multiplicity for ORCA (black squares) and ARCA (blue squares) from data and expectations for CCSN signal at 10 kpc for the 27 M⊙ (yellow filled) and the 11 M⊙ (green filled) progenitors. On the right, ORCA rates before (black dots) and after the filter (red squares).

Atmospheric muons are another prominent background producing Cherenkov tracks over long distances with high multiplicity events on multiple DOMs. Suppression of M≥4 correlated coincidences between two or more DOMs on a few μs time scale has been proven to be an effective strategy to reject the muon background. A muon filter has been developed and optimized for ORCA.

On Fig. 1 (left), it can be seen the multiplicity rates per KM3NeT DOM for signal simulations and background for both KM3NeT detectors, ARCA and ORCA. The effect of the muon filter on the ORCA background rates above M=4 is shown on the right plot.

3 Significance results

The number of signal and background events when applying the multiplicity cut and the muon filtering are used to compute the detection significance as a function of the distance to the source.

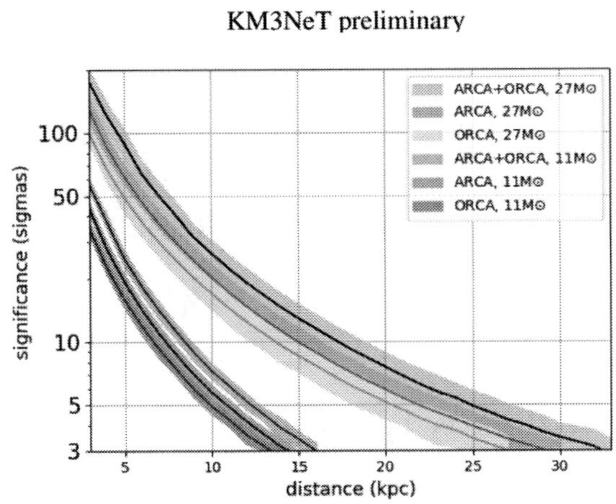

Figure 2. KM3NeT significance to a CCSN neutrino signal as a function of the distance to the source.

The results for ORCA, ARCA and the combination of both detectors are shown in Fig. 2 for the two stellar progenitors considered in this study. The more relevant systematics, coming mostly from the uncertainties on the models, and the statistical uncertainties have been taken into account. A combined significance of 5σ is achieved at 25 kpc for the 27 M⊙ progenitor, ensuring the coverage of the full Galaxy. In the case of the 11 M⊙ progenitor, a significance of 5σ is reached beyond the Galactic Centre with the full ORCA detector.

4 Trigger performance

An approach for the real-time triggering of a CCSN neutrino burst, applicable to both ORCA and ARCA, has been evaluated and will be operational from the first phases of the detector deployment.

Data from the deployed ARCA and ORCA lines have been extensively analyzed by sampling the background on a τ = 100 ms time scale in order to identify patterns in the DOM response that could mimic the CCSN signal. A robust trigger is implemented by defining the *trigger level* as the number of DOMs that detect a multiple coincidence in the multiplicity cut over a larger sliding time window of n times τ, with n an integer allowing to cover the signal duration.

Given ρ the average background rate in the multiplicity cut (after muon filtering), the false rate for a given trigger level X can be estimated and related to the distance to the source in the following way:

$$R_B(X) = \tau^{-1} \cdot P(N_{lines} \cdot n\rho\tau, X) \quad (4)$$

$$X = N_{lines} \cdot X_{10kpc}(n\tau) \cdot \left(\frac{D}{10kpc}\right)^2 \quad (5)$$

Estimated false trigger rates are shown for the ORCA detector in Fig. 3 as a function of the CCSN distance, where we highlight the $\leq 1/week$ fake trigger requirement to participate in the global SNEWS alert network [7], for the two progenitors studied. A CCSN of 27 M⊙ could be triggered with full ORCA up to 18.6 kpc and above the Galactic Centre for the 11 M⊙ CCSN.

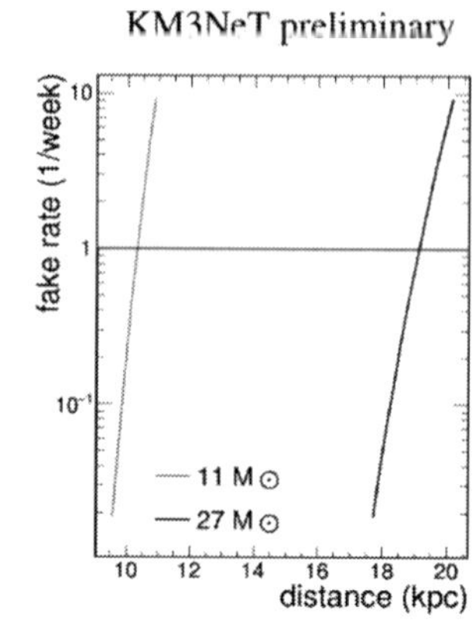

Figure 3. Fake trigger rate in ORCA as a function of the distance to the CCSN for the two progenitors studied. The black line indicates the SNEWS threshold.

5 Sensitivity to fast time variations on the neutrino light-curve.

Latest 3D simulations predict anisotropic hydrodynamical instabilities during the CCSN accretion phase, which may play an important role in the efficient neutrino heating [8] [9]. We focus here on the Standing Accretion Shock Instability (SASI) that could explain the observed energetic explosion. It leaves as a footprint fast time variations in the neutrino light-curve around 200 ms after the core bounce, with a characteristic oscillation frequency [4].

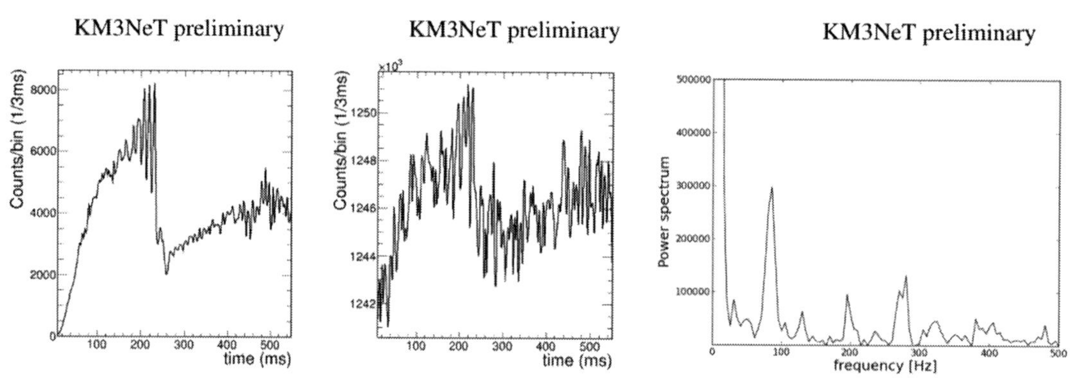

Figure 4. On the left, the expected light-curve for CCSN at 5kpc, where SASI oscillations are seen at around 200 ms. On the middle, the expected light-curve in one KM3NeT block when adding background. On the right, the power spectrum for background+signal at 5 kpc, where the peak is recovered at the SASI oscillation frequency.

The detected CCSN neutrino light-curve has been simulated for a the 27 M⊙ progenitor at 5 kpc, including all hits. Background has been included assuming a Poisson distribution, which is a preliminary an quite optimistic assumption. As in [10], a Fourier analysis has been used to recover the SASI frequency and the significance to the SASI peak has been estimated through MC pseudo-experiments. Results are given in Fig. 4. A 3σ sensitivity to the SASI oscillation peak is obtained at 5 kpc for the 27 M⊙ progenitor model with one KM3NeT detection block. A more detailed description of the background will follow in a more advance analysis.

6 Conclusions

The response of the KM3NeT neutrino detectors to core-collapse supernova neutrinos has been estimated by means of a complete Monte Carlo simulation and an exhaustive study of the background from the first KM3NeT datasets.

The KM3NeT combined sensitivity for a future CCSN detection has been estimated to be of 5σ at 25 kpc (coverage of the full Galaxy) for a 27 M$\odot$ stellar progenitor. Assuming an 11 M$\odot$ stellar progenitor, a significance of 5σ is reached above the Galactic Centre with the full ORCA detector.

As for the online triggering capabilities, evaluated for the ORCA detector, the maximum triggering distance meeting the SNEWS false alert rate requirement has been estimated to be well beyond the Galactic Centre for the two progenitors studied.

7 Acknowledgements

This project is partly supported by the LabEx UnivEarthS exploratory project E9: Low energy astrophysics with KM3NeT.

References

[1] KM3NeT Collaboration, *"The prototype detection unit of the KM3NeT detector"*, *(arXiv:1510.01561)* (2015)

[2] KM3NeT Collaboration, *"Letter of intent for ARCA and ORCA"*, *(arXiv:1601.07459v2)* (2016)

[3] I. Tamborra et al., *"Neutrino emission characteristics and detection opportunities based on three-dimensional supernova simulations"*, *Phys. Rev. D 90, 045032 (arXiv e-prints 1406.0006)* (2014)

[4] I. Tamborra et al., *"Neutrino signature of supernova hydrodynamical instabilities in three dimensions"*, *Phys. Rev. Lett. 111,121104 (arXiv e-prints 1307.7936)* (2013)

[5] A. Strumia and F. Vissani, *Phys. Lett. B* 564 (2003) 42

[6] G. Radel and R. Beyer, *Mod. Phys. Lett. A*, 08 (1993) 12

[7] K.Scholberg, *"The Supernova Early Warning System"*, *(arXiv:0803.0531)* (2008)

[8] A. Marek, H.-T. Janka and E. Müller, *"Equation-of-state dependent features in shock-oscillation modulated neutrino and gravitational-wave signals from supernovae"*, *Astron. Astrophys. 496, 66* (2009)

[9] H.T.Janka, *"Neutrino-driven Explosions"*, *arXiv:1702.08825* (2017)

[10] J.Miganda, *"Detecting Fast Time Variations in the Supernova Neutrino Flux with Hyper-Kamiokande"*, *arXiv:1609.04286* (2015)

Upgrading Plan Towards Multi-messenger Observation with LHAASO

Z. Cao[1,2,a], M.J. Chen[1], H.C. Li[1], Z.G. Yao[1], for LHAASO Collaboration

[1] *Institute of High Energy Physics*
19B, Yuquan St., Shijingshan, Beijing, China, 100049
[2] *Chinese Academy of Science University*
19A, Yuquan St., Shijingshan, Beijing, China, 100049

Abstract. LHAASO is planning to enhance its sensitivity at energies around 100 GeV by utilize MCP staffed 20" PMT in the Water Cherekove Detector Array. The effective area for gamma ray detection will reach to 1800 m^2 and differential sensitivity to 0.2 CU at 50 GeV. It will be the very useful survey detection for transient phenomena at 50 GeV in the northern sky. LHAASO is expected to play an important role in the multi-messenger observation with the upgrading.

1 Introduction

The first multi-messenger observation of gravitational wave (GW) event GW170817 by LIGO and VIRGO GW observatories together with many other Electro-magnetic (EM) wave observations, such as FERMI[1] and the multi-wavelength campaign on the possible EM partner of the very high energy neutrino IceCube-170922A detected by IceCube experiment[2] are very significant progresses in astroparticle physics in the past year. They opened new windows for exploring the high energy phenomena in the universe. This, however, becomes a challenge to Large High Altitude Air Shower Observatory (LHAASO) experiment with its original proposal which is designed to target the high energy (above several hundred GeV) gamma ray sources and charged particles at even higher energies up to a few EeV. At energies lower than 300 GeV, the gamma ray detection sensitivity is not sufficient to detect those sources which is typically faint. In order to enhance the sensitivity below 100 GeV, we proposed to enlarge the sensitive area of the photo cathode of the water Cherenkov detector (WCD) in the LHAASO array. In this paper, we are going to briefly describe the LHAASO experiment and its WCD Array as well as the upgrading plan in particular in the second section, and the performance of WCDA in gamma ray astronomy with the the enhancement at low energy region in third section. The whole upgrading plan is summarized in the forth section.

2 LHAASO/WCDA Experiment and the Upgrading Plan

LHAASO is a multipurpose complex of EAS detection consisting of four major detector arrays[3], ie. 5195 scintillation counters (ED) and 1171 muon detectors (MD) covering an area of 1.3 km^2, 78,000 m^2 water Cherenkov detector (WCD) with 3120 gap-less detecting units, and 18 wide field of view Cherenkov telescopes watching over the sky above the whole complex with a coverage of 4608 square degree. As shown in Figure 1, the WCDA in the center of the array is divided into 3 components as 3 independent water pools, namely two smaller pools with the area of 150 $m \times 150$ m each and the larger one with the area of 300 $m \times 110$ m.

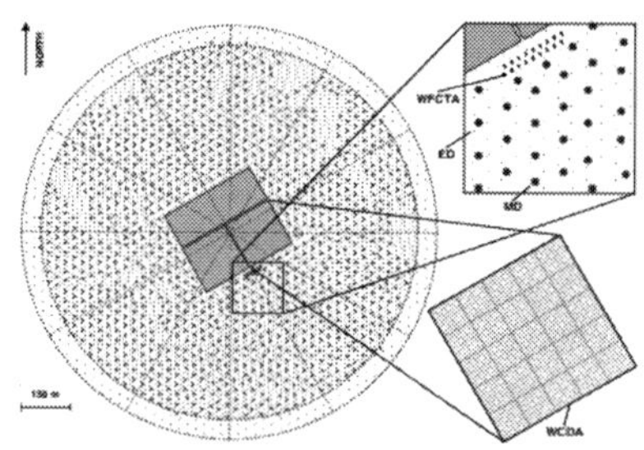

Figure 1. The layout of the scintillator counter (small dots) array, muon counter (big dots) array, water Cherenkov detector (rectangle in the center) array and 18 wide field of view Cherenkov telescopes (small rectangles) in the LHAASO complex of multi-detector array of 1.3 km^2.

The firstly built pool in south-west has 900 WCD units, 25 m^2 each, equipped by a large (8") PMT for timing and a small (1.5") PMT for pulse size at the center of each unit 4 m beneath the water surface, and measures shower directions with a resolution better than 0.2° above 10 TeV and 1.0° above 600 GeV. Only direct Cherenkov light generated by the shower particles is collected by the upwards watching PMTs. To suppress the cross talking effect and improve the timing resolution, black plastic divisions are

[a] e-mail: caozh@ihep.ac.cn

installed between units. The Front End Electronics (FEE) of the large PMTs is designed to have the timing resolution of 0.5 ns. The dynamic range of the detector is enlarged very much by using the small PMT. This enables the measurement of the detailed particle density distribution in the shower cores without significant saturation even for energetic showers up to 10 PeV and achievement of the core location resolution better than 3 m over a wide energy range. This is designed for the identification of the primary particle species in the cosmic ray composition and spectrum measurements. It is also very useful in locating the shower inside the pool with minimal loss of good detected events. The pool is planned to be turned on for operation early 2019.

Low energy showers are small in terms of total number of particles that reach to the pools, therefore the total Cherenkov signal generated by those secondary particles in every detector unit is faint, even for units being near the cores of the showers. In order to enhance the gamma ray detecting sensitivity at low energies, enlarging the sensitive photo-cathode area of the PMT in the same size unit could be one effective way to catch the faint signals. LHAASO's upgrading plan is along with this approach, namely to replace the 8" PMTs by 20" PMTs in the rest two pools of 55,500 m^2 in total. The customized design of the PMTs using multi-channel-plate (MCP) instead of the traditional dynodes enables good uniformity between PMTs as well as the Transit Time Spreads (TTS) less than 7 ns, Cathode Transit Time Distribution (CTTD) less than 2 ns and long lifetime. The photo cathode is a factor of 6.25 larger than the 8" tube, therefore the dynamic range is also shrunk by the same factor. In order to compensate the loss, a 3" PMT is installed beside the large PMT in each unit, read out only for the pulse size by a simplified version of FEE covering 4 orders of amplitudes in number of photo-electrons.

3 Performances and Prospects for Gamma Ray Astronomy

Gamma ray induced showers are different from showers induced by charged CR nuclei in terms of the hits distribution in the pool. In general, the later is more spread out than the former ones as shown in Figure 2, where two MC simulated events due to 1 TeV gamma ray and 2 TeV proton are compared with each other. Even more significantly, the CR events have many "hot spots", the populated hits, outside the core region, while the gamma ray events are much cleaner beyond some distance from the core. This enables us to identify the gamma events out of the CR background, nevertheless they are much more ($10^{4\sim5}$) than gamma ray signals even if within a very small angular region defined by the point-spread-function (PSF) near the sources. By eliminating the events which have the most populated hit in the outer region, 45 m from the core, greater than certain number of photoelectrons, N_{th}, the CR background will be suppressed to very low in the nearby region of sources. Making balancing between the elimination of the background CR events and the loss the gamma-like signal events to maximize the sensitivity, it is found that N_{th} increases with the number of hits that are involved in the shower front fit, denoted as N_{fit} which measures the shower energy, i.e. proportional to $N_{fit}/logN_{fit}$. Other parameters characterize the distribution of hits in the whole pool area, such as hit density in the outer region, are used in the identification of gamma ray events. This results an effective area for gamma ray detection of about 230 m^2 at 50 GeV and 30,000 m^2 at 1 TeV, respectively, if all 3 pools were equipped by 8" PMTs. The corresponding sensitivity of the gamma ray point-like source detection is plotted in Figure 3 as a function of gamma ray energy.

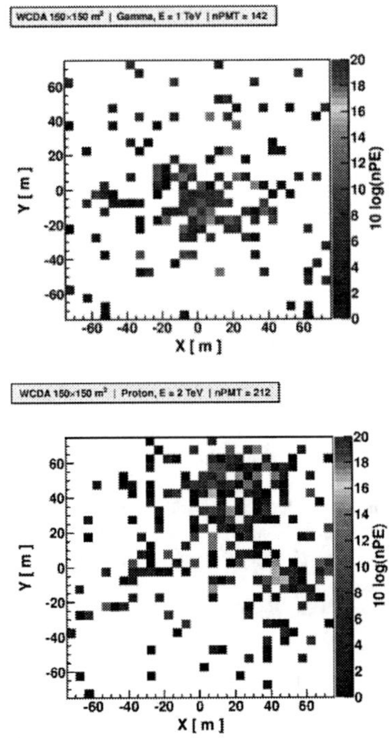

Figure 2. Two simulated events recorded by one of the pools of WCDA induced by 1 TeV gamma ray (upper) and 2 TeV proton (lower), respectively.

With this configuration, LHAASO has a survey power for discovering all sources that are brighter than 7 mini crab unit above 1 TeV. Taking into account the wide field of view of ~ 1/7 of the entire sky and the constant exposure time of 24 hours, LHAASO is very significant in finding new sources. It is actually estimated that about 40 new AGNs[4] could be discovered within one year after it is fully operated. It is also expected that LHAASO will make a deeper survey inside our galaxy comparing with what has been done by HAWC experiment[5].

3.1 Enhancement at Low Energy

With the upgraded configuration using 20" PMTs in the other two pools, the effective area at energies below 300 GeV is significantly enlarged, i.e. reaches to 1,800 m^2 at

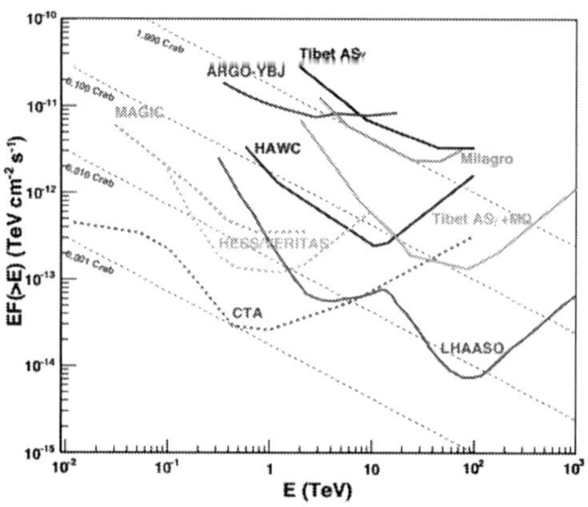

Figure 3. The integrated sensitivity of LHAASO (in pink) comparing with other experiments. The part of the curve below 10 TeV is the contribution of WCDA optimized with the cuts described in text.

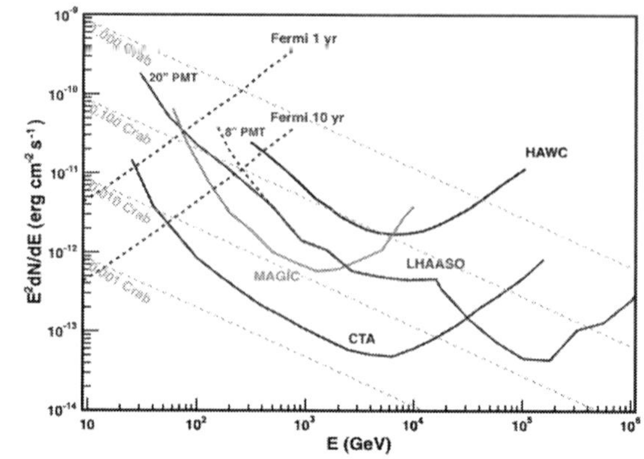

Figure 4. The differential sensitivity of LHAASO (in red) comparing with other experiments. Below 300 GeV, both estimates with 8" PMT configuration (dashed line) and 20" PMT configuration (solid line) are plotted. It is noticed that WCDA with 20" PMTs is almost as same sensitive as FERMI/LAT at 70 GeV.

50 GeV and 44,000 m^2 at 1 TeV. The corresponding differential sensitivity around 50 GeV is expected to be 0.2 crab unit per a quart decade of energy which is compatible with the space borne detector FERMI/LAT, as shown in Figure 4. The difference, however, is that the effective area is a factor of 1,800 larger than the later. This means that more than 1000 photons are expected to be recorded if the gamma ray burst event GRB090510 happened again in the field of view of LHAASO. In the event, FERMI/LAT recorded a single gamma photon at 95 GeV[6]. This opens a window for the multi-wavelength campaign in a much convenient way because of the clock-round operation of LHAASO. With any global alarm for transient phenomena, such as GBR, it is easy for LHAASO to recall the data in the window in which the alarm was ringing. Not only the status of the source at T_0 can be observed, but also it is in principle possible to find any pre-emission of gamma rays if there were.

With such a sensitivity, LHAASO will be a transient phenomenon finder as well. An alarm trigger algorithm is going to be operated to constantly watch all interested AGNs in LHAASO's FoV for any excess in various time windows, e.g. from few seconds to hours. It is useful for monitoring any AGN flare, e.g. if its emission level rises to be greater than 1 crab unit within a hour, an alarm will be broadcasted to the whole community.

3.2 Prospects for Multi-messenger Exploring

Investigating sources with multi-messengers is very powerful in viewing of inside mechanism of high energy phenomena in the universe, particularly for possible common origins of the messengers, such us neutrinos, gamma rays, charged particles and gravitational waves. However, to identify the sources and verify the association, all corresponding detectors are required for sufficient sensitivities. As an example, we investigated the possible association between the ultra high energy muon neutrino event IC-170922A detected by IceCube experiment[2] and the blazer TXS 0506+056 which had a faint flare within 20 days after the neutrino event in multiple wavelength bands, including]X-ray (SWIFT), gamma ray (FERMI-LAT) and very high energy gamma ray (MAGIC). The SED of the blazer during the flare is reported in Ref. [2] and is quoted here in Figure 5 over a very wide energy range. Also shown in the figure, sensitivity curves of several experiments, including HAWK, HESS, VERITAS and upgraded LHAASO. According to this, LHAASO will play a significant role in such multi-messenger observation by covering an important energy range starting from as low as 30 GeV to few hundred TeV.

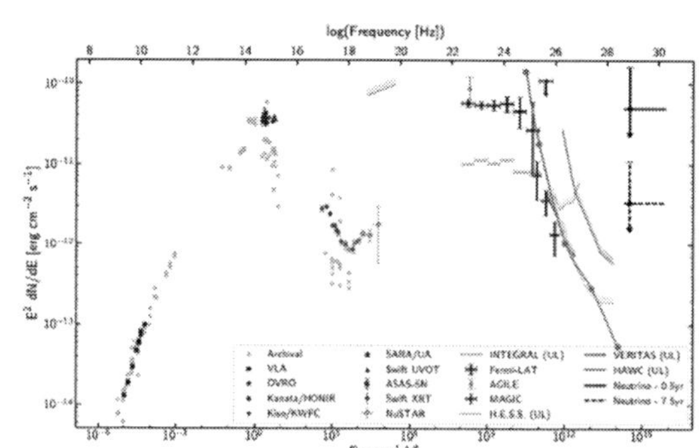

Figure 5. The SED of TXS 0506+056 during the flare within 20 days after the neutrino event IC-170922A. It is reported in Ref. [2] including sensitivity curves of HAWC, HESS and VERITAS experiments. The upgraded LHAASO sensitivity is also plotted in the same figure (light blue) from 30 GeV to 100 TeV.

4 Summary

LHAASO has made its upgrading plan by replacing the 8" PMTs in 70% of area of the central water Cherenkov detector with 20" PMTs. The first part, 30% of the total area of the detector, is under construction and going to be operated in early 2019. The whole array, including the scintillator counter array and muon detector array and Cherenkov telescopes, will be built up by 2021. With the upgraded configuration, the gamma ray detecting sensitivity below 300 GeV will be boosted to be about 0.2 crab unit per quarter decades of energy around 50 GeV and as same sensitive as FERMI at 70 GeV. LHAASO therefore will play a significant role in surveying for new sources brighter than 7 mini-crab-unit above 1 TeV in the northern sky and monitoring for transient phenomena in its FoV of the size of 1/7 sky at any moment. The LHAASO effective area of 1800 m^2 around 50 GeV is going to be useful tool in the multi-messenger observation involving ultra high energy neutrino or gravitational wave detections.

5 acknowledgement

We acknowledge the support to the author (CZ) for his attendance of the conference by the National Science Foundation of China under contract No. 11761141001.

References

[1] Abbott, B. P., *et al.* (LIGO and VIRGO Collaboration), PRL **119**, 141101 (2017)
[2] Aartsen, M.G., *et al.*, Science **361**, eaat 1378 (2018) (arXiv:1807.08816)
[3] Z. Cao (for LHAASO Coll.), Chin.Phys. **C34** 249-252 (2010); H. He (for LHAASO Coll.), Radiation Detection Technology and Methods 2-7, (2018)
[4] Y. Zhao, *et al.*, int.J.mode.Phys. **D25** 160, (2016)
[5] Abeysekara, A. U. *et al.*, Astrophysical Journal **843** 1, (2017)
[6] Ackermann, M *et al.*, Science **343** 6166, 42-47 (2014)

EPJ Web of Conferences **209**, 01011 (2019)
RICAP-18
https://doi.org/10.1051/epjconf/201920901011

e-μ discrimination at high energy in the JUNO detector

Giulio Settanta[1,2,a], *Stefano Maria* Mari[1,2], *Cristina* Martellini[1,2] and *Paolo* Montini[1,2]

[1] *Dipartimento di Matematica e Fisica, Università degli Studi Roma Tre, Via della Vasca Navale 84, 00146 Rome, Italy*

[2] *INFN Sezione di Roma Tre, Via della Vasca Navale 84, 00146 Rome, Italy*

Abstract. Cosmic Ray and neutrino oscillation physics can be studied by using atmospheric neutrinos. JUNO (Jiangmen Underground Neutrino Observatory) is a large liquid scintillator detector with low energy detection threshold and excellent energy resolution. The detector performances allow the atmospheric neutrino oscillation measurements. In this work, a discrimination algorithm for different reaction channels of neutrino-nucleon interactions in the JUNO liquid scintillator, in the GeV/sub-GeV energy region, is presented. The atmospheric neutrino flux is taken as reference, considering $\overset{(-)}{\nu}_\mu$ and $\overset{(-)}{\nu}_e$. The different temporal behaviour of the classes of events have been exploited to build a time profile-based discrimination algorithm. The results show a good selection power for $\overset{(-)}{\nu}_e$ CC events, while the $\overset{(-)}{\nu}_\mu$ CC component suffers of an important contamination from NC events at low energy, which is under study. Preliminary results are presented.

1 Introduction

The origin and properties of Cosmic Rays (CRs) are still matter of study by a number of experiments. A fraction of the energy of an air shower, resulting after a CR interaction in the atmosphere, is carried by neutrinos. The atmospheric ν flux is almost entirely composed of $\overset{(-)}{\nu}_\mu$ and $\overset{(-)}{\nu}_e$ and spans many decades in energy, from the MeV up to the PeV scale. Neutrinos travelling across the Earth can be also used as a probe to study flavor oscillation effects in matter. After the experimental results in past years [1–4], further contributions to the atmospheric ν spectrum will come from the next generation of neutrino detectors, operating in the next decade. JUNO is a large liquid scintillator detector with low energy threshold and excellent energy resolution ($\sim 3\%/\sqrt{E\,[MeV]}$), under construction in China [5]. Its core consists of a ~36 m diameter acrylic sphere, which will be filled with 20kt of liquid scintillator. The light produced in ν interactions will be collected by a double-system of photosensors: 17.000 20" PMTs and 25.000 3" PMTs. Given the large fiducial volume and the excellent energy resolution, JUNO will be able to detect atmospheric neutrinos.

A crucial part in the atmospheric ν study is the correct identification of the original ν flavor. Above ~100 MeV neutrinos can be assumed to interact with nucleons only. This implies that hadronic particles are always produced in the final state. In the case of a charged-current (CC) interaction, the corresponding charged lepton is also produced (either $\nu_\mu N \rightarrow \mu X$ or $\nu_e N \rightarrow eX$). In the case of a

[a] e-mail: giulio.settanta@uniroma3.it

neutral-current (NC) interaction, only secondary hadrons and the scattered neutrino are in the final state.

2 The Monte Carlo simulation

Since the JUNO detector is still under construction, all the analysis so far relies on Monte Carlo (MC) simulations only. The MC production proceeds in several steps:

- Generation of the atmospheric ν flux at the JUNO location, according to the predictions from [6]. The model provides also the zenith- and azimuth-angle flux dependance.
- Interaction of neutrinos with the JUNO liquid scintillator, by means of the *GENIE Neutrino Monte Carlo Generator* [7]. The events have been considered up to 20 GeV and consist of 10^6 $\overset{(-)}{\nu}_\mu$ + $\overset{(-)}{\nu}_e$ events (fig. 1 - left). The output of the simulation is the full list of secondaries and their associated properties (ID, momentum, direction ...).
- Propagation of secondaries in the scintillator by means of a *GEANT4*-based simulation [8], which includes energy losses, photon production and propagation, photon-PMT cathode interaction and photo-electrons (PE) generation. This last step includes 50k events only. Since JUNO includes an active volume of 20000 tons and more than 40000 PMTs, running a full simulation is both CPU and storage consuming, which is a limitation in the number of simulated events.

At the end of the simulation, a map of the hits on the detector is available, distributed in time and position (fig. 1 - right).

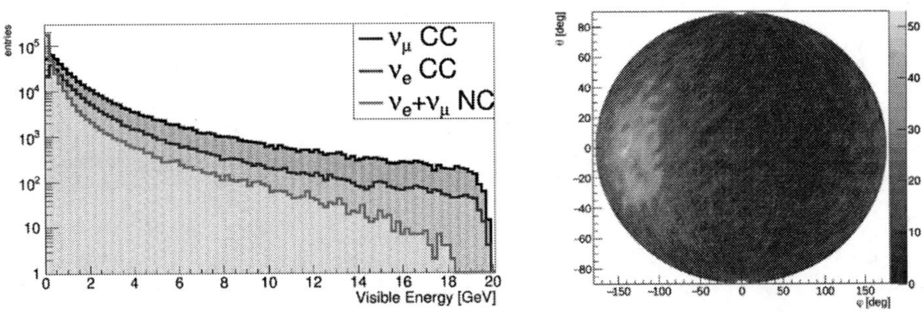

Figure 1. Left: visible energy of the MC sample, for ν_μ CC (blue), ν_e CC (red) and NC events (green); right: map of the hits on the 3" PMT system after a E = 378 MeV ν_μ CC interaction at positions X = 4.9 m, Y = 7.2 m, Z = -0.05 m, assuming the detector centre at (0,0,0).

3 Flavor identification

Since in CC interactions there is always an electron or a muon in the final state, they constitute the preferred channel for the neutrino flavor identification. Muons with energy > 1 GeV travel in general for a longer distance inside the detector with respect to electrons. Additionally, muons can decay inside the scintillator volume, giving a delayed energy release at low energy too. The above differences make ν_μ CC events more elongated in time with respect of ν_e CC events, which indeed appear more point-like. Hadronic particles, which constitute NC events, have in general a long-living

energy release, because of interactions and decays. Furthermore, since hadrons add in general late energy releases to all events, they also make more similar ν_μ and ν_e CC event topologies.
The different temporal behaviour of the classes of events has been exploited to build a time profile-based discrimination algorithm. The 3" PMT system has been used in the algorithm, since it is more accurate in time measurement. To reproduce realistic Time-Transit Spread (TTS) effects, an artificial smearing has been applied to the true hit time over every PMT. A gaussian function with $\sigma = 4\ ns$ has been used. A further selection on the smeared vertex is applied, in order to avoid edge effects, requiring it to be be within 16 m from the detector centre. The hit time-residual is defined for each PMT as:

$$t^i_{res} = t^i_{hit} - \left(\frac{c}{n \cdot R^i_V}\right) \quad (1)$$

where c/n is the speed of light inside the scintillator and R^i_V is the distance between the i-th PMT and the interaction vertex. In order to reproduce conservatively the uncertainty on the vertex reconstruction at the energy considered, a smearing with a $\sigma = 1\ m$ gaussian function has been applied to all events. Since ν_μ and ν_e CC events result in different light production duration, the RMS of the t_{res} is used as a discrimination variable (here called $\sigma(t_{res})$). In fig. 2 the $\sigma(t_{res})$ distribution is reported for the three populations: ν_μ CC, ν_e CC and NC events. The variable is also reported for 4 different bins of number of PEs (NPE) collected by all 20" PMTs, selected in order to have equal statistics in each bin. From fig. 2, a good separation of the ν_e CC component is evident, in all the energy bins.

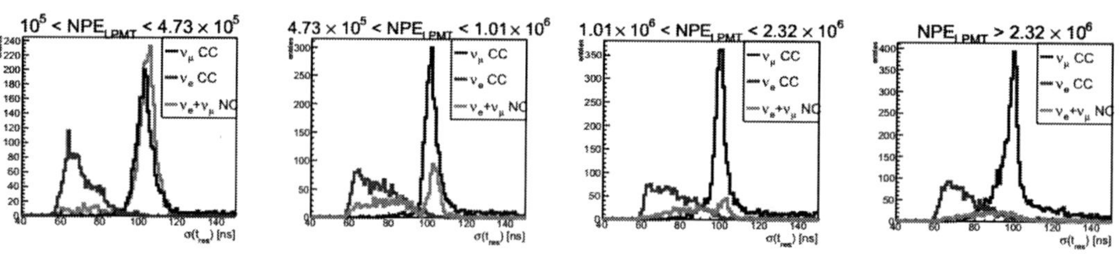

Figure 2. Distribution of $\sigma(t_{res})$ for ν_μ CC (blue), ν_e CC (red) and NC events (green). The NPE range is reported above the figures.

On the contrary, the ν_μ CC and the NC components are almost overlapped on the whole energy range. The NC component contribution, however, becomes less significant at high energy. The spectral distribution of the MC sample, according to the interaction channel, is shown in fig. 3 and reflects the sub-dominance of the NC component at high energy.

Given the distributions reported in fig. 2, a sharp cut on the $\sigma(t_{res})$ is applied to select the populations of interest: $\sigma(t_{res}) < 80\ ns$ for ν_e CC and $\sigma(t_{res}) > 90\ ns$ for ν_μ CC. In fig. 4 the NPE distribution is reported, after the above cuts. Preliminary studies on the effects of the cuts have been evaluated on an independent MC sample of 10k events. Considering a range between 10^5 and 10^7 NPE, the $\sigma(t_{res}) < 80\ ns$ cut keeps 70% of the ν_e CC events with a contamination around 17%. In the range above 10^6 NPE, the $\sigma(t_{res}) > 90\ ns$ cut keeps 93% of the ν_μ CC events with a contamination around 17%. These performances represent a preliminary result in the separation of the samples and are under study for further optimization.

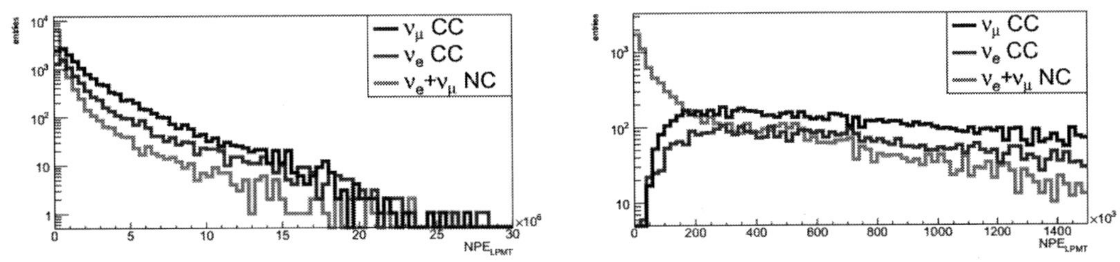

Figure 3. Left: distribution of the NPE collected by the 20" PMT system, according to the the interaction channel (same legend of fig. 2); right: same distribution, in the lower energy band.

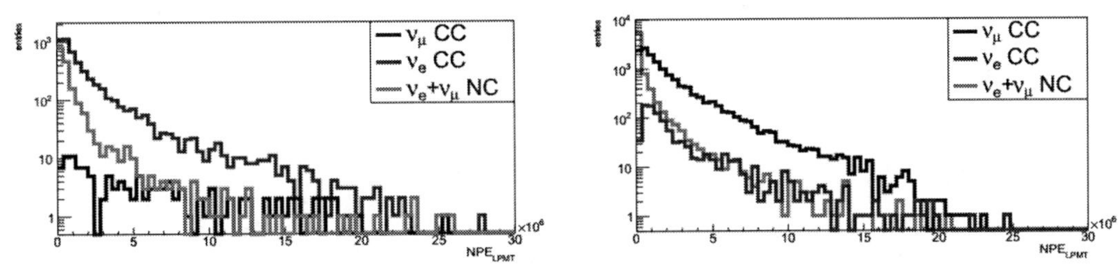

Figure 4. Left: distribution of the NPE collected by the 20" PMT system, of events with $\sigma(t_{res}) < 80\ ns$, according to the the interaction channel (same legend of fig. 2); right: same distribution, for $\sigma(t_{res}) > 90\ ns$.

Conclusions

An event time-profile algorithm is used for the e/μ separation after atmospheric neutrinos interaction in the JUNO detector. Preliminary results are reported on the poster and show a good selection power for ν_e CC events. The ν_μ CC is still contaminated by an important fraction of NC events at low energy and is under study. By evaluating the cuts on an independent MC sample, in a range between 10^5 and 10^7 NPE, the $\sigma(t_{res}) < 80\ ns$ cut keeps 70% of the ν_e CC events with a contamination around 17%. In a range above 10^6 NPE, the $\sigma(t_{res}) > 90\ ns$ cut keeps 93% of the ν_μ CC events with a contamination around 17%. The above results are still preliminary and are under optimization.

References

[1] K. Daum et al. (Frejus Collaboration), Z. Phys. C 66, 417 (1995).
[2] E. Richard et al. (Super-Kamiokande Collaboration), Phys. Rev. D 94, 052001 (2016).
[3] M. G. Aartsen et al. (IceCube Collaboration), Phys. Rev. D 91, 122004 (2015).
[4] M. G. Aartsen et al. (IceCube Collaboration), Eur. Phys. J. C 77, 692 (2017).
[5] F. An et al., Journ. of Phys. G: Nucl. and Part. Phys. 43, 030401 (2016).
[6] M. Honda, M. Sajjad Athar, T. Kajita, K. Kasahara, and S. Midorikawa, Phys. Rev. D 92, 023004 (2015).
[7] C.Andreopoulos et al., Nucl. Instrum. Meth. A614 87-104 (2014).
[8] F. An et al., J. Phys.: Conf. Ser. 664 072053 (2015).

EPJ Web of Conferences **209**, 01012 (2019) https://doi.org/10.1051/epjconf/201920901012
RICAP-18

Towards a reconstruction of Supernova Neutrino Spectra in JUNO

Cristina Martellini[1,2,a], *Stefano* Maria Mari[1,2], *Paolo* Montini[1,2] and *Giulio* Settanta[1,2]

[1] *Dipartimento di Matematica e Fisica, Universita' degli Studi di Roma Tre, Via della Vasca Navale 84, 00146 Rome, Italy*
[2] *INFN Sezione di Roma Tre, Via della Vasca Navale 84, 00146 Rome, Italy*

Abstract. Observation of supernovae (SN) through their neutrino emission is a fundamental point to understand both SN dynamics and neutrino physical properties. JUNO is a 20kton liquid scintillator detector, under construction in Jiangmen, China. The main aim of the experiment is to determine neutrino mass hierarchy by precisely measuring the energy spectrum of reactor electron antineutrinos. However due to its properties, JUNO has the capability of detecting a high statistics of SN events too. Existing data from SN neutrino consists only of 24 events coming from the SN 1987A,the detection of a SN burst in JUNO at $\sim 10kpc$ will yield $\sim 5x10^3$ inverse beta decay (IBD) events from electron antineutrinos, about 1500 from proton elastic scattering (pES) above the threshold of 0.2 MeV, about 400 from electron elastic scattering (eES), plus several hundreds on other CC and NC interaction channels from all neutrino species.

1 Introduction

The Jiangmen Underground Neutrino Observatory (JUNO)[1] is a 20kton multi-purpose underground liquid scintillator (LS) detector, which is located at Kaiping, Jiangmen in South China, is design primarily to determine the neutrino mass hierarchy (MH) using reactor antineutrinos.

The detector features make it suitable for a whole series of measurements in the ν Physics such as atmospheric neutrinos, geoneutrinos, solar neutrinos and supernova and diffuse supernova neutrinos, as a natural accessible source and other exotic searches[1].The JUNO central detector consists of a $\sim 36m$ diameter acrylic sphere, filled with 20kton of liquid scintillator (LS). The light will be collected by a double-system of photosensors: 18.000 20" PMTs and 25.000 3" PMTs. As a LS calorimeter it has a very low energy threshold and can measure supernova neutrinos with unprecedent performaces and excellent energy resolution. In this work, we present simulations studies on Core Collapse Supernova (CCSN) neutrino event for different detection channels involving different flavours of SN neutrinos. Using the supernova flux model from Nakazato *et al* [2] and assuming the galactic supernova explosion occurring at three different distances, a statistic sample can be simulated to be able to build an unfolding method to get back to the original SN parameters and hence reconstruct the SN neutrino energy spectra.

[a] e-mail: cristina.martellini@uniroma3.it

2 Core Collapse Supernova Neutrinos

A massive star of mass above $\sim 8M_\odot$ is expected to experience a core collapse under its own gravity and then a violent explosion, where 99% of the gravitational binding energy of the core-collapse supernova (CCSN) will be carried by the intense burst of neutrinos. Because of this they play a decisive role during all the stages of such an event. While the neutrinos detected from the SN 1987A were useful to give basic confirmation of the theory of neutron star (NS) formation, their event statistics was too poor to highlight some useful information about the dynamics of the explosion mechanism of the CCSN[3]. High-statistics measurement of neutrinos from the next Galactic SN will therefore be of major importance for astrophysics, but also neutrino and nuclear physics. As the largest new generation LS detector, JUNO will be superior in its high statistics, in the energy resolution ($3\%/\sqrt{MeV}$) and flavour informations. Since the average SN distance is around 10 kpc, JUNO will register about 5000 from inverse beta decay (IBD) caused by the interaction of electron antineutrinos with the LS, ($\bar{\nu}_e + p \rightarrow n + e^+$), around 1500 events from all-flavour elastic neutrino-proton elastic scattering, ($\nu + p \rightarrow \nu + p$), more than 300 events from neutrino-electron elastic scattering, ($\nu + e^- \rightarrow \nu + e^-$), as well as other charged current (CC) and neutral current (NC) interactions on the ^{12}C nuclei.

Table 1. Numbers of neutrino events in JUNO for a SN at a typical distance of 10 kpc. Three representative values of the average neutrino energy $\langle E_\nu \rangle$ = 12, 14 and 16 MeV are taken for illustration. For the elastic neutrino-proton scattering a threshold of 0.2 MeV for the proton recoil energy is considered.

Channel	Type	Events for different $\langle E_\nu \rangle$ values		
		12 MeV	14 MeV	16 MeV
$\bar{\nu}_e + p \rightarrow e^+ + n$	CC	4.3×10^3	5.0×10^3	5.7×10^3
$\nu + p \rightarrow \nu + p$	NC	0.6×10^3	1.2×10^3	2.0×10^3
$\nu + e \rightarrow \nu + e$	ES	3.6×10^2	3.6×10^2	3.6×10^2
$\nu +^{12} C \rightarrow \nu +^{12} C^*$	NC	1.7×10^2	3.2×10^2	5.2×10^2
$\nu_e +^{12} C \rightarrow e^- +^{12} N$	CC	0.5×10^2	0.9×10^2	1.6×10^2
$\bar{\nu}_e +^{12} C \rightarrow e^+ +^{12} B$	CC	0.6×10^2	1.1×10^2	1.6×10^2

The numbers of neutrino events at JUNO[1] for a SN at a typical distance of 10 kpc are presented in Table 2. With these measurements of SN neutrinos, JUNO may have the capability to get unique informations to measure the initial SN neutrino fluxes, to constrain the neutrino's mass scale and ordering, to test the scenario of collective neutrino oscillations and even to probe the neutrino electromagnetic properties.

3 Monte Carlo Supernova Simulations

For the simulation the Supernova Generator implemented in the JUNO Software has been used, setting a sample based on different solar masses ($M_\odot$), metallicity (Z) and revival time (τ_{rev}), setting our simulation supposing normal hierarchy (NH), we generated SN at three different reference distances, building the statistic sample shown in Table2. We simulated therefore 18 SN bursts, which provided us with different event rates and then we run those through the detector to evaluate its response. We plotted the energy spectrum from each of the different channels first as a function of the visible energy and subsequently as a function of the number of photoelectrons (PE) detected by the two PTMs optical systems in JUNO. In Figure 1 below the distributions for all the different channels are shown.

Table 2. Different SN fluxes generated by different progenitor star at three different distances and with two values of the metallicity.

Mass	Metallicity (Z)	Revival time τ (ms)	Distance D (kpc)
13 $M_\odot$	0.02 0.04	300	2 10 20
20 $M_\odot$	0.02 0.04	300	2 10 20
50 $M_\odot$	0.02 0.04	300	2 10 20

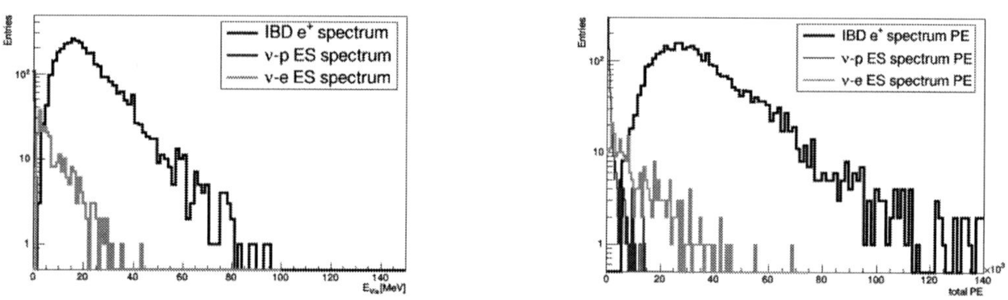

(a) Visible\Energy (b) PEdistribution

Figure 1. Left: Distribution of the energy spectra as a function of the visible energy for IBD (blue),$\nu - p$ ES (red) and $\nu - e$ ES (green). Right: distribution of the NPE for the IBD (blue), $\nu - p$ ES (red) and $\nu - e$ ES (green).

4 Unfolding Method

Following the analysis of the simulation the main aim of this work is to build an unfolding method to get back to the original parameters of the SN and reconstruct the original neutrino spectra. Taking into consideration the contributions given to their event rate in the detector, the three main channels have been taken into account for the development of this work, which are the IBD event, the $\nu - p$ ES and the $\nu - e$ ES event. Depending on the channel, the probability of having a $\nu - \bar{\nu}$ of any flavour of a given energy have been identified as follow:

$$
\begin{aligned}
P_{IBD}(E_{\bar{\nu}_e}) &\propto \int_{E_{min}}^{E_{max}} P_{IBD}(E_{\bar{\nu}_e}|E_0) \cdot P_{IBD}(E_0)\mathrm{d}E_0 \\
P_{pES}(E_\nu) &\propto \sum_{flavour=1}^{3} \int_{E_{min}}^{E_{max}} P_{pES}^{flavour}(E_\nu|E_0) \cdot P_{pES}^{flavour}(E_0)\mathrm{d}E_0 \qquad (1) \\
P_{eES}(E_\nu) &\propto \sum_{flavour=1}^{3} \int_{E_{min}}^{E_{max}} P_{eES}^{falvour}(E_\nu|E_0) \cdot P_{eES}^{flavour}(E_0)\mathrm{d}E_0
\end{aligned}
$$

where we consider the probability of measuring the energy E_0 in the detector $P_{IBD}(E_0)$, $P_{pES}^{flavour}(E_0)$ and $P_{eES}^{flavour}(E_0)$ as the observables, the $P_{IBD}(E_{\bar{\nu}_e})$, $P_{pES}^{flavour}(E_\nu)$ and $P_{pES}^{flavour}(E_\nu)$ as the unfolded spectra which are given by the convolution of the prior probabilities and the conditional probabilities that given the visible energy E_0 measured by the JUNO detector, this has been generated by $\nu/\bar{\nu}$ of energy E_ν. These last ones are obtained from the Bayes Theorem. The expressions are then integrated within an energy range (E_{min}, E_{max}) and for the two elastic scattering also sum on the three different neutrino flavours.

5 Conclusions

This work illustrates a preliminary study on the capability of the JUNO experiment to detect SN neutrinos and to act together with other neutrino experiment as an alert system for astrophysical object. How reconstructing the SN energy neutrino spectra gives us the chance to learn useful informations about the SN evolution mechanism has been appointed. JUNO should be competitive and improve our knowledge not just on the neutrino oscillation parameters but also with a unique potential for SN neutrinos.

References

[1] F. An et al. (JUNO), J. Phys. **G43**, 030401 (2016), 1507.05613

[2] K. Nakazato, E. Mochida, Y. Niino, H. Suzuki, Astrophys. J. **804**, 75 (2015), 1503.01236

[3] A. Mirizzi, I. Tamborra, H.T. Janka, N. Saviano, K. Scholberg, R. Bollig, L. Hudepohl, S. Chakraborty, Riv. Nuovo Cim. **39**, 1 (2016), 1508.00785

Neutrino Sources from a Multi-Messenger Perspective

Markus Ahlers[a]

[1] *Niels Bohr International Academy & Discovery Centre, Niels Bohr Institute, University of Copenhagen, Blegdamsvej 17, DK-2100 Copenhagen, Denmark*

Abstract. The field of high-energy neutrino astronomy is undergoing a rapid evolution. After the discovery of a diffuse flux of astrophysical TeV-PeV neutrinos in 2013, the IceCube observatory has recently found first compelling evidence for neutrino emission from blazars. In this brief review, I will summarize the status of these neutrino observations and highlight the strong role of multi-messenger astronomy for their interpretation.

1 Introduction

Neutrinos are unique cosmic messengers. Their weak interaction with matter allows to probe the interior of dense sources, which are otherwise opaque to photons. Unlike γ-rays, which can suffer from strong absorption in the cosmic microwave background above TeV energies, high-energy neutrinos arrive at Earth from distant and therefore early sources of the Universe. And unlike cosmic rays, which are observed at Earth as a diffuse flux after repeated scattering in Galactic or extra-galactic magnetic fields, neutrinos point back to their emitters.

High-energy neutrino emission is expected from the interaction of cosmic ray (CR) nucleons with gas or radiation. These interactions produce pions, that subsequently decay via $\pi^+ \to \mu^+ + \nu_\mu$ followed by $\mu^+ \to e^+ + \nu_e + \bar{\nu}_\mu$ and the charge-conjugated processes. Typically, a neutrino from this production channel receives about 5% of the energy from the initial CR nucleon. At the same time, CR interaction also produce neutral pions that decay as $\pi^0 \to \gamma+\gamma$, where the individual γ-ray have an average energy fraction of 10% with respect to the nucleon. Neutrinos are therefore tightly connected to other cosmic messengers and their sources need to be understood in the context of multi-messenger observations.

Presently, the most sensitive neutrino telescope in the TeV-PeV energy range is the IceCube Observatory located at the South Pole. The detector consists of 5,160 digital optical modules (DOMs) that fill a volume of about one cubic-kilometer of clear glacial ice. The DOMs are organized along 86 read-out and support cables ("strings") at a depth between 1,450 and 2,450 meters below the surface. Most strings follow a triangular grid with a width of 125 meters, evenly spaced over the volume. The IceCube detector is fully instrumented since 2011, following a seven-year construction phase with incrementally increasing number of strings.

High-energy neutrinos are detected in IceCube via the Cherenkov light emission of charged particles produced by neutrino interactions in the vicinity of the detector. The most valuable events for neutrino astronomy are charged current interactions of muon neutrinos that can produce high-energy

[a]e-mail: markus.ahlers@nbi.ku.dk

muons traversing the detector. These events allow for a good angular resolution of $\lesssim 0.4°$ at the highest energy ($\gtrsim$ 100 TeV). Charged current interactions of tau neutrinos can also be visible by muon tracks in 18% of the cases. All other neutrino interactions can be visible via the charged particles created in electro-magnetic cascades of the secondary electrons or in hadronic cascades from the breakup of the struck nucleons or the hadronic decay of taus. Whereas these cascade events have a poorer angular resolution in IceCube, at about $10° - 15°$, they allow for a good estimator of the initial neutrino energy with a resolution of better than 15 %.

Cosmic ray interactions in the atmosphere produce a large background of muons and neutrinos that require a careful selection of astrophysical neutrino candidates. One possibility to eliminate the background of atmospheric muons is the selection of up-going muon track events. The remaining background of atmospheric neutrinos follows a steep spectrum that can be distinguished form a hard astrophysical neutrino flux via a statistical analysis. Another possibility consists of the selection of high-energy starting events (HESE). These events are required to have an interaction vertex inside a virtual outer veto layer of optical modules. The advantage of this selection is that atmospheric neutrinos from above the detector (Southern Hemisphere) are accompanied by atmospheric muons that trigger the veto [1, 2].

In the following we will summarize the status of diffuse neutrino observations with IceCube and recent evidence of neutrino emission from blazars.

2 Diffuse Astrophysical Neutrinos

In 2013, the IceCube Observatory reported first evidence of TeV-PeV astrophysical neutrinos in an analysis of two years of data based on the HESE selection [3, 4]. First evidence of this astrophysical flux in the up-going $\nu_\mu + \bar{\nu}_\mu$ sample was found shortly afterwards [5]. With continued observation, both analyses have now reached a significance of more than 5σ [6–8]. The diffuse flux inferred from the two event selections is shown in Fig. 1. The HESE selection allows to study the astrophysical neutrino emission at lower energies due to the self-veto of atmospheric backgrounds. Above 100 TeV both results are consistent with a hard power-law spectrum, whereas at lower energies the HESE selection indicates a rising flux that deviates from a naive power-law behavior.

Both analyses are consistent with a power-law extrapolation of the neutrino emission above 1 PeV. Neutrinos at these energies are associated with primary CR nuclei with an energy of the order of 20 PeV in the case of protons or 1 EeV in the case of iron. These CRs are therefore located in the CR spectrum between the "knee" (2 – 5 PeV) and the "ankle" (3 – 5 EeV), which is the transition region between the dominance of Galactic and extragalactic sources. Both types of sources have been considered in the literature as possible neutrino production sites and a summary can be found in the recent review [9]. However, the arrival directions of neutrinos are consistent with an isotropic distribution, after correcting for the angular acceptance of the detector and Earth absorption. This suggests that the signal is likely to originate from a population (or even multiple populations) of relatively weak extragalactic sources.

Figure 1 shows the diffuse neutrino flux ϕ in terms of the product $E^2\phi$, which is a measure of its local energy density. The energy density in cosmic neutrinos is comparable to that of the isotropic γ-ray background (IGRB) observed with the Fermi satellite [10] (blue data in Fig. 1) and to that of ultra-high-energy (UHE) CRs (above 10^9 GeV) observed, *e.g.*, by the Auger observatory [11] (green data in Fig. 1). We will highlight in the following that this might indicate a common origin of these signals, which provides excellent conditions for multi-messenger studies.

The simultaneous production of neutral and charged pions in CR interactions suggests that the sources of high-energy neutrinos could also be strong TeV-PeV γ-ray emitters. For extragalactic

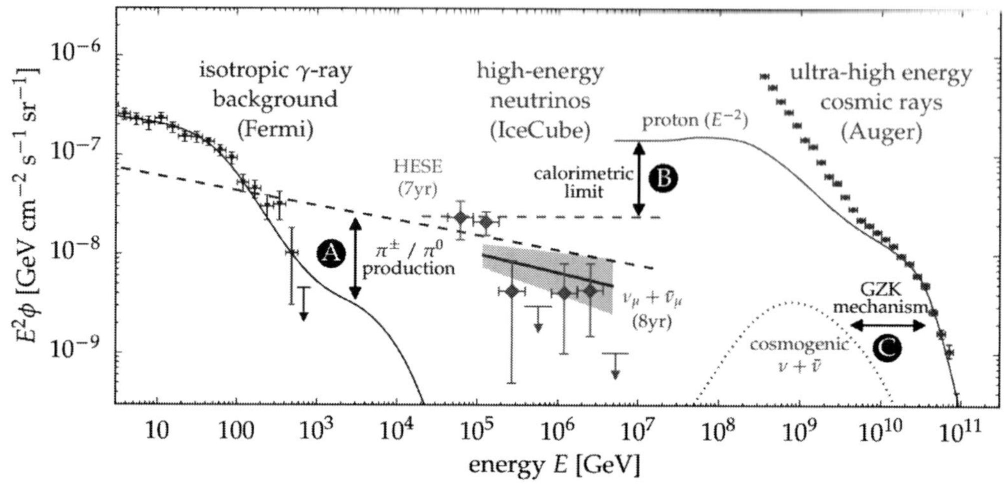

Figure 1. The spectral flux (ϕ) of neutrinos inferred from the eight-year upgoing track analysis (red fit) and preliminary results of the seven-year HESE analysis [8] (magenta data) compared to the flux of unresolved extragalactic γ-ray sources [10] (blue data) and ultra-high-energy cosmic rays [11] (green data). The $\nu_\mu + \bar{\nu}_\mu$ spectrum is indicated by the best-fit power-law (solid line) and 1σ uncertainty range (shaded range). We highlight the various multimessenger relations: **A:** The joined production of charged pions ($\pi^\pm$) and neutral pions (π^0) in cosmic-ray interactions leads to the emission of neutrinos (dashed blue) and γ-rays (solid blue), respectively. **B:** Cosmic ray emission models (solid green) of the most energetic cosmic rays imply a maximal flux (calorimetric limit) of neutrinos from the same sources (green dashed). **C:** The same cosmic ray model predicts the emission of cosmogenic neutrinos from the collision with cosmic background photons (GZK mechanism).

scenarios, this γ-ray emission is not directly observable because of the strong absorption of photons by e^+e^- pair production in the extragalactic background light (EBL) and the cosmic microwave background. The high-energy leptons initiate electromagnetic cascades of repeated inverse-Compton scattering and pair production in the CMB that eventually yield photons that contribute to the Fermi γ-ray observations in the GeV-TeV range.

The corresponding γ-ray and neutrino fluxes of a simple power-law emission model are shown as solid and dashed blue lines, respectively, in Fig. 1. The γ-ray flux is normalized to the Fermi data to indicate the maximal contribution of neutrinos. The observed neutrino flux above 100 TeV is very close to the corresponding upper limit (dashed blue line), which would imply a large contribution of the underlying source population to the γ-ray background. The neutrino flux below 100 TeV slightly overshoots this bound and poses a challenge for this type of emission models [12]. This suggests that the accompanying γ-rays produced via cosmic-ray interactions needs to be "hidden" by absorption in the neutrino sources [13]. Intriguingly, the mild tension between the power-law fits of the HESE and $\nu_\mu + \bar{\nu}_\mu$ data [8], respectively, might indicate that a second population of "hidden" sources dominate the HESE data below 100 TeV.

The extragalactic γ-ray background observed by Fermi [10] has contributions from identified point-like sources on top of the IGRB shown in Fig. 1. This IGRB is expected to consist mostly of emission from the same class of γ-ray sources that are individually below Fermi's point-source detection threshold (see, e.g., Ref. [14]). These underlying source populations can be deciphered via one-point fluctuation analyses [15–17] and the remaining isotropic γ-ray background allows to put stronger limits on neutrino emission models [18]. Alternatively, a significant contribution of γ-rays associated with IceCube's neutrino observation would have the implication that many extragalactic γ-ray sources are also neutrino emitters.

Another intriguing observation is that the high-energy neutrinos observed at IceCube could be related to the sources of UHE CRs. The simple argument is as follows: UHE CR sources can be embedded in environments that act as "storage rooms" for CRs with energies far below the "ankle". This energy-dependent trapping can be achieved via CR diffusion in magnetic fields. While these CRs are trapped, they can produce γ-rays and neutrinos via collisions with gas. If the conditions are right, this mechanism can be so efficient that the total energy stored in low-energy CRs is converted to that of γ-rays and neutrinos. The contribution of UHE CR sources to the diffuse neutrino flux is therefore bounded by a "calorimetric limit" [19, 20], which is indicated as green dashed line in Fig. 1

Interestingly, the observed diffuse neutrino flux is just below this calorimetric limit. It is therefore feasible that UHE CRs and neutrinos observed with IceCube have a common origin. If this is the case, the neutrino spectrum should reflect the energy-dependent release of CRs from the calorimeters. Ultra-high energy CRs are strongly attenuated by resonant interactions with background photons, as first pointed out by Greisen, Zatsepin and Kuzmin [21, 22] (GZK). This GZK mechanism is responsible for the suppression of the UHECR proton flux at the highest energies (green solid line) and predicts a detectable flux of cosmogenic neutrinos [23] (green dotted line). The observation of these cosmogenic neutrinos would provide a complementary measure of UHE CR composition models and the corresponding calorimetric limit. Future measurements of the diffuse PeV-EeV neutrino spectrum can provide supporting evidence for the UHE CR connection.

3 Neutrino Emission from Blazars

On September 22nd, 2017 IceCube observed a high-energy muon that entered the detector from a direction a few degrees below the Horizon (see left panel of Fig. 2). These rare type of events trigger alerts that are sent out to partner observatories to find hints of electromagnetic emission from the same part of the Universe. After one week the Fermi satellite reported that the neutrino alert IC-170922A was spatially coincident with a known γ-ray source, the blazar TXS 0506+056, that was undergoing a period of enhanced γ-ray emission. The chance correlation of the neutrino alert with the γ-ray source is at the level of 3σ and provides evidence that the blazar was responsible for the event [24].

The blazar's spectral energy distribution at the time of the outburst can be well modeled by lepto-hadronic and proton-synchrotron models [25–30]. It was pointed out that in these models predict a low neutrino flux ($\ll 1$ event) for the 2017 outburst that is limited by the observed level of the X-ray emission [25, 26] and the theoretically feasible proton luminosity, unless one considers enhanced pion production efficiencies in alternative models [31–34].

However, the low model predictions are not surprising. The emission of TXS 0506+056 has to be discussed in the larger context of the blazar population [35]. The collection of blazar flares might individually only contribute with a low neutrino flux, but their sum can still provide observable event numbers. The question, which flare will eventually be responsible for the observed neutrino, becomes then simply a matter of chance. This is the essence of the *Eddington bias*, *i.e.* the trend to overestimate the intrinsic neutrino flux of a source observed with one neutrino: TXS 0506+056 was just the lucky winner of the cosmic lottery [35].

Motivated by the correlation seen by Fermi, IceCube investigated archival data for evidence of past neutrino emission of TXS 0506+056. Figure 3 shows the local p-values of past neutrino outbursts of this analysis using a Gaussian (orange) and box-shaped (blue) neutrino time profile. This analysis revealed that during a period from September 2014 to March 2015 the sources showed a prolonged

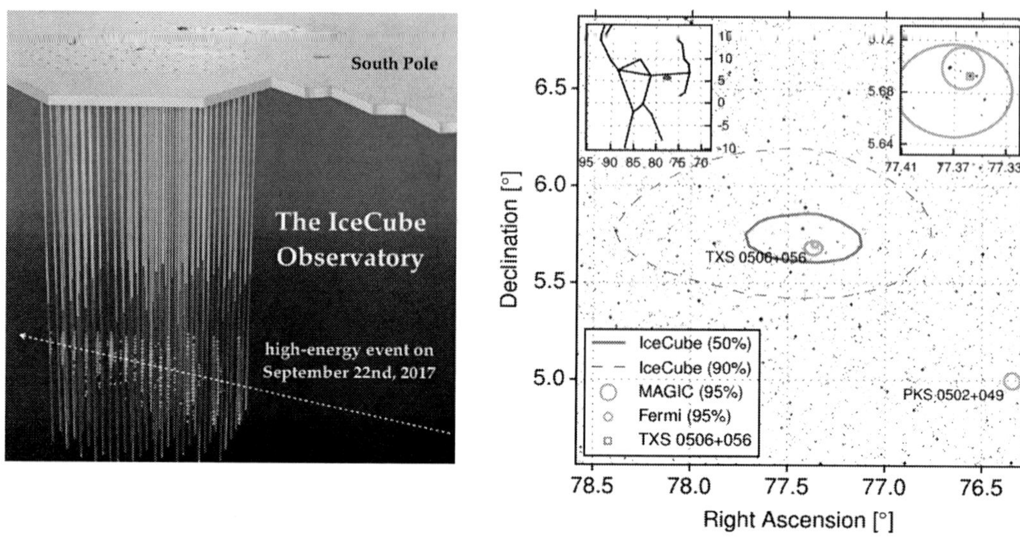

Figure 2. Left panel: A sketch of the IceCube Observatory at the South Pole. The detector consists of 5,160 digital optical modules (DOMs) that are attached along 86 support and read-out cables at a depth of 1,450 to 2,450 meters below surface. The alert IC-170922A is shown by colored DOMs, with color scale ranging from red to blue for early and late Cherenkov light detection, respectively. **Right panel:** The blazar TXS 0506+056 located just off the left shoulder of the constellation Orion (top-left inset plot). The contours indicate the localization in γ-rays (Fermi & MAGIC) and neutrinos (IceCube). (Figure from Ref. [24].)

outburst with an estimated 13 ± 5 neutrino candidates. A chance correlation of this type of neutrino outburst can be excluded at a confidence level of 3.5σ [36]. Together with the earlier observation, this provides compelling evidence that this blazar is a source of high-energy neutrinos.

The 2014/15 outburst of neutrinos is not accompanied by a flare in the archival Fermi data. The corresponding neutrino luminosity is about four times larger than the γ-ray luminosity in the quiescent state. There is a statistically weak hint that the Fermi spectrum during the time of the neutrino outburst shows a hard spectral index, that could indicate an unobserved outburst of TeV γ-rays [37]. In any case, the different photon and neutrino emissions in the 2014/15 and 2017 observations, respectively, are presently the greatest challenge in finding a unified model of the neutrino emission mechanism.

Are blazars then also responsible for the diffuse TeV-PeV neutrino emission? This question was already addressed in an earlier study by IceCube [38] looking for the combined time-integrated emission of blazars observed in γ-rays. This study could not find evidence for neutrino emission and placed an upper limit on the relative contribution of blazars to the diffuse neutrino flux. For an $E^{-2.5}$ spectrum this upper limit is at the level of 27%. In light of the recent results, it is therefore plausible that blazars have a large contribution, but it is unlikely that they dominate the observed diffuse neutrino flux. Future analyses with IceCube data can further probe this hypothesis by looking for time-dependent neutrino emission from blazar populations.

4 Conclusions

The future of neutrino astronomy is promising. Recent observations with IceCube have revealed a diffuse TeV-PeV neutrino flux of unknown origin. The intensity of the neutrino flux is comparable

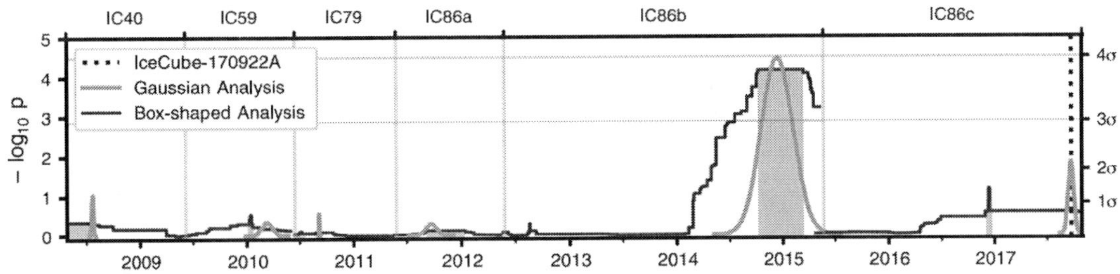

Figure 3. The local p-values of past neutrino outbursts in the direction of TXS 0506+056. The orange and blue line shows results using a Gaussian and box-shaped neutrino time-profile. The vertical dotted line shows the time of the alert IC-170922. (Figures from Ref. [36].)

to that of γ-rays and ultra-high energy cosmic rays, which provides excellent conditions for multi-messenger studies. We now also have first compelling evidence that blazars are sources of high-energy neutrinos, but they are unlikely the sole contributors of the diffuse flux. With the help of future optical Cherenkov telescopes – KM3NeT [39], Baikal-GVD [40], and IceCube-Gen2 [41] – we will be able to shed more light on these recent observations and establish high-energy neutrinos as an essential component for multi-messenger astronomy.

Acknowledgements

I would like to thank the organizers of the *7th Roma International Conference on Astroparticle Physics* for their invitation to present this work. I would also like to thank my colleagues in the IceCube collaboration for their support. This work was supported by Danmarks Grundforskningsfond (project no. 1041811001) and Villum Fonden (project no. 18994).

References

[1] S. Schonert, T.K. Gaisser, E. Resconi, O. Schulz, Phys. Rev. **D79**, 043009 (2009), 0812.4308
[2] T.K. Gaisser, K. Jero, A. Karle, J. van Santen, Phys. Rev. **D90**, 023009 (2014), 1405.0525
[3] M.G. Aartsen et al. (IceCube), Phys. Rev. Lett. **111**, 021103 (2013), 1304.5356
[4] M.G. Aartsen et al. (IceCube), Science **342**, 1242856 (2013), 1311.5238
[5] M.G. Aartsen et al. (IceCube), Phys. Rev. Lett. **115**, 081102 (2015), 1507.04005
[6] M.G. Aartsen et al. (IceCube), Phys. Rev. Lett. **113**, 101101 (2014), 1405.5303
[7] M.G. Aartsen et al. (IceCube), Astrophys. J. **833**, 3 (2016), 1607.08006
[8] M.G. Aartsen et al. (IceCube) (2017), 1710.01191
[9] M. Ahlers, F. Halzen, Prog. Part. Nucl. Phys. **102**, 73 (2018), 1805.11112
[10] M. Ackermann et al. (Fermi-LAT), Astrophys. J. **799**, 86 (2015), 1410.3696
[11] A. Aab et al. (Pierre Auger) (2015), 1509.03732
[12] K. Murase, M. Ahlers, B.C. Lacki, Phys. Rev. **D88**, 121301 (2013), 1306.3417
[13] K. Murase, D. Guetta, M. Ahlers, Phys. Rev. Lett. **116**, 071101 (2016), 1509.00805
[14] M. Di Mauro, F. Donato, Phys. Rev. **D91**, 123001 (2015), 1501.05316
[15] M. Ackermann et al. (Fermi-LAT), Phys. Rev. Lett. **116**, 151105 (2016), 1511.00693
[16] M. Lisanti, S. Mishra-Sharma, L. Necib, B.R. Safdi, Astrophys. J. **832**, 117 (2016), 1606.04101

[17] M. Di Mauro, S. Manconi, H.S. Zechlin, M. Ajello, E. Charles, F. Donato, Astrophys. J. **856**, 106 (2018), 1711.03111
[18] K. Bechtol, M. Ahlers, M. Di Mauro, M. Ajello, J. Vandenbroucke, Astrophys. J. **836**, 47 (2017), 1511.00688
[19] E. Waxman, J.N. Bahcall, Phys. Rev. **D59**, 023002 (1999), hep-ph/9807282
[20] J.N. Bahcall, E. Waxman, Phys. Rev. **D64**, 023002 (2001), hep-ph/9902383
[21] K. Greisen, Phys. Rev. Lett. **16**, 748 (1966)
[22] G.T. Zatsepin, V.A. Kuzmin, JETP Lett. **4**, 78 (1966), [Pisma Zh. Eksp. Teor. Fiz.4,114(1966)]
[23] V.S. Berezinsky, G.T. Zatsepin, Phys. Lett. **28B**, 423 (1969)
[24] M.G. Aartsen et al. (Liverpool Telescope, MAGIC, H.E.S.S., AGILE, Kiso, VLA/17B-403, INTEGRAL, Kapteyn, Subaru, HAWC, Fermi-LAT, ASAS-SN, VERITAS, Kanata, IceCube, Swift NuSTAR), Science **361**, eaat1378 (2018), 1807.08816
[25] S. Gao, A. Fedynitch, W. Winter, M. Pohl (2018), 1807.04275
[26] A. Keivani et al., Astrophys. J. **864**, 84 (2018), 1807.04537
[27] H. Zhang, K. Fang, H. Li (2018), 1807.11069
[28] M. Cerruti, A. Zech, C. Boisson, G. Emery, S. Inoue, J.P. Lenain (2018), 1807.04335
[29] A. Gokus, S. Richter, F. Spanier, M. Kreter, M. Kadler, K. Mannheim, J. Wilms (2018), 1808.05540
[30] N. Sahakyan, Astrophys. J. **866**, 109 (2018), 1808.05651
[31] S. Ansoldi et al. (MAGIC), Astrophys. J. Lett. (2018), [Astrophys. J.863,L10(2018)], 1807.04300
[32] C. Righi, F. Tavecchio, S. Inoue (2018), 1807.10506
[33] K. Murase, F. Oikonomou, M. Petropoulou, Astrophys. J. **865**, 124 (2018), 1807.04748
[34] R.Y. Liu, K. Wang, R. Xue, A.M. Taylor, X.Y. Wang, Z. Li, H. Yan (2018), 1807.05113
[35] N.L. Strotjohann, M. Kowalski, A. Franckowiak (2018), 1809.06865
[36] M.G. Aartsen et al. (IceCube), Science **361**, 147 (2018), 1807.08794
[37] P. Padovani, P. Giommi, E. Resconi, T. Glauch, B. Arsioli, N. Sahakyan, M. Huber, Mon. Not. Roy. Astron. Soc. **480**, 192 (2018), 1807.04461
[38] M.G. Aartsen et al. (IceCube), Astrophys. J. **835**, 45 (2017), 1611.03874
[39] S. Adrian-Martinez et al. (KM3Net), J. Phys. **G43**, 084001 (2016), 1601.07459
[40] A.D. Avrorin et al. (Baikal-GVD), EPJ Web Conf. **191**, 01006 (2018), 1808.10353
[41] M.G. Aartsen et al. (IceCube) (2014), 1412.5106

The AMS-02 detector on the ISS - Status and highlights, after the first 7 years on orbit

Matteo Duranti[1,a] *on behalf of the* AMS Collaboration

[1] *INFN Sezione Perugia*

Abstract. The Alpha Magnetic Spectrometer, AMS-02, detector is operating on the International Space Station (ISS) since May the 19th, 2011. More than 120 billion events have been collected by the instrument in the first 7 years of data taking, providing detailed insight on the features of different species of cosmic rays. This contribution reviews the recent AMS-02 results based on 7 years of operations in space and their contribution to the advances in the understanding of cosmic ray origin, acceleration and propagation physics.

1 The AMS experiment and its scientific objectives

The Alpha Magnetic Spectrometer, AMS-02, is a general purpose high energy particle physics detector. It was launched to space with the Space Shuttle STS-134 mission and installed onboard the ISS in 2011, on May 19*th*, to conduct a unique long duration mission (up to the lifetime of the ISS) of fundamental physics research in space. More than 120 billion events have been collected by the instrument in the first 7 years of data taking.
The scientific objectives of the AMS collaboration cover a wide range of measurements in the field of the fundamental physics and of the Charged Cosmic Rays (CCR) physics in general:

- direct search of antimatter of primordial origin, through the detection of antinuclei;
- indirect search of "exotic" sources, through the precise measurement of the rare antimatter components (positrons, antiprotons and antideuterons);
- strong constraints on the models of origination and acceleration of the CCR, through the accurate and high statistics measurement of the primary nuclear components (proton, Helium, Carbon, Oxygen, etc... fluxes);
- strong constraints on the models of propagation in the Galaxy and in the Inter-Stellar Medium (ISM), through the accurate and high statistics measurement of the secondary components and of the secondary-to-primary ratios (Boron/Carbon (B/C), and Boron flux, Lithium flux, etc...).

To reach its scientific goals, the AMS-02 instrument, whose schematic layout is reported in Fig.1 (left), has been conceived with a large redundancy in the measurement of particle properties with complementary techniques by different sub-detectors.
The detector core is the magnetic spectrometer, composed of a 0.14 T dipole permanent magnet and

[a] e-mail: matteo.duranti@pg.infn.it

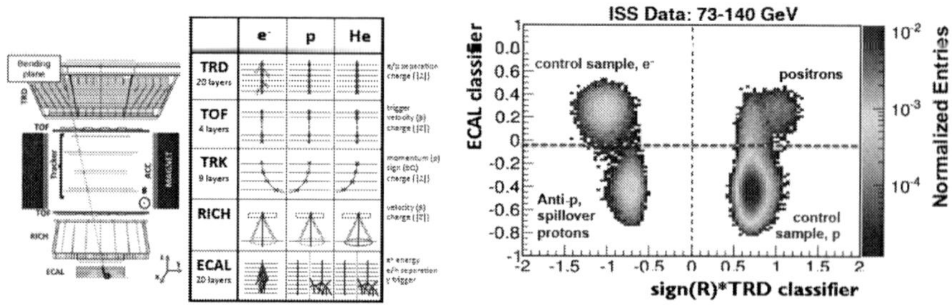

Figure 1. Left - Schematic view of the AMS-02 spectrometer (bending view). Center - A summary sketch of the measurement redundancy and complementarity of the sub-detectors of the AMS-02 apparatus. Right - Distribution of the events, in the energy range 73-140 GeV, in the *ECAL classifier - TRD classifier* plane. Negative from positive events are distinguished by means of the sign of the Rigidity in the spectrometer, *sign(R)*. The separation of the various components of e^-, e^+, p and $\bar{p}$ is clearly visible.

9 layers of double sided Si microstrip tracker detectors. The particle rigidity, R, is measured over a lever arm up to 3 m, the maximum detectable rigidity is 2 TV for $|Z| = 1$ particles and 3.2 TV for $|Z| \geq 2$ particles. The direction of the curvature inside the magnetic field identifies the sign of the charge and consequently separates matter from the rare antimatter CR component. The charge resolution of the inner tracker amounts to $\Delta Z/Z = 3.5\%$ for He and improves up to 1.5% for O nuclei. Four Time-Of-Flight (TOF) planes trigger the readout of the detector and measure the particle flight direction and velocity. The 17 X0 ECAL imaging calorimeter provides the measurement of the $e^{\pm}$ energy (E), and the topology of the shower is used to separate $e^{\pm}$ from hadrons. The energy deposit in the 20 layers of the TRD are used to further differentiate between $e^{\pm}$ and protons. The analysis of the radiation in the Ring Imaging Cherenkov detector (RICH) provides accurate information on nuclei charges and velocity. Details about the performances of the different sub-detectors can be found in the first published result paper [1] and in the references therein.

As sketched in Fig.1 (center) the large number of sub-detectors results, in terms of the measured particle's properties, in a large redundancy (as for example for the measurement of the charge) and, moreover, guarantees a full complementarity that allows for a full Particle Identification (matter from antimatter distinction, electromagnetic from hadronic separation, etc...) at the single event level.

The identification of the various nuclear components is granted by the measurement of the charge of the impinging particles, all along the detector. Essentially all the sub-detectors have some charge measurement capability: the most accurate is the inner tracker, providing a $\sim$ 0.1 c.u. charge resolution, with its 7 layers (each layer, alone, has a resolution of about 0.3 c.u.); the RICH and the TRD have a resolution of $\sim$ 0.3 c.u., while each pair of planes of the TOF has a resolution of $\sim$ 0.16 c.u.. This permits to follow the particle traversing the detector and to search for fragmentation that could spoil the charge reconstruction and constitute a source of background in the less abundant nuclear components (for example the secondaries). The dynamic range of the various detectors permits, with a resolution slightly degrading with Z, the charge measurement up to the higher charges: Iron and above (the Zinc peak is clearly visible) for Tracker, TOF and RICH, and up to $Z \sim 6$ for the TRD.

The discrimination power to the rare electromagnetic components, e^+ and e^-, from the overwhelming background of protons, and from the even rarer antiprotons component is sketched in Fig.1 (right). The two key detectors to distinguish e^+ and e^- from p and $\bar{p}$ are the ECAL (via *classifiers* exploiting the 3D imaging capability of the detector and built using multi-variate techniques such as BDT [2] or *log-likelihood* [3]), and the TRD (via a *log-likelihood* to the energy deposit in its straw tubes). The discrimination between matter and anti-matter (i.e. e^+ from e^- or $\bar{p}$ from p) is provided by the spectrometer. The redundancy is also a key capability that permits to build, for example using the TRD classifier and the sign of the rigidity, almost pure control samples of e^- and p, to study the performances of the ECAL.

2 Experimental results

The analysis of the enormous amount of data being collected by the experiment permitted the publication of high statistics and very accurate measurement of the more important CCR components. In particular, in 2018, the Collaboration published four papers:

- Nitrogen flux in the range 2.2-3300 GV [4], based on the first 5 years of the collected data, corresponding to 2.2 million nitrogen nuclei;
- Li, Be and B flux in the range 1.9-3300 GV [5], based on the first 5 years of the collected data, corresponding to 1.9 million lithium, 0.9 million beryllium and 2.6 million boron nuclei;
- p and He fluxes as a function of rigidity (up to 60 GV) and time [6], based on the first 6 years of collected data, corresponding to 79 Bartels [7] rotations;
- e^- and e^+ fluxes as a function of rigidity (up to 50 GV) and time [8], based on the first 6 years of collected data, corresponding to 79 Bartels [7] rotations;

The results about Lithium, Beryllium, Boron and Nitrogen [5] [4] are summarized in Fig.2: as He-C-O were found to have a nearly identical rigidity dependance at high energies, being all primaries [9], also Li-Be-B are found to have a nearly identical dependance, being all secondaries. Their dependance is distinctly different to the one of primaries. The Nitrogen, expected to have both primary and secondary origin, is found to have a distinct rigidity dependance that is well fitted as a weighted sum of the primary and secondary dependences.

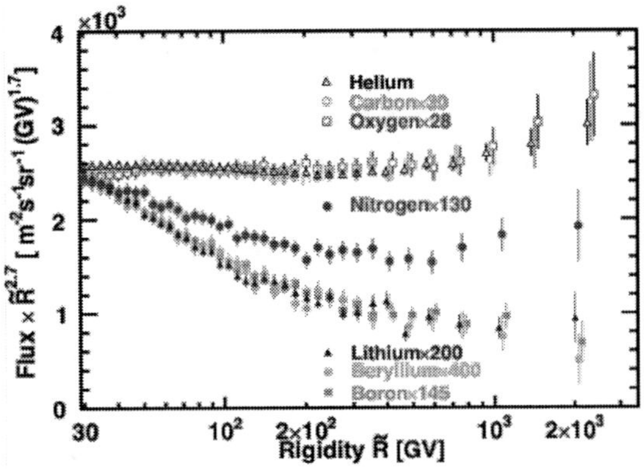

Figure 2. He-C-O, N and Li-Be-B fluxes, each one scaled to superimpose the other ones with similar rigidity dependence. The rigidity dependance of primaries (He-C-O), secondaries (Li-B-B) and nuclei with both the origins (N) are distinctly different.

The results about proton and helium [6] are summarized in Fig.3 (left): the proton and the helium fluxes show nearly identical short-term and long-term structures in both time and relative amplitude. The amplitudes of the flux structures decrease with increasing rigidity and vanish above 40 GV. The electron and positron fluxes [8], as shown in Fig.3 (right), present short-term structures on the timescale of months coincident in both the fluxes. These structures are not visible in the e^+/e^- flux ratio. Differently from p and He, the e^+ and e^- fluxes present also long-term structures that, even if coherent between the two species, have amplitudes markedly different, with the difference that is a smooth function of time.

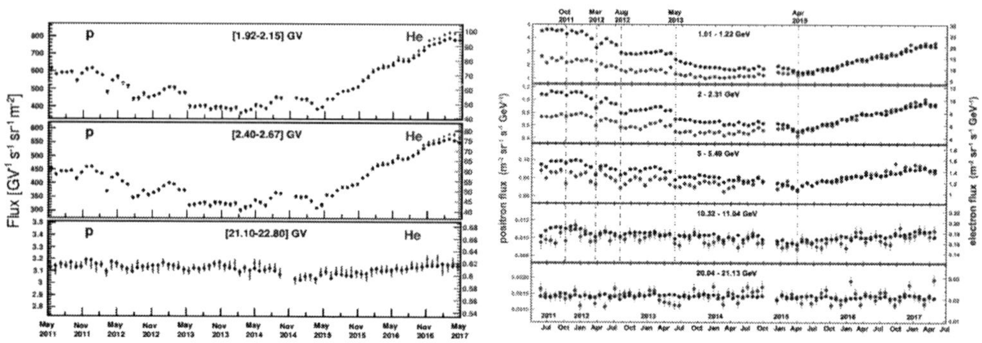

Figure 3. Left: proton (blue, left axis) and helium (red, right axis) fluxes as function of time for 3 rigidity bins. Detailed structures are clearly present below 40 GV and are nearly identical for the two species. Right: fluxes of positrons (red, left axis) and electrons (blue, right axis) as functions of time, for five energy bins. Prominent and distinct time structures visible in both the positron spectrum and the electron spectrum and at different energies are marked by dashed vertical lines. Long-term trends, instead, are markedly different between the two species.

These new results published in 2018 are part of the systematic campaign of measurement of the AMS-02 physics program:

- $e^+/(e^+ + e^-)$ in the range 0.5-500 GeV [1, 10], based on 30 months and ~ 10.9 million positron and electron events;
- positron flux in the range 0.5-500 GeV and electron flux in the range 0.5-700 GeV [11], based on 30 months and ~ 0.6 million positrons and ~ 9.2 million electrons ;
- *all-electron*, $e^+ + e^-$, flux, in the range 0.5-1000 GeV [12], based on 30 months and ~ 10.6 million positron and electron events;
- p flux in the range 1-1800 GV [13], based on 30 months and ~ 300 million protons;
- He flux in the range 1.9-3000 GV [14], based on 30 months and ~ 50 million helium events;
- antiproton flux and $\bar{p}/p$ ratio in the range 1-450 GV [15], based on the first 4 years of the collected data, corresponding to ~ 350000 antiproton and ~ 2.4 billion proton events;
- B/C, ratio in the range 1.9-2600 GV [16], based on the first 5 years of the collected data, corresponding to 2.3 million boron and 8.3 million carbon nuclei;
- He, C and O fluxes in the range ~ 2-3000 GV [9], based on the first 5 years of the collected data, corresponding to ~ 90 million helium, 8.4 million carbon and and 7.0 million oxygen nuclei.

All these measurements, together, can be seen in Fig.4, where each flux is scaled differently, just for display purpose, to superimpose the other ones with a similar trend as a function of the energy. The resulting 'scheme' is populated, so far, by five different 'families' of fluxes: electrons, Li-Be-B,

Nitrogen, protons-antiprotons and positrons, He-C-O.

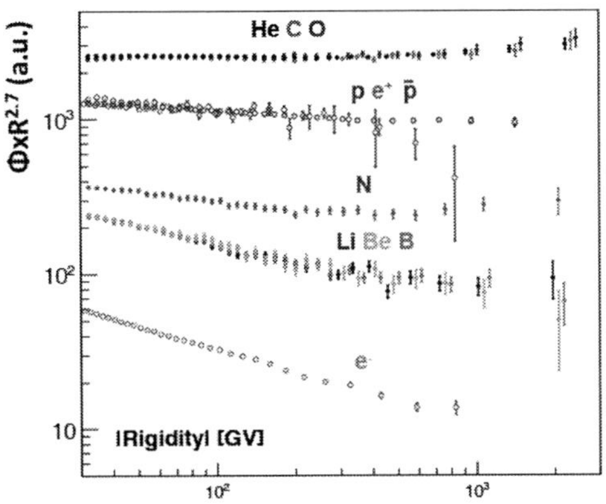

Figure 4. The fluxes measured by the AMS-02 experiment up to 2018. Each flux is scaled differently, just for display purpose, to superimpose the other ones with a similar trend as a function of the energy.

The big amount of collected data is being analyzed for the measurements of the other nuclear components (F, Ne, Na, Mg, Al, Si, etc..) and for the study of the other characteristics of the already published spectra (for example the variability of the C and O fluxes as a function of time). Of particular interest are the extension, in both statistics (factor 3) and energy reach (factor 2) of the $e^{\pm}$ fluxes and ratios, to be submitted for publication by the end of 2018. In 2019 the measurements of nuclei will be broadened by the first isotopic results (such as $^3He/^4He$, $^6Li/^7Li$ and deuteron). In the next couple of years also the antimatter search (i.e. anti-deuteron and anti-Helium) analyses should be mature and deep enough for publication.

3 Conclusions

Data recorded in the first $\sim$ 7 years of mission are being analyzed to cover the physics program of the AMS-02 experiment: all the measurements performed so far show interesting features, extending previous observations and adding new findings whose experimental accuracy challenges the actual theoretical predictions. AMS-02 represents the first instrument with the capability to simultaneously study all the different nuclear CR species, including the less abundant ones and up to Iron, electrons and anti-matter with unprecedented accuracy and in an extended energy range. These unique features will allow AMS-02 to shed light on new phenomena and improve the current understanding of CR origin and propagation.

Acknowledgments

This work has been supported by acknowledged persons and institutions in the AMS-02 publications [1, 4–6, 8–16], as well as by the Italian Space Agency under ASI-INFN Agreement No. 2013-002-R.0.

References

[1] M. Aguilar, G. Alberti, B. Alpat, A. Alvino, G. Ambrosi, K. Andeen, H. Anderhub, L. Arruda, P. Azzarello, A. Bachlechner et al. (AMS Collaboration), Phys. Rev. Lett. **110**, 141102 (2013)

[2] B.P. Roe, H.J. Yang, J. Zhu, Y. Liu, I. Stancu, G. McGregor, Nuclear Instruments and Methods in Physics Research Section A: Accelerators, Spectrometers, Detectors and Associated Equipment **543**, 577 (2005)

[3] A. Kounine, Z. Weng, W. Xu, C. Zhang, Nuclear Instruments and Methods in Physics Research Section A: Accelerators, Spectrometers, Detectors and Associated Equipment **869**, 110 (2017)

[4] M. Aguilar, L. Ali Cavasonza, B. Alpat, G. Ambrosi, L. Arruda, N. Attig, S. Aupetit, P. Azzarello, A. Bachlechner, F. Barao et al. (AMS Collaboration), Phys. Rev. Lett. **121**, 051103 (2018)

[5] M. Aguilar, L. Ali Cavasonza, G. Ambrosi, L. Arruda, N. Attig, S. Aupetit, P. Azzarello, A. Bachlechner, F. Barao, A. Barrau et al. (AMS Collaboration), Phys. Rev. Lett. **120**, 021101 (2018)

[6] M. Aguilar, L. Ali Cavasonza, B. Alpat, G. Ambrosi, L. Arruda, N. Attig, S. Aupetit, P. Azzarello, A. Bachlechner, F. Barao et al. (AMS Collaboration), Phys. Rev. Lett. **121**, 051101 (2018)

[7] J. Bartels, Terrestrial Magnetism and Atmospheric Electricity **39**, 201 (1934), `https://onlinelibrary.wiley.com/doi/pdf/10.1029/TE039i003p00201`

[8] M. Aguilar, L.A. Cavasonza, G. Ambrosi, L. Arruda, N. Attig, S. Aupetit, P. Azzarello, A. Bachlechner, F. Barao, A. Barrau et al. (AMS Collaboration), Phys. Rev. Lett. **121**, 051102 (2018)

[9] M. Aguilar, L. Ali Cavasonza, B. Alpat, G. Ambrosi, L. Arruda, N. Attig, S. Aupetit, P. Azzarello, A. Bachlechner, F. Barao et al. (AMS Collaboration), Phys. Rev. Lett. **119**, 251101 (2017)

[10] L. Accardo, M. Aguilar, D. Aisa, B. Alpat, A. Alvino, G. Ambrosi, K. Andeen, L. Arruda, N. Attig, P. Azzarello et al. ((AMS Collaboration)), Phys. Rev. Lett. **113**, 121101 (2014)

[11] M. Aguilar, D. Aisa, A. Alvino, G. Ambrosi, K. Andeen, L. Arruda, N. Attig, P. Azzarello, A. Bachlechner, F. Barao et al. ((AMS Collaboration)), Phys. Rev. Lett. **113**, 121102 (2014)

[12] M. Aguilar, D. Aisa, B. Alpat, A. Alvino, G. Ambrosi, K. Andeen, L. Arruda, N. Attig, P. Azzarello, A. Bachlechner et al. ((AMS Collaboration)), Phys. Rev. Lett. **113**, 221102 (2014)

[13] M. Aguilar, D. Aisa, B. Alpat, A. Alvino, G. Ambrosi, K. Andeen, L. Arruda, N. Attig, P. Azzarello, A. Bachlechner et al. (AMS Collaboration), Phys. Rev. Lett. **114**, 171103 (2015)

[14] M. Aguilar, D. Aisa, B. Alpat, A. Alvino, G. Ambrosi, K. Andeen, L. Arruda, N. Attig, P. Azzarello, A. Bachlechner et al. (AMS Collaboration), Phys. Rev. Lett. **115**, 211101 (2015)

[15] M. Aguilar, L. Ali Cavasonza, B. Alpat, G. Ambrosi, L. Arruda, N. Attig, S. Aupetit, P. Azzarello, A. Bachlechner, F. Barao et al. (AMS Collaboration), Phys. Rev. Lett. **117**, 091103 (2016)

[16] M. Aguilar, L. Ali Cavasonza, G. Ambrosi, L. Arruda, N. Attig, S. Aupetit, P. Azzarello, A. Bachlechner, F. Barao, A. Barrau et al. (AMS Collaboration), Phys. Rev. Lett. **117**, 231102 (2016)

EPJ Web of Conferences **209**, 01015 (2019) https://doi.org/10.1051/epjconf/201920901015
RICAP-18

Baikal-GVD: first results and prospects

A.D. Avrorin[1], A.V. Avrorin[1], V.M. Aynutdinov[1], R. Bannash[9], I.A. Belolaptikov[2], V.B. Brudanin[2], N.M. Budnev[3], G.V. Domogatsky[1], A.A. Doroshenko[1], R. Dvornicky[2,7], A.N. Dyachok[3], Zh.-A.M. Dzhilkibaev[1], L. Fajt[7,8], S.V. Fialkovsky[5], A.R. Gafarov[3], K.V. Golubkov[1], T.I. Gress[3], Z. Honz[2], K.G. Kebkal[9], O.G. Kebkal[9], E.V. Khramov[2], M.M. Kolbin[2], K.V. Konischev[2], A.V. Korobchenko[2], A.P. Koshechkin[1], A.V. Kozhin[4], V.F. Kulepov[5], D.A. Kuleshov[1], M.B. Milenin[5], R.A. Mirgazov[3], E.R. Osipova[4], A.I. Panfilov[1], L.V. Pan'kov[3], E.N. Pliskovsky[2], M.I. Rozanov[6], E.V. Rjabov[3], V.D. Rushay[2], G.B. Safronov[2], B.A. Shaybonov[2], M.D. Shelepov[1], I. Stekl[8,2], F. Simkovic[7,2], A.V. Skurihin[4], A.G. Solovjev[2], M.N. Sorokovnikov[2], O.V. Suvorova[1,a], V.A. Tabolenko[3], B.A. Tarashansky[3], S.A. Yakovlev[9], A.V. Zagorodnikov[3], and V.L. Zurbanov[3]

[1]*Institute for Nuclear Research, Moscow, 117312 Russia*
[2]*Joint Institute for Nuclear Research, Dubna, 141980 Russia*
[3]*Irkutsk State University, Irkutsk, 664003 Russia*
[4]*Institute of Nuclear Physics, Moscow State University, Moscow, 119991 Russia*
[5]*Nizhni Novgorod State Technical University, Nizhni Novgorod, 603950 Russia*
[6]*St. Petersburg State Marine Technical University, St. Petersburg, 190008 Russia*
[7]*Comenius University, Mlynská dolina F1, SK-842 48 Bratislava, Slovakia*
[8]*Czech Technical University in Prague, 12800 Prague, Czech Republic*
[9]*EvoLogics, Germany*

Abstract. Next generation cubic kilometer scale neutrino telescope Baikal-GVD is currently under construction in Lake Baikal. The detector is specially designed for search for high energies neutrinos whose sources are not yet reliably identified. Since April 2018 the telescope has been successfully operated in complex of three functionally independent clusters i.e. sub-arrays of optical modules (OMs) where now are hosted 864 OMs on 24 vertical strings. Each cluster is connected to shore by individual electro-optical cables. The effective volume of the detector for neutrino initiated cascades of relativistic particles with energy above 100 TeV has been increased up to about 0.15 km^3. Preliminary results obtained with data recorded in 2016 and 2017 are discussed.

1 Introduction

Construction and continuous operation of gigaton volume neutrino telescopes in the Northern hemisphere (Baikal-GVD and KM3Net) is highly relevant for acquiring complementary astrophysical neutrino observations for fluences identified by the IceCube detector since 2013 [1], including the first observed PeV-scale events of unidentified origins. Moreover, the 300 TeV neutrino fluence from BL LAC galaxy TX-0506-056 (event IC170922A [2]) is expected to be confirmed by comparable detectors. Experimental search for high energy astrophysical fluences provides the most successful and

[a]e-mail: suvorova@cpc.inr.ac.ru

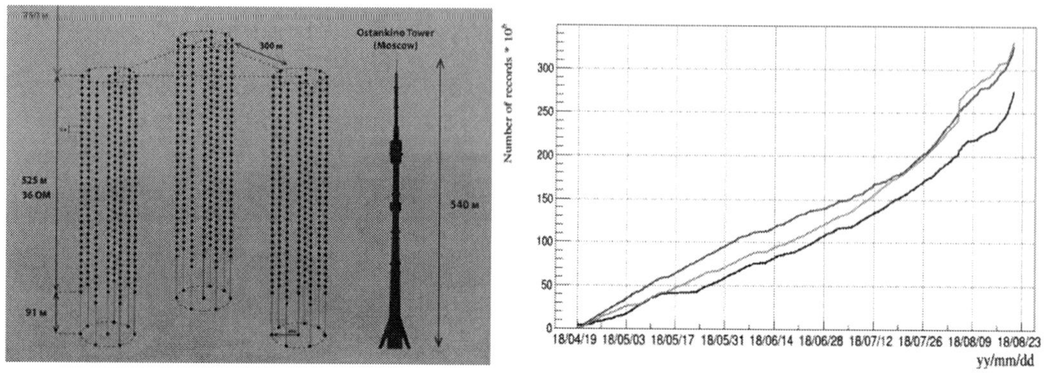

Figure 1. Left: Schematic drawing of the Baikal-GVD three clusters operated in 2018. Right: Integrated number of events recorded by the clusters in period of April 10–September 2 of 2018.

reliable results when based on multimessenger approach and a multi-domain alert system, as exemplified by the gravitation wave event GW170817A [3]. Furthermore, the combined analysis of data acquired at multiple wavelengths provides a better understanding of physical processes of particles production with energies above hundreds of TeV and PeV inside astrophysical sources and allows us to discover new sources of this kind.

Deep underwater neutrino telescope Baikal-GVD (Gigaton Volume Detector) is currently under construction in the southern basin of Lake Baikal [4], about four kilometres away from the shore. Lake depth at the site is $\sim$ 1360m. The site was chosen because of an optimal combination of hydrological, hydro-physical and landscape factors for deployment and operation of a neutrino telescope. Baikal-GVD has a modular structure and consists of functionally independent sub-arrays (clusters). Each cluster is an array of photodetectors (optical modules, OMs), i.e. photomultiplier tubes housed in transparent pressure spheres [5]. In a baseline cluster configuration there are 288 OMs installed on 8 vertical load-bearing cables (strings) with instrumented length of 525 meters (active depths 735–1260 m). Each cluster is connected to the shore via an individual electro-optical cable used for transmitting electrical power and data. Staring April 2018, three clusters of the telescope totalling 864 OMs and 24 strings have been acquiring data. As shown on Fig. 1 (left), the cluster centers are spaced at 300 m. The effective volume of the telescope in present configuration is about 0.15 km^3 for incoming particle energies above 100 TeV.

2 Baikal-GVD detector and the data processing

The first configuration of the Baikal-GVD was the demonstration cluster "Dubna" deployed in April 2015 [7]. It consisted of eight 345 m long strings of optical modules spaced at 40 m (192 OMs in total). Each string comprised two sections of OMs. Due to successful operation of "Dubna" and the registration efficiency 72% in period of April 2015 - February 2016, in next winter expedition 2016 the cluster has been extended by additional third top sections at each string and by prolongation of radius to 60 meters for 7 side strings from central one, according to the baseline configuration of GVD cluster. The upgraded cluster comprises 288 OMs operated at active depths of 735–1260 meters. Instrumented volume of cluster encloses 5.9 Megatons. During 2017 and 2018, each year, one new cluster had been deployed. In total 3 clusters are operated currently.

Basic element of the Baikal-GVD structure is a section of 12 OMs distributed vertically along the string, spaced by 15m, and a central electronics module (CeM). There are 3 sections of 36 OMs per string and 8 strings in cluster, each is an independently working sub-array of the Baikal-GVD. One photomultiplier tube Hamamatsu R7081-100 with a 10-inch hemispherical photocathode and quantum efficiency up to 35% is used as light sensor of OM [8]. The signals arriving from the OMs are digitized in ADC boards located at the CeM (12 channels with sampling frequency 200 MHz). When digitized, the signal from the ADC is transferred into programmable gate array (FPGA), which logic has been configured for generating a trigger, reading data from the ADC channels, and transmitting them over the Ethernet channel into the data acquisition center. A basic trigger mode is a signal amplitude in 4 FADC counts exceeding low and high thresholds in two adjacent OMs within 100 ns in a section: for thresholds 1.5 p.e. and 4.5 p.e. the trigger rate is in range of 60 -110 Hz. When the trigger condition is fulfilled, a request signal is generated to the cluster center module (CC) through 1 km long line. CC generates an acknowledgement signal to all CMs in a cluster, when the request signal is received. When CM is caught the signal, a timestamp is defined and the CM starts to form data. The FPGA memory buffer allows to acquire the OM signal waveform for a time interval as long as 5 μs, this information then is transferred to the shore station and stored as raw data in binary files. Such a trigger system approach allows to have all signal waveforms on each channel in an event from only one triggered pair of adjacent channels, while the data arriving from the sections can be processed in real time mode. Additionally, there are raw data from the acoustic positioning system and from OM monitoring. Details of data acquisition, basic controls, methods of calibrations, hard-and soft-ware triggers can be found in [9] and [10].

All raw data, for an amount as large as 40 Gbyte per cluster per day, are transferred to the shore station, stored there, and then transmitted at 5Mbit/s on a 40km channel to Baikalsk (Irkutsk region), where data are archived and transferred through a high-bandwidth internet to the central storage and processing facility in the JINR in Dubna (Moscow region) every 4 hours. The cumulative number of section records generated by array in period of April – September of 2018 is shown in Fig. 1 (centre). Preliminary results with data sample of 2016 and 2017 years are presented in section 4 and some more in Ref. [6].

3 Two reconstruction modes

There are two developed modes in the reconstruction of Cherenkov radiation being emmitted by cascade particles and straight-line moving muons generated in the HE neutrino interactions with Baikal water. A long-term experience of the Baikal collaboration in the search for diffuse neutrino flux with NT200 array using cascade mode [11, 12] was presently applied to a search for high-energy neutrino flux of astrophysical origin with energies higher 100 TeV by the selection of cascade events with the vertex interaction within the sensitive volume of the Baikal-GVD array [13].

After a precise determination of the OMs position and after the calibration of the OMs timing and pulse amplitude, a study of the OM rates is performed in order to reject the noise due to PMT dark current and to chemiluminescence background in the Baikal water. Rates of noise impulses vary between 20 and 100 kHz depending on season and depth. The noise is suppressed by causality requirement: $|t_i - t_j| \leq \frac{\Delta R_{ij}}{c_w} + t_s$, where t_i, t_j are impulse times, ΔR_{ij} is distance between modules, c_w is speed of light in water and $t_s = 10$ ns. In cascades reconstruction the cascade hypothesis is tested and the quality cuts are applied. First the shower vertex coordinates are reconstructed by χ^2_t minimization using the time information from the GVD clusters's triggered OMs. Based on the reconstructed shower coordinates the shower energy and the direction are reconstructed by applying the maximum-likelihood. For each event as shower energy (E_{sh}) and direction (zenith (θ) and azimuth (ϕ) angles) are taken the

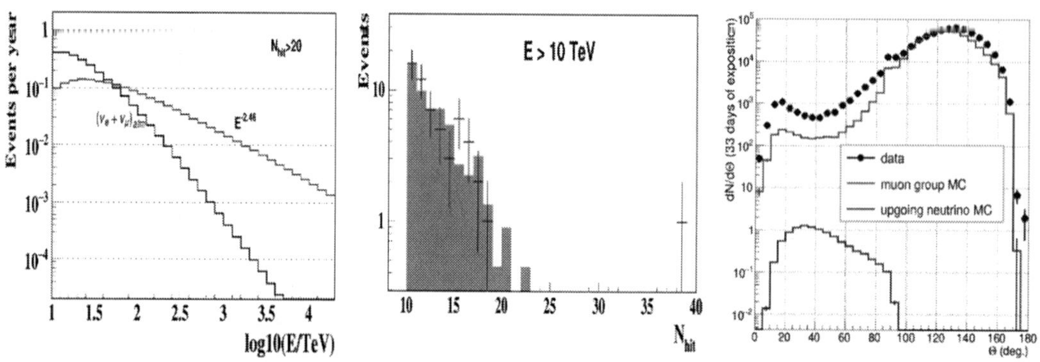

Figure 2. Left: Energy distributions of events expected for one year observation from astrophysical fluxes with $E^{-2.46}$ spectra and IceCube normalization, and from atmospheric neutrinos. Centre: Measured distribution of hit OMs multiplicity in cascade-like events with energies above 10 TeV and expected one from atmospheric muons (green histogram). Right: Zenith angle distribuitions reconstructed with data samlpe of muon-like events (dots) and with MC atmospheric muons (red) and neutrinos (blue).

values that maximize the event likelihood. Finaly, events with a multiplicity of hit OMs N_{hit} >20 are selected as high-energy neutrino events. The accuracy of the cascade energy reconstruction is about 30%, the accuracy of direction reconstruction is about 4.5° (median value) and vertex resolution is about 2 m [5]. Shown in Fig. 2 (left) are distributions of the expected number of events per year from astrophysical flux with $E^{-2.46}$ spectrum and the IceCube normalization and from atmospheric muons. An estimate of number of astrophysical events in one year per cluster with energies above 100 TeV is about 0.6 and for background events about 0.08. In muon reconstruction χ_t^2 minimization is done first in simple track position and direction estimation for each causally-connected group. OM hits which do not fall into model of muon propagation and direct Cherenkov light emission are excluded in gradually tightening set of cuts on hit residuals. The noise contamination to the sample of hits selected with the described procedure amounts to 1-2% depending on the elevation of muon track. The resulting median resolution of the procedures measured in up-going neutrino Monte Carlo sample is at the level of 1 degree [6].

4 Selected results with neutrino candidates

The first neutrino candidate with energy 107 TeV have been found with data sample of 2015 of the cluster "Dubna" [7]. The data collected with first of three Baikal-GVD clusters are presented here. Total number of accumulated events between April 2016 and January 2017 is 6.86×10^8 for effective time 182 days and conditions of low/high thresholds 1.5/4 ph.el. and hit charges ≥1.5 ph.el. on ≥ 3 strings. After applying causality cut and requirement of N ≥ 4 hit OMs data sample reduced to 3.3×10^8 events. Finally, 57 events with reconstructed energies E_{sh} > 10 TeV. Hit OM multiplicity distribution of these events is shown in Fig. 2 (center); also shown is expected distribution of background events from atmospheric muons. Five events are reconstructed with energy E_{sh} > 100 TeV. While four of them have hit OM multiplicities consistent with expected distribution of background events from atmospheric muons, one event with 38 hit OMs was reconstructed as downward moving contained cascade. For a more precise reconstruction of cascade parameters, this event was reanalysed including hits with charges lower than 1.5 ph.el. A total of 53 hits is consistent with a cascade hypothesis

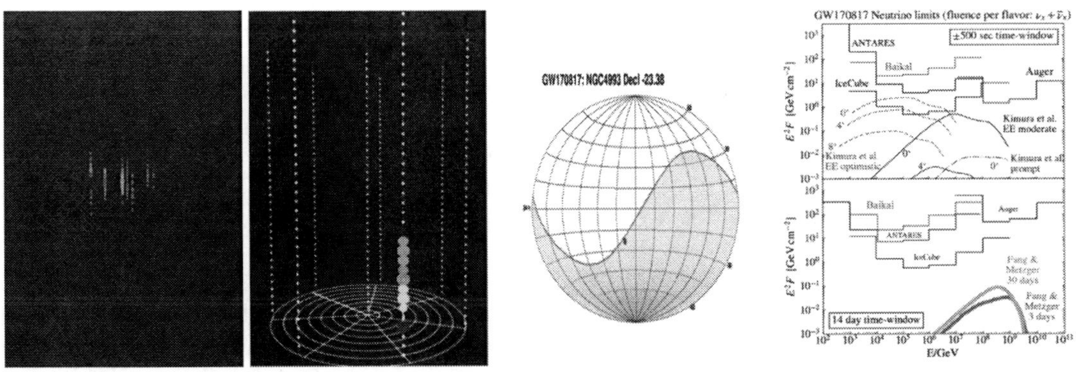

Figure 3. Left1: Reconstructed cascade event with energy 154.9 TeV. Left2: Reconstructed vertical event with 10 hit OMs. Centre: Horizon of the Baikal-GVD in alert time of the GW170817A (blue point). Right: The Baikal-GVD upper limits on neutrino fluences from direction of the GW170817A event (see text).

and reconstructed cascade parameters are as follows: cascade energy E_{sh} = 154.9 TeV, zenith angle $\theta = 57.3°$ and azimuthal angle $\phi = 249.4°$, distance from array axis $\rho = 44.7$ m. Found neutrino candidate in the cascade event is shown in Fig. 3 (left 1) after noise hit rejection. Since 2015 we have selected two cascade events with high energy above 100 TeV. The coordinates of the first cascade event in equatorial system are $\delta = 5.56°$, $\alpha = 139.7°$ for MJD 57342. For the second cascade: $\delta = 13.95°$, $\alpha = 173.35°$ for MJD 57507.

For the search for upward going muon neutrinos a set of 70 runs was chosen within the data collected by the first cluster in 2016. Total exposition time of the analysed sample is close to 33 days. Plot in Fig. 2 (right) shows reconstructed angular distribution in data compared to the prediction of Monte Carlo simulation of the detector response to atmospheric muon flux. Good agreement in shape of distributions is achieved, although muon rate in data is 1.5 times larger than in MC. That discrepancy requires our additional study to reject contribution from down going muon bundles, which could most likely be mis-reconstructed as upward going muons from neutrinos. First iteration in the selection of neutrino events was developed by the procedure based on boosted decision tree (BDT) as implemented in TMVA framework [14]. BDT was trained on events reconstructed as upward going in MC samples of atmospheric upward going neutrinos (signal) and atmospheric muons (background). Then BDT value was reconstructed for the data events. Those events with BDT value larger than found value for good discrimination of signal and backgorund were selected. Overall, 23 neutrino candidate events for 33 live days were found in data sample, while 42 events are expected from up-going neutrino MC. Number of expected background from atmospheric muons is about 6. Improvements in quality of track-like reconstruction is the next iteration of the data sample analysis.

Separately, a data analysis approach was developed for nearly vertical upward going muons, that takes into account particular features of single string event reconstruction. Within the whole data sample of 182 live days in 2016 there have been selected 5674 neutrino candidates. In this sample, there was one candidate event with 10 hits, 6 events with 8 hits, 15 events with 7 hits and 144 events with 6 hits. No candidate events with 9 hits were found. The one event with 10 hits is shown in Fig. 3 (left 2). Expected near vertical neutrino fluence can exceed number of background atmoshperic neutrino events as long as there can be an addtional flux arising from the Earth's core. For the hypothesis of dark matter in form of weakly interacting massive particles, WIMPs have been accumulated inside the Earth during a time comparable to the Universe age and currently weakly annihilate into ordinary matter, producing high energy pairs of neutrinos and antineutrinos through decays of particles.

Off-line analysis has been done in cascade mode with two cosmic events over fixed direction and alert time: GW170817A, gravitation wave origined in galaxy NGC4993 from merged double neutron star, and IC170922A, arising activity of BL LAC galaxy TX-0506+056. Two GVD-clusters were operational in 2017. In looking for fluence in prompt emission (±500 seconds) from GW170817A the source was located slightly below the horizon for Baikal-GVD (zenith angle 93°, horizon is shown in Fig. 3 in center), while in case of the IC170922A the source was above the horizon ($\theta = 57°$). No neutrino events associated with these sources have been found in cascade search mode both in the prompt emissions and in the emission delayed on 14 days after alert times. More details of analysis are presented in Ref. [15]. Assuming E^{-2} spectral behavior and equal fluence in neutrino flavors, upper limits at 90% c.l. have been derived on the neutrino fluence from GW170817A for each energy decade and are shown in Fig. 3 (on right).

In summary, Baikal-GVD comprises total of 864 OMs arranged at 24 strings and becomes, at present, the largest underwater neutrino telescope. The first two promising high-energy cascade events have been selected as candidates for events from astrophysical neutrinos in collected data sample of 2015-2016 years as well the first neutrino-like events are reconstructed in muon track mode. The modular structure of Baikal-GVD design allows studies of neutrinos of different origin at early stages of Baikal-GVD construction, as was done in search for neutrinos assosiated with GW170817A. By 2020, it is planned the GVD to be consisted of 2304 cumulative number of OMs on 8 clusters with a total volume of about 0.4 km^3, that aims on discovery of non-atmospheric neutrinos.

This work was supported by the Russian Foundation for Basic Research (Grants 16-29-13032, 17-02-01237). O.V.Suvorova acknowledges support by RSCF grant 17-12-01547 in part of works on analysis of GW event and events incoming from the Earth core direction.

References

[1] M.G.Aartsen et al. , Phys. Rev. Lett. 113 101101. Science 342 1242856 (2013).
[2] M.G.Aartsen et al., IceCube Coll., Science, 361, eaat1378 (2018)
[3] B. Abbott et al., Phys. Rev. Lett., **119**, 161101 (2017).
[4] V.Aynutdinov et al., NIM A742 82-88 (2014).
[5] A.D.Avrorin et al., Phys.Part.Nucl. 46 no.2, 211-221 (2015).
[6] A.D.Avrorin et al., EPJ Web Conf. 191 (2018) 01006 (2018).
[7] A.D.Avrorin et al., EPJ Web Conf. EPJ Web Conf. 136 (2017) 04007.
[8] A.D.Avrorin et al., EPJ Web Conf. 116 , 01003 (2016).
[9] A.V.Avrorin et al., Instr. and Exp. Tech. 57, issue 3, 262-273 (2014).
[10] A.Avrorin et.al., PoS (ICRC2017)**1046**, (2017)
[11] V. Aynutdinov et al., Astropart.Phys. **25**, 140 (2006).
[12] A. Avrorin et al., Astronomy Letters **35**, 651 (2009).
[13] A.D. Avrorin et al., PoS (ICRC2017)**962**, (2017).
[14] A.Hoecker et al., arXiv:physics/0703039 (2009).
[15] A.D. Avrorin et al., arXiv:1810.10966[astro-ph.HE] submitted to Pis'ma v ZhETF.

Giant Molecular Clouds as probes of of Galactic Cosmic Rays with *Fermi*-LAT

Giada Peron[1,a], Felix Aharonian[1,2,3,b], Sabrina Casanova[1,4,c], Ruizhi Yang[1,d], and Roberta Zanin[1,e]

[1]*Max-Planck-Institut für Kernphysik P.O. Box 103980, 69029 Heidelberg, Germany*
[2]*Dublin Institute for Advanced Studies, 31 Fitzwilliam Place, Dublin 2, Ireland*
[3]*Gran Sasso Science Institute, Viale Francesco Crispi 7, 67100, L'Aquila, Italy*
[4]*Institute of Nuclear Physics, Radzikowskiego 152, 31-342 Krakow, Poland*

Abstract. Direct measurements of Cosmic Rays (CRs) can not extend much further than the Solar System. In order to probe the so-called "sea" of Galactic Cosmic Rays, one should rely on secondary emission. Gamma rays produced in the interstellar gas, when hit by high energy particles, trace the energy distribution of the parent CRs. Giant Molecular Clouds, being huge reservoirs of hydrogen, serve as perfect targets for interaction with CRs. The high particle density in these objects enables to have enhanced γ emission from small isolated location, and hence to derive information about single spots of the Milky Way. We analyzed more than 9 years data of *Fermi*-LAT from 18 molecular clouds, located in different regions from 100 pc to more than 10 kpc from us, allowing us to have the most comprehensive study on Galactic Cosmic Rays from Molecular Clouds ever.

Cosmic Rays (CRs) in the Galaxy diffuse through the interstellar magnetic fields and spread in the Galactic Disk. The confinement time of these relativistic particles is approximately of 10^7 years for GeV-particles and decreases with energy as $E^{-\delta}$, with $\delta \sim 0.5$. The typical lifetime of accelerators is much shorter, so that CRs have time to propagate and mix uniformly in the Milky Way. For this reason, we expect to find the same energy density and spectrum, essentially a *sea*, of CRs, almost everywhere in the Galaxy. Exceptions may occur in the vicinity of an accelerator, where freshly injected particles can cause an excess over the 'sea' typical density.

The diffuse γ-ray emission in the Galactic Disk is produced by interaction of CRs with the interstellar medium and therefore contains information about the spatial and energy distributions of these particles. γ-ray photons reach the Earth undeflected from the region where they were produced and their detection allows us to trace back their parent particles distribution. In particular in the energy range from 1 GeV and above, one of the main channels of production of γ-s comes from decay of neutral pions, produced by interaction of high energy protons with protons of the interstellar gas [1, 2]. Remarkably, where the proton density is high, as in the case of molecular clouds ($\sim$100 cm^{-3}), this

[a]e-mail: μgiada.peron@mpi-hd.mpg.de
[b]e-mail: μfelix.aharonian@mpi-hd.mpg.de
[c]e-mail: μsabrina.casanova@mpi-hd.mpg.de
[d]e-mail: μruizhi.yang@mpi-hd.mpg.de
[e]e-mail: μroberta.zanin@mpi-hd.mpg.de

process dominates [3]. Detection of this γ radiation provides direct information about the parent proton spectrum as $F_\gamma \propto n_H \cdot \rho_{CR}$. Different studies on the diffuse emission of the Galactic Disk [4, 5] showed that the CR density is not constant throughout the plane, but instead is enhanced in the ring within 4 and 6 kpc from the Galactic Center (GC). That effect could be explained by considering that the analyzed emission is an average on a region that contains a higher number of potential accelerator, or that it could be enhanced because of the contribution of numerous weak unresolved sources. To have a comprehensive understanding of the CR distribution, and to actually test the CR sea, one should probe different specific locations in the Galaxy. Giant Molecular Clouds (GMCs), being huge reservoirs of protons ($M \sim 10^5 - 10^7$ $M_\odot$), concentrated in a small radius (10-100 pc) enable this kind of investigation. In these regions the particle density is by order of magnitudes higher than the average density of the interstellar medium. For this reason, the probability of interaction of cosmic protons with the gas is bigger in GMCs and consequently the γ-ray flux from these objects is enhanced. On the other hand MCs are often the dominant objects, in terms of gas, in one specific direction. In some cases the emission of a single cloud could represent even the 80%, or more, of the entire emission of the line of sight and it is possible then to identify, with a certain precision, the origin of the emission with the position of the cloud. Measuring γ-rays from isolated MCs, permits to derive the CR distribution in the cloud ambient and therefore they are special, if not the sole, targets that could serve as CR barometers throughout the Galaxy [6, 7].

In the case of MCs the emitted flux in γ-rays depends on the column density of the cloud, usually expressed in terms of the ratio M_5/d_{kpc}^2 (with M_5 and d_{kpc} the mass and the distance expressed in units of 10^5 $M_\odot$ and kpc, respectively). Consequently, detection is possible if clouds are sufficiently massive, or very close. A value of $M/d_{kpc}^2 \sim 1$ is ideal for *Fermi*-LAT, but this threshold could be significantly decreased, when considering small objects in uncrowded regions. So far, this study has been successfully applied to clouds of the near (few hundred parsecs) Gould Belt region [8–10], and to Sagittarius B2 [11], a very massive ($\sim$150 $M_\odot$) molecular cloud, close to the Galactic Center. In order to probe the CR sea, one needs to have the highest possible number of targets, and preferably located in many different regions. In this work we focused on different environments from the near Gould Belt region to the inner Galaxy . We considered the clouds listed in ref. [12] as they provide a new improved estimation of the distance of these objects . Among all we chose to reanalyze three clouds of the Gould Belt, as they are representative of the local ($\sim$ 200 pc) environment, and cross-checked our methods with former results. We included in the analysis Lupus, that was not studied before, and we performed a new analysis on Taurus and Orion A, by taking advantage of the new improved statistics. From the same list we picked three other objects that stand outside the Gould Belt system, at a distance from 0.8 kpc to more than 2 kpc, We then looked at the recent catalog of ref. [13] that contains 1064 objects spread in the entire Galactic Disk, several of them with masses exceeding 10^6 $M_\odot$. From this catalog we selected 10 good candidates for the *Fermi*-LATanalysis located from 600 pc to 12 kpc from us. All chosen clouds have a parameter $M_5/d_{kpc}^2 \gtrsim 0.4$, they are the dominant object on the gas L.o.S and they are not overlapping with known γ-ray sources. This allows us to obtain high quality results and to precisely localize the origin of the gamma radiation with the position of the cloud.

We analyzed *Fermi*-LAT data accumulated for more than nine years, from MET 239557417 (4th August 2008) to MET 533045411 (22nd November 2017). We put a lower energy threshold at 800 MeV, to take advantage of the better angular resolution and to reduce the source confusion. We considered PASS8 data and selected ‘FRONT+BACK’ events (`evtype=3`). We cut in zenith angles (z_{max}=90°), to avoid the Earth limb events and imposed `DATA QUAL==1 && LAT CONFIG==1`. The considered ROI was a 10°x10° square, around the cloud centre. For each molecular cloud under analysis we created a spatial template derived from the CO radio-maps of ref. [14]. As spectral model, we assumed a PowerLaw and left normalization and spectral index free. As template for the background, we could not make use of the standard galactic background model provided by the *Fermi*-LAT collaboration [5], since the emission from the molecular clouds is included there as background. We built then a customized background model, by considering the main channels of production of γ rays above 1 GeV: π^0-decay emission, inverse Compton scattering and extra-galactic diffuse. We created a spatial template of the π^0-decay emission by considering the gas map of HI [15] and CO [14], from which we excluded the cloud; we assumed a PowerLaw spectrum and freed normalization and index. For what concerns the Inverse Compton scattering, we took the map of `galprop v.54` [16] and we derived the isotropic extra-galactic component by fitting a region at high latitude in the outer galaxy, where the galactic contribution is minimum. We included in the model also the sources from the 3FGL catalog [17].

For the first time CR spectra has been successfully extracted from specific regions of the Milky Way. The entire discussion of the results is left to the main paper [18]. Preliminary results from many different locations are in good agreement with theoretical assumption that the CRs embedded in the clouds regions are characterized by the same spectrum as the locally measured one, reported by the AMS collaboration [19]. Deviations are measured only in the inner Galaxy, where we see in general an enhanced flux, as observed in ref. [4, 5]. The heterogeneity of the derived parameters anyhow suggests that flux variation depends on local factors as the vicinity to a specific accelerator. On the other hand, CR spectra measured in the outer Galaxy perfectly agrees with AMS02 data [19]. Our independent analysis based on 9 years of *Fermi*-LAT data of the Gould Belt clouds confirms what found in previous work [9, 10], but our wider sample of clouds give new strength to the hypothesis of a general 'sea' level of galactic cosmic rays . In particular the spectral parameter extracted in regions beyond 2 kpc from the Earth, are in good agreement with the locally measured parameters, suggesting that the behaviour of CRs has a more general character, and is not influenced by local factors. This remarkable constancy of the derived densities and the spectral indices of CRs suggests that we are dealing with the sea of CRs.

Acknowledgments

S.C. acknowledges the Polish Science Centre under contract number UMO-2017/27/B/ST9/ 02272

References

[1] F.A. Aharonian, A.M. Atoyan, **362**, 937 (2000), `astro-ph/0009009`
[2] A.W. Strong, I.V. Moskalenko, O. Reimer, The Astrophysical Journal **613**, 962 (2004)
[3] S. Gabici, F.A. Aharonian, P. Blasi, **309**, 365 (2007), `arXiv:astro-ph/0610032`
[4] R. Yang, F. Aharonian, C. Evoli, **93**, 123007 (2016), `1602.04710`
[5] F. Acero (Fermi-LAT collaboration), The Astrophysical Journal Supplement Series **223**, 26 (2016)
[6] F. Aharonian, Space Science Reviews **99**, 187 (2001)
[7] S. Casanova, F.A. Aharonian, Y. Fukui, S. Gabici, D.I. Jones, A. Kawamura, T. Onishi, G. Rowell, H. Sano, K. Torii et al., Publications of the Astronomical Society of Japan **62**, 769 (2010)
[8] A. Neronov, D.V. Semikoz, A.M. Taylor, Physical Review Letters **108**, 051105 (2012), `1112.5541`
[9] R.z. Yang, E. de Oña Wilhelmi, F. Aharonian, Astronomy & Astrophysics **566**, A142 (2014)
[10] A. Neronov, D. Malyshev, D.V. Semikoz, **606**, A22 (2017), `1705.02200`

[11] R.z. Yang, D.I. Jones, F. Aharonian, Astron. Astrophys. **580**, A90 (2015), 1410.7639
[12] E. Schlafly, G. Green, D. Finkbeiner, H.W. Rix, E. Bell, W. Burgett, K. Chambers, P. Draper, K. Hodapp, N. Kaiser et al., The Astrophysical Journal **786**, 29 (2014)
[13] T.S. Rice, A.A. Goodman, E.A. Bergin, C. Beaumont, T.M. Dame, The Astrophysical Journal **822**, 52 (2016)
[14] T.M. Dame, D. Hartmann, P. Thaddeus, The Astrophysical Journal **547**, 792 (2001)
[15] N.B. Bekhti, L. Flöer, R. Keller, J. Kerp, D. Lenz, B. Winkel, J. Bailin, M. Calabretta, L. Dedes, H. Ford et al., Astronomy & Astrophysics **594**, A116 (2016)
[16] A.E. Vladimirov, S.W. Digel, G. Jóhannesson, P.F. Michelson, I.V. Moskalenko, P.L. Nolan, E. Orlando, T.A. Porter, A.W. Strong, Computer Physics Communications **182**, 1156 (2011), 1008.3642
[17] F. Acero (Fermi-LAT collaboration), **218**, 23 (2015), 1501.02003
[18] F. Aharonian, G. Peron, R. Yang, S. Casanova, R. Zanin, arXiv preprint arXiv:1811.12118 (2018)
[19] M. Aguilar et al. (AMS), Phys. Rev. Lett. **114**, 171103 (2015)

Status of the KM3NeT project

Carla Distefano[1,a] for the KM3NeT Collaboration
[1] *INFN Laboratori Nazionali del Sud, Via S.Sofia 62, 95123, Catania, Italy*

Abstract. The KM3NeT Collaboration is constructing a research infrastructure hosting a next generation underwater neutrino observatory in the Mediterranean Sea. It will be a distributed infrastructure with two nodes: ORCA (Mton scale 2500 depth offshore Toulon (France)) and ARCA (Gton scale, 3500 m offshore Capo Passero (Italy)). The ORCA detector, optimised to detect neutrinos of tens of GeV, aims at determining the neutrino mass hierarchy. ARCA is the high-energy component and is dedicated to the search for extraterrestrial neutrino sources in the TeV-PeV range. The KM3NeT Collaboration has started the construction of both detectors. The first detection units were deployed and the data analysis is ongoing to validate the detector performances. In this paper we give a general overview of the two detectors. We also report the present status of the detector and the first results of data analysis.

1 Introduction

KM3NeT is a research infrastructure hosting neutrino telescopes in the Mediterranean Sea. Two detectors with the same technology but different granularity are planned: ORCA (Oscillation Research with Cosmics in the Abyss) at a depth of 2500 m offshore Toulon, France and ARCA (Astroparticle Research with Cosmics in the Abyss) at a depth of 3500 m offshore Capo Passero, Italy.

The ORCA detector is optimised to measure neutrinos with energies at the GeV scale. Its primary goal is resolving a fundamental question of particle physics: the relative ordering of the neutrino masses, i.e. the question whether the third neutrino mass eigenstate is heavier (normal hierarchy, NH) or lighter (inverted hierarchy, IH) than the first and second states. The mass hierarchy influences the distortion of neutrino oscillation pattern due to the mass effects in the Earth. Looking at atmospheric neutrinos, ORCA has the possibility to measure the oscillation pattern distortion in a wide range of neutrino energies (GeV - PeV) and zenith angles, the latter corresponding to a wide range of baselines (50 - 12800 km).

The ARCA telescope is designed to observe high-energy neutrinos (TeV ÷ PeV) of cosmic origin. It will be the next generation of the underwater kilometre-cube scale neutrino telescopes. The main physics cases for ARCA are the search for diffuse neutrino fluxes, in particular the investigation from a complementary field of view of IceCube [1, 2], and the search for point-like neutrino sources, with a particular focus on the galactic ones.

[a] e-mail: distefano_c@lns.infn.it

2 The KM3NeT detector design

The KM3NeT detectors are three dimensional arrays of optical sensors detecting the Cherenkov light induced in water by the secondary charged particles produced in neutrino interactions.

The basic unit of the detectors is the Digital Optical Module (DOM) [3]. The DOM is a pressure-resistant glass sphere of 17-inch diameter containing 31 3-inch photomultiplier tubes (PMTs) with their associated electronics for digitisation and data transmission as well as the calibration instrumentation. Nineteen of these PMTs are located in the lower hemisphere, and are thus looking downwards, while 12 in the upper hemisphere are looking upwards (Fig.1). This novel design offers significant improvements with respect to optical modules with a single large-area PMT: the total photocathode area is about three times larger, the field of view covers almost the full solid angle, pixelisation allows for high-purity photon counting and directional sensitivity.

A vertical sequence of 18 DOMs forms a Detection Unit (DU) [3]. Two thin parallel ropes hold the DOMs in place. The DU is anchored on the seabed and kept taut by a submerged buoy at its top. A vertical "backbone" cable, running along the full length of the DU, provides connectivity for power and data transmission. This cable is an oil-filled plastic tube, in equi-pressure with sea water, containing electrical wires and optical fibres, with a breakout box at each DOM. A sketch of the DU is shown in Fig.1.

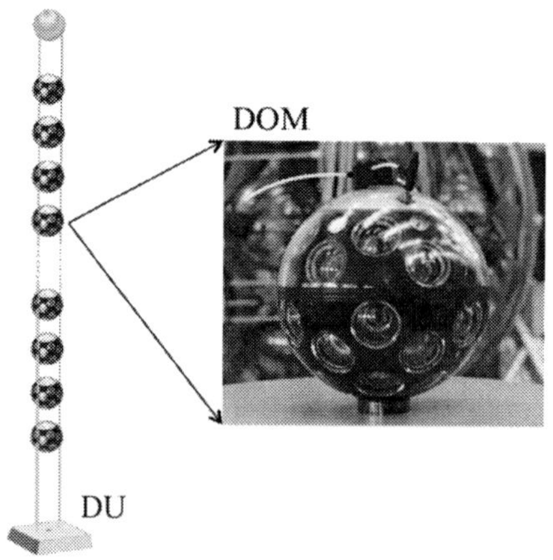

Figure 1. Schematic outline of a KM3NeT/ARCA building block.

A set of 115 DUs, arranged with a roughly circular footprint, forms a "building block". The DUs are connected to junction boxes and finally to shore by means of a main electro-optical cable. The architecture is intrinsically modular allowing for a staged implementation of the telescope. For the deployment the DU is coiled on a spherical aluminum frame with 2 m diameter, called Launcher of Optical Modules (LOM). In this compact configuration the DU is deployed on the seabed and connected to the seafloor network. The LOM is then released, unfurling the DU while rotating upwards. It floats to the surface where it is recovered for further use [3].

3 The KM3NeT/ORCA detector

The ORCA detector is located at the KM3NeT-France site off-shore Toulon and at a depth of 2450 m (see Fig. 2). The DOM spacing along the DU is 9 m and the average distance among the DUs is 23 m. It will consists of one building block, as shown in Fig. 2, with an instrumented mass of 8 Mton.

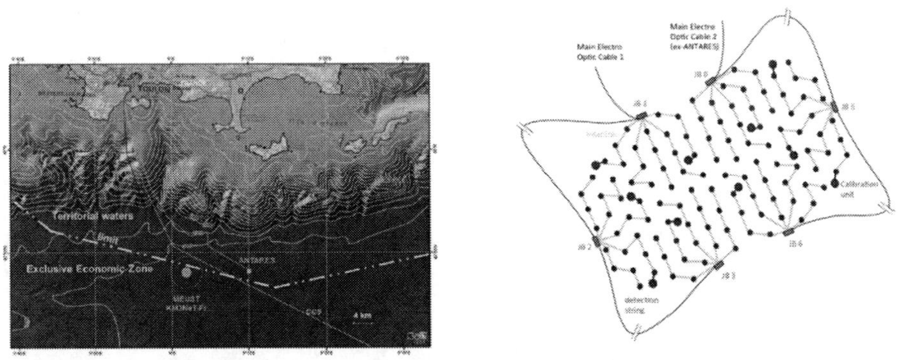

Figure 2. Left: map of the Mediterranean Sea south of Toulon, France. The location of the KM3NeT-France and ANTARES installations are indicated. Right: layout of the ORCA detector, depicting the 115 DUs, cables and connection devices of the full array.

The arrival times of the light collected by optical sensors and their positions are used to reconstruct the secondary particles. From these the neutrino arrival directions can be inferred and their energies estimated. An angular resolution of 5° and and energy resolution of 30% at 10 GeV, for both track and shower channels, are estimated for ORCA with Monte Carlo simulations. An algorithm based on the Random Decision Forest technique is used to perform particle identification. Monte Carlo simulations show that 90% of electron neutrinos and 70% of muon neutrinos are correctly identified at 10 GeV (Fig. 3). The algorithm is also able to suppress atmospheric muon background and noise to 3% level, preserving 95% of the neutrino signal. With a LLR test statistic is then possible to estimate the expected sensitivity to exclude the wrong hierarchy. The ORCA detector could be able to resolve the question of the neutrino mass ordering after only 3 years of data taking, as shown in Fig. 3.

4 The KM3NeT/ARCA detector

The ARCA telescope is located at the KM3NeT-Italy site, about 100 km off-shore Portopalo di Capo Passero (Sicily), at a depth of 3500 m (see Fig. 4). The DOM spacing along the DU is 36 m with an average distance among DUs of 90 m. ARCA will consists of two building blocks as shown in Fig. 4 and will have an instrumented volume of 1 km^3.

The golden channel is the detection of long track muon produced in CC muon neutrino interaction. For this kind of events an angular resolution better than 0.1° and an energy resolution of 30% in log-energy are reached for neutrino energies greater than 10 TeV. Shower events, induced by all flavour neutrino interaction in NC and electron neutrinos in CC, are also detectable. Monte Carlo simulations show that showers events will be reconstructed with an angular resolution better than 2° and energy resolution better $\approx$5% at 1σ.

The identification of point-like sources is one of the main goals of the ARCA telescope, for which the sensitivity to a generic E^{-2} neutrino flux is shown in Fig. 5. Thanks to its geographical location,

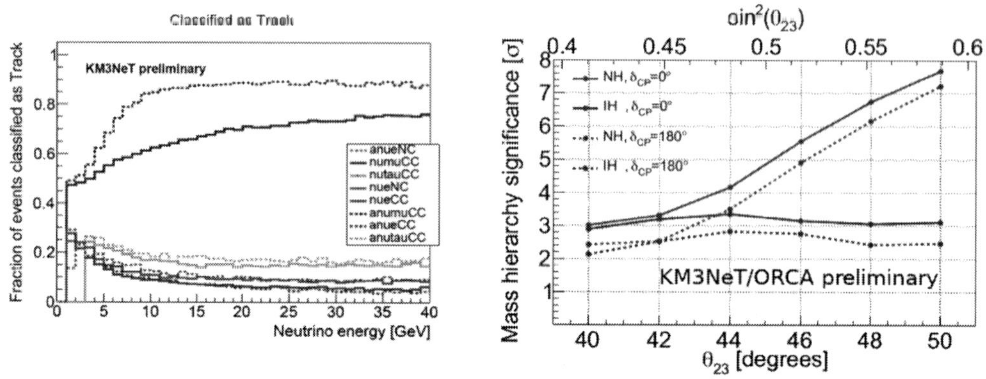

Figure 3. Fraction of events classified as tracks with the ORCA detector (left) and detector sensitivity to reject the wrong hierarchy after 3 years of data taking (right).

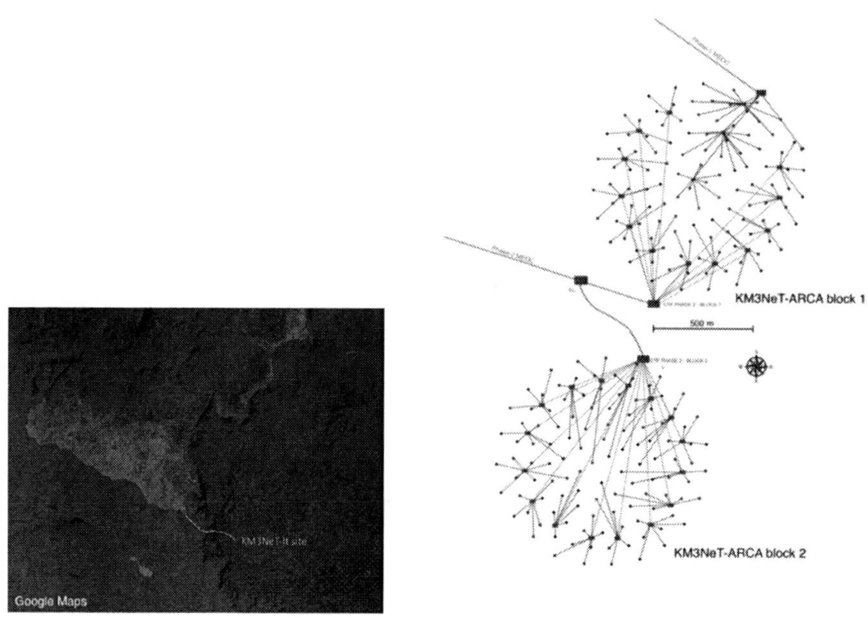

Figure 4. Left: map of the Mediterranean Sea south of Sicily, Italy. The location of the KM3NeT-Italy is indicated. Right: layout of the ARCA detector, depicting the two building blocks, cables and connection devices.

the telescope will allow for surveying a large part of the Galactic Plane, including the Galactic Centre. The sensitivity for specific galactic sources are shown in the right panel of Fig. 5.

5 First results

The KM3NeT detectors are currently under construction. The first two strings of ARCA were deployed in December 2015 and in May 2016 and the first string of ORCA was deployed in September

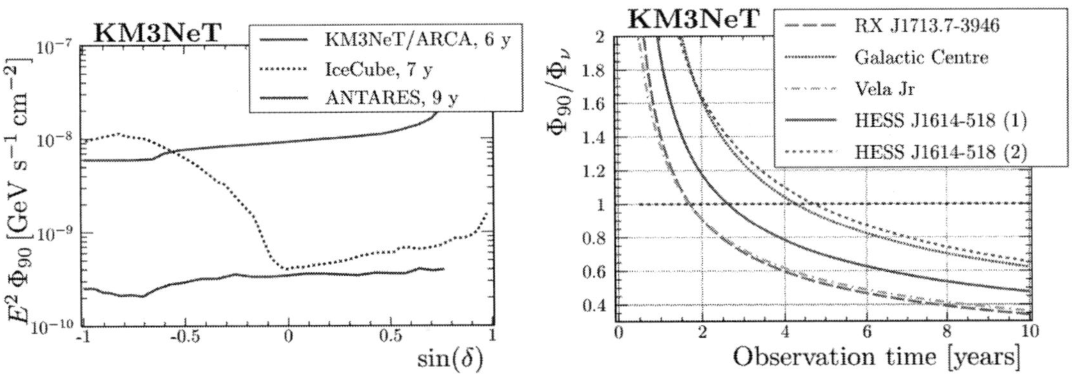

Figure 5. Left: ARCA sensitivity for a generic, unbroken E^{-2} neutrino flux as a function of the source declination. For comparison, the corresponding IceCube [4] and ANTARES [5] results are also shown. Right: ratio of ARCA sensitivity to the expectation flux as a function of the observation time for some galactic sources. See ref. [6] for details.

2017. Data collected since their deployment were analysed to perform *in-situ* calibrations to determine the relative PMT efficiencies and PMT-specific transit times ('intra-DOM calibration') and to measure the DOM specific optical fibre lengths ('inter-DOM calibration') [7].

A measurement of the depth dependence of the coincidence rates induced by atmospheric muons looking at the coincidence multiplicity occurring in the different DOMs is on going [8]. Results for the two ARCA DUs are shown in Fig.6. The good agreement between the data and the Monte Carlo simulations confirms that the measured coincidence rate as a function of the depth matches the expected behaviour due to the depth dependence of the atmospheric muon flux.

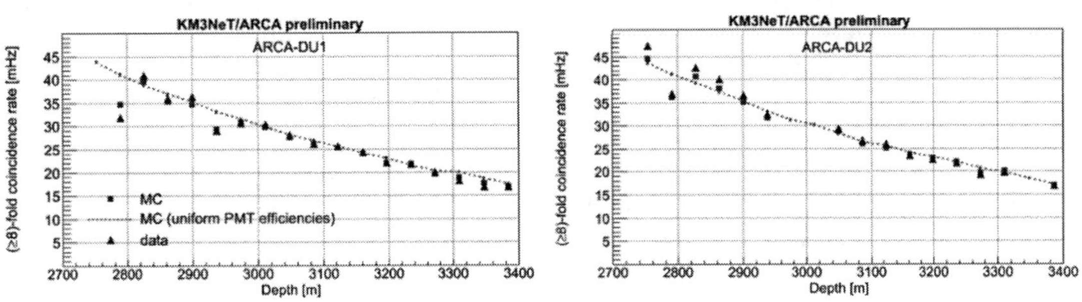

Figure 6. ≥8-fold coincidence rate as a function of the depth for the two ARCA DUs. The black points are the data. The red dashed lines are the base expectation from an ideal Monte Carlo with uniform PMT efficiencies which neglects the differences between individual PMTs as determined from in-situ calibration. The blue points are the Monte Carlo simulations where PMT efficiencies have been taken into account.

Data collected are also under analysis to reconstruct atmospheric muons and neutrinos in order to check the reconstruction algorithms and to optimise the event selection criteria for atmospheric muon rejection. The angular distribution of the events reconstructed with the 82 days live time data acquired with the ORCA DU are shown in Fig. 7.

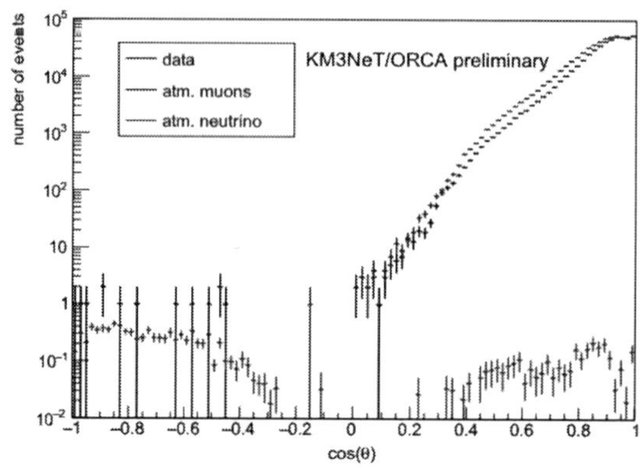

Figure 7. Angular distribution of events detected by the ORCA DU in a live time of 82 days, compared with Monte Carlo simulations.

6 Conclusions

The construction of the KM3NeT detectors has begun in the Mediterranean sea. The ORCA detector, devoted to study the atmospheric neutrino oscillations, will be able to determine the neutrino mass hierarchy in a few years of data taking. The ARCA telescope, with an instrumented volume of 1 km^3, will offer independent confirmation of the IceCube flux and it will be optimised to study point-like astrophysical neutrino sources. In particular, it will profit of the advantage of its location to observe the Galactic Plane and the Galactic Centre.

The first two ARCA DUs and the first ORCA DU were deployed between 2015 and 2017 and data collected since their deployment are under analysis. A measurement of the depth dependence of the PMT coincidence rates induced by atmospheric muons is on-going. Data are also under analysis to reconstruct atmospheric muons and neutrinos in order to check the reconstruction algorithms and to optimise the event selection criteria for atmospheric muon rejection. First results show a good agreement with Monte Carlo simulations and then the functionality of the detector.

References

[1] M.G. Aartsen et al. (IceCube), Science **342**, 1242856 (2013)
[2] M.G. Aartsen et al. (IceCube), Science **361**, 147 (2018)
[3] S. Adrián-Martínez et al. (KM3NeT), J. Phys. G: Nucl. Part. Phys **43**, 084001 (2016)
[4] M.G. Aartsen et al. (IceCube), Astroph. J. **835**, 151 (2017)
[5] A. Albert et al. (ANTARES), Phys. Rev. D **96**, 082001 (2017)
[6] S. Aiello et al. (KM3NeT), submitted to Astrop. Phys. (2018), arXiv:1810.08499
[7] K. Melis (KM3NeT), Proc. of Science, PoS(ICRC2017)1059 (2017)
[8] M. Jongen (KM3NeT), Proc. of Science, PoS(ICRC2017)1018 (2017)

Ultra High Energy Cosmic Rays: Origin, Composition and Spectrum

Roberto Aloisio[1,2,a]

[1] *Gran Sasso Science Institute, L'Aquila, Italy*
[2] *INFN - Laboratori Nazionali del Gran Sasso, Assergi (L'Aquila), Italy.*

Abstract. The physics of Ultra High Energy Cosmic Rays will be reviewed, discussing the latest experimental results and theoretical models aiming at explaining the observations in terms of spectra, mass composition and possible sources. It will be also discussed the emission of secondary particles such as neutrinos and gamma rays produced by the interaction of Ultra High Energy Cosmic Rays with astrophysical photon backgrounds. The content of the present proceeding paper is mainly based on the review papers [1, 2].

1 Introduction

Ultra High Energy Cosmic Rays (UHECR) are the most energetic particles observed in nature, with energies exceeding 10^{20} eV. The study of these particles dates back to 1962 when J. Linsley in the Volcan Ranch experiment observed the first extensive air shower generated by a cosmic particle of estimated energy around 10^{20} eV (see [1, 2] and references therein).

Nowadays, the most advanced experiments to detect UHECR are the Pierre Auger Observatory in Argentina [3], far the largest experimental setup devoted to the study of UHECR, and the Telescope Array (TA) experiment [4], placed in the United States, with roughly 1/10 of the Auger statistics.

Both detectors exploit the hybrid concept, combining an array of surface detectors (SD) to sample Extensive Air Showers (EAS, shower) when they reach the ground and telescopes, overlooking the surface array, to collect the fluorescence light of the atmospheric nitrogen excited by the EAS (fluorescence detectors, FD). The advent of the hybrid approach has been a major breakthrough in the detection of UHECR since, combining the measurements of the SD and FD, the tiny flux of UHECRs is studied with a 100% duty cycle and with an almost calorimetric estimation of the shower energies.

The experimental study of UHECR clarified few important characteristics of these particles: (i) UHECR are charged particles with a limit on photon and neutrino fluxes around 10^{19} eV at the level of few percent and well below respectively (see [1, 2] and references therein), (ii) the spectrum observed at the Earth shows a slight flattening at energies around 5×10^{18} eV (called the ankle) with (iii) a steep suppression at the highest energies (see [1, 2] and references therein), (iv) the arrival directions of particles with $E > 5 \times 10^{18}$ eV show a strong (at the level of 5.4σ) dipole anisotropy signalling the extragalactic origin of these particles [5].

Together with the number of particles per unit energy arriving at the Earth (flux) and their arrival direction (anisotropy) the third observable that characterises cosmic rays phenomenology is the mass

[a] e-mail: roberto.aloisio@gssi.it

composition. In the case of UHECR, mass composition is still matter of some debate. Before the advent of Auger the experimental evidences were all pointing toward a light composition with a proton dominated flux until the highest energies (see [1, 2] and references therein). The measurements carried out by the Auger observatory [6] have shown that the mass composition, from prevalently light at $\sim 10^{18}$ eV, becomes increasingly heavier towards higher energies. If confirmed, these findings would represent a change of paradigm respect to the picture of ten years ago. On the other hand, the TA experiment, even if with 1/10 of the Auger statistics, collected data that seem to confirm the pre-Auger scenario [7], the mass composition is compatible with being light for energies above 10^{18} eV. Nevertheless, the analysis performed by the joint working group Auger-TA showed that the results of the two detectors are compatible within the systematic uncertainties [8].

A clear experimental determination of mass composition is a difficult task as it is performed in an indirect way through the observation of the EAS development in the atmosphere and its comparison with the results of Monte Carlo simulations. These simulations are affected by the uncertainties in the hadronic interaction model between primary CR and the nuclei of the atmosphere. Direct measurements of the interaction cross sections are performed (at the LHC accelerator in CERN) only till (center of mass) energies around 10 TeV and therefore extrapolated at the interesting energies for UHECR (up to 500 TeV). As discussed by several authors (see [1, 2] and references therein), the cross section extrapolation, being model dependent, introduces a certain level of uncertainty, particularly at the highest energies, that could even spoil a clear determination of the mass composition itself.

In order to interpret the observations at the Earth, trying to constrain the possible sources of UHECR, it is very important a detailed modelling of UHECR propagation in the intergalactic medium, which is mainly conditioned by the interaction with astrophysical photon backgrounds and intergalactic magnetic fields, being the presence of these fields still matter of debate we will not touch here the effect of magnetic fields on the propagation of UHECR. The interactions with astrophysical photon backgrounds (CMB, Cosmic Microwave Background, and EBL, Extragalactic Background Light) shape the spectrum of UHECR observed at the Earth and are also responsible for the production of secondary (cosmogenic) particles: photons and neutrinos. This secondary radiation can be observed through ground-based or satellite experiments and brings important informations about the mass composition of UHECR and, possibly, on their sources.

The study of UHECR is a vital and intriguing field of research with a wide impact also in fundamental physics and cosmology, as it involves particles with extreme energies not attainable in any Earth based laboratory. This is the case of theories with Lorentz invariance violations or super heavy dark matter that in principle can be tested through UHECR observations, we will not discuss this issues here interested readers can refer to [1, 9] and reference therein.

In the forthcoming sections we will briefly review the main theoretical models on UHECR sources compatible with observations (section 2) and the production of secondary cosmogenic gamma rays and neutrinos (section 3). We will conclude in section 4.

2 Source models

Following the discussion presented in [1, 2], to constrain the basic characteristics of UHECR sources we adopt a purely phenomenological approach in which sources are homogeneously and isotropically distributed with the basic parameters: γ_g injection power law index, E_{max} maximum energy at the source (rigidity dependent), $\mathcal{L}_S$ emissivity and relative abundances of different elements at injection. These parameters are fitted to experimental data (both spectrum and mass composition) with as little as possible *a priori* theoretical prejudice on what the values should be.

The interactions suffered by UHECR with astrophysical photons backgrounds are pair-production, photo-pion production and, only for nuclei heavier than protons, photo-disintegration. Taking into

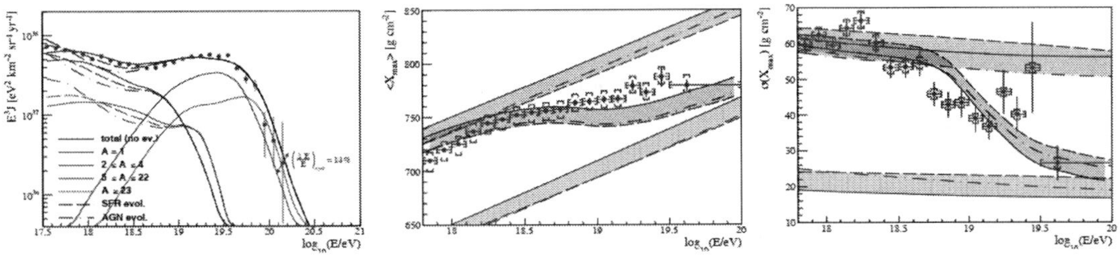

Figure 1. Comparison of the flux (left panel), elongation rate (central panel) and its root mean square (right panel) as observed by Auger and computed assuming the model with two classes of different extragalactic sources [10].

account these channels of energy losses, together with the adiabatic energy losses due to the expansion of the universe, solving the transport equations for UHECR we can determine the theoretical flux, to be compared with observations. Mass composition is inferred from the mean value of the depth of shower maximum $\langle X_{max} \rangle$ and its dispersion (RMS) $\sigma(X_{max})$, computed as shown in [11] and compared with observations too. The combined analysis of $\langle X_{max} \rangle$ and $\sigma(X_{max})$, even if not conclusive allows to obtain less model dependent information on the mass composition.

The qualitative new finding of Auger that mass composition might be mixed has served as a stimulus to build models that can potentially explain the phenomenology of Auger data. These models all show that the Auger spectrum and mass composition at $E \geq 5 \times 10^{18}$ eV can be fitted at the same time only at the price of requiring very hard injection spectra for all nuclei ($\propto E^{-\gamma_g}$ with $\gamma_g = 1 \div 1.6$) and a maximum acceleration energy $E_{max} \leq 5Z \times 10^{18}$ eV (see [1, 2] and references therein). The need for hard spectra can be understood taking into account that the low energy tail of the flux of UHECR reproduces the injection power law. Therefore, taking $\gamma \geq 2$ cause the low energy part of the spectrum to be polluted by heavy nuclei thereby producing a disagreement with the light composition observed at low energy.

One should appreciate here the change of paradigm that these findings imply: while in the case of a pure proton composition it is needed to find sources and acceleration mechanisms able to energise CR protons up to energies larger than 10^{20} eV with steep injection ($\gamma_g \simeq 2.5 \div 2.7$) [1, 2], the Auger data require that the highest energy part of the spectrum ($E > 5 \times 10^{18}$ eV) has a flat injection ($\gamma_g \simeq 1.0 \div 1.6$) being dominated by heavy nuclei with maximum energy not exceeding a few $\times$ $Z \times 10^{18}$ eV [1, 2, 12].

By accepting the new paradigm, it follows that the Auger spectrum at energies below 5×10^{18} eV requires an additional component that, composed by protons and helium nuclei, could be, in principle, of galactic or extra-galactic origin [1, 12]. Nevertheless, the anisotropy expected for a galactic light component extending up to 10^{18} eV exceeds by more than one order of magnitude the upper limit measured by Auger [13]. This observation, just restricting the analysis to Auger data, would constrain the transition between galactic and extra-galactic CR at energies below 10^{18} eV [1, 14], thus excluding a galactic component at the highest energies.

In order to reproduce the Auger observations, the additional (light and extra-galactic) contribution to the flux at energies below 5×10^{18} eV should exhibit a steep power law injection at the source with $\gamma_g \simeq 2.6 \div 2.7$ and a maximum acceleration energy not exceeding a few $\times$ 10^{18} eV, as for the heavier component (see [1, 2] and references therein). The possible origin of this radiation can be modelled essentially in two ways: (i) assuming the presence of different classes of sources: one injecting also heavy nuclei with hard spectrum and the other only proton and helium nuclei with soft spectrum or (ii) identifying a peculiar class of sources that could provide at the same time a steep light component and a flat heavy one (see [1, 2] and references therein). The second approach is based on a specific hypothesis on the sources that should be surrounded by an intense radiation field that, through photo-disintegration of heavy nuclei in the source neighbourhood, can provide a light component of (secondary) protons with a steep spectrum together with a hard and heavier component. In figure 1 we plot the flux and mass composition as observed by Auger and as reproduced theoretically assuming two classes of extragalactic sources with different injection characteristics as discussed above.

3 Secondary neutrinos and gamma rays

The propagation of UHECR in the intergalactic space, through the interactions with CMB and EBL, gives rise to the production of several unstable particles that in turn produce high energy photons, electrons/positrons and neutrinos. The possible detection of these signal carriers, realised already at the end of sixties, is extremely important to constrain models for UHECR sources, composition and the details of propagation (see [1, 2] and references therein). In the following we will separately discuss the case of neutrinos and gamma rays.

There are two processes by which neutrinos can be emitted in the propagation of UHECR: (i) the decay of charged pions, produced by photo-pion production, $\pi^{\pm} \rightarrow \mu^{\pm} + \nu_{\mu}(\bar{\nu}_{\mu})$ and the subsequent muon decay $\mu^{\pm} \rightarrow e^{\pm} + \bar{\nu}_{\mu}(\nu_{\mu}) + \nu_e(\bar{\nu}_e)$; (ii) the beta-decay of neutrons and nuclei produced by photo-disintegration: $n \rightarrow p + e^- + \bar{\nu}_e$, $(A, Z) \rightarrow (A, Z-1) + e^+ + \nu_e$, or $(A, Z) \rightarrow (A, Z+1) + e^- + \bar{\nu}_e$. These processes produce neutrinos in different energy ranges: in the former the energy of each neutrino is around a few percent of that of the parent nucleon, whereas in the latter it is less than one part per thousand (in the case of neutron decay, larger for certain unstable nuclei). This means that in the interaction with CMB photons, which has a threshold Lorentz factor of about $\Gamma \geq 10^{10}$, neutrinos are produced with energies of the order of 10^{18} eV and 10^{16} eV respectively. Interactions with EBL photons contribute, with a lower probability than CMB photons, to the production of neutrinos with energies of the order of 10^{15} eV in the case of photo-pion production and 10^{14} eV in the case of neutron decay (see [10] and reference therein).

The flux of secondary neutrinos is very much sensitive to the composition of UHECR. In figure 2 we plot the flux of cosmogenic neutrinos expected in the case of a pure proton composition (left panel) and in the case of mixed composition (central panel). Comparing the two panels of figure 2 it is evident the huge impact of the composition on the expected neutrino flux: heavy nuclei provide a reduced flux of neutrinos because the photo-pion production process in this case is subdominant.

The production of cosmogenic neutrinos is almost independent of the variations in sources' distribution because the overall universe, up to the maximum red-shift, could contribute to the flux. Typically, the maximum red-shift of astrophysical structures is expected around $z_{max} \simeq 10$, which is the redshift of the first stars (pop III). Once produced at these cosmological distances neutrinos travel toward the observer almost freely, except for the adiabatic energy losses and flavour oscillations, the opacity of the universe to neutrinos being relevant only at redshifts $z \gg 10$. This is an important point that makes neutrinos a viable probe not only of the mass composition of UHECR but also of the cosmological evolution of sources. In figure 2 (left and central panels) three different hypothesis on the cosmological evolution of sources are taken into account: no cosmological evolution (red bands), evolution typical of the star formation rate (green band) and of active galactic nuclei (blue band).

There is a solid consensus about the light composition of UHECR in the low energy part of the observed spectrum. This assures a flux of cosmogenic neutrinos in the PeV energy region, produced by the protons' photo-pion production on the EBL photons. Coloured bands in figure 2 show the uncertainties connected with the EBL background (see [1, 2, 10] and references therein). Another

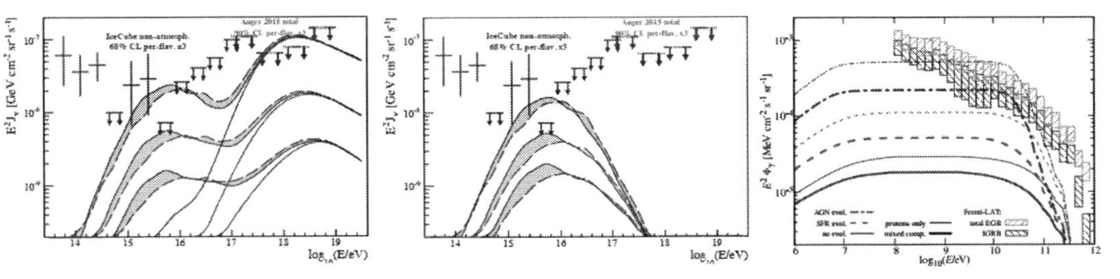

Figure 2. [Left panel] Fluxes of neutrinos in the case of a pure proton composition. The three different fluxes correspond to different assumptions on the cosmological evolution of sources (from bottom to top): no evolution (red), star formation rate evolution (green) and active galactic nuclei evolution (blue), coloured bands show the uncertainties due to the EBL model (see [1, 2] and references therein). Thin solid lines are neutrino fluxes obtained taking into account the sole CMB field. [Central Panel] Neutrino fluxes in the case of mixed composition, as shown in figure 1, with the same color code of left panel. Experimental points are the observations of IceCube on extra-terrestrial neutrinos [15] and the Auger limits on neutrino fluxes [16]. Figures taken from [10]. [Right Panel] Diffuse extragalactic gamma ray background expected from UHECR propagation in the two cases of a pure proton composition (thin lines) and mixed composition (thick lines) compared with Fermi-LAT observations [17].

important uncertainty in the expected neutrino flux comes from the contribution of UHECR sources at high red-shift. Given the energy losses suffered by UHE protons and nuclei, sources at red-shift larger than $z > 1$ can be observed only in terms of cosmogenic neutrinos [1, 10].

While neutrinos reach the observer without being absorbed, high energy photons and electrons/positrons colliding with astrophysical photon backgrounds (CMB and EBL) produce Electromagnetic Cascades (EMC) through the processes of pair production (PP, $\gamma + \gamma_{CMB,EBL} \to e^+ + e^-$) and Inverse Compton Scattering (ICS, $e + \gamma_{CMB,EBL} \to \gamma + e$). While PP is characterised by a threshold the ICS process does not. From this simple observation follows that once a cascade is started by a primary photon/electron/positron it develops since the energy of photons produced by ICS are still above the PP threshold. The final output of the cascade, i.e. what is left behind when the cascade is completely developed, is a flux of low energy photons all with energies below the PP threshold.

The two astrophysical backgrounds CMB and EBL against which the EMC develops are characterised by typical energies $\epsilon_{CMB} \simeq 10^{-3}$ eV and $\epsilon_{EBL} \simeq 1$ eV. Hence, the typical threshold energy scale for pair-production will be respectively[1] $\mathcal{E}_{CMB} = m_e^2/\epsilon_{CMB} = 2.5 \times 10^{14}$ eV and $\mathcal{E}_{EBL} = m_e^2/\epsilon_{EBL} = 2.5 \times 10^{11}$ eV. The radiation left behind by the cascade will be restricted to energies below $\mathcal{E}_{EBL}$ and will pile up into the extragalactic gamma ray background.

In figure 2 (right panel) we plot the expected gamma ray background produced by the propagation of UHECR assuming a pure proton composition (thin lines) or a mixed composition (thick lines). The case of pure protons maximises the number of secondary photons produced and, therefore, it is more constrained by the extragalactic gamma rays background, as observed by Fermi-LAT [17]. In the right panel of figure 2 we have also considered different hypothesis on the cosmological evolution of sources, as for neutrinos, apart from the case of no cosmological evolution (solid lines), we considered the evolution typical of star formation rate (dashed lines) and of active galactic nuclei (dot dashed lines). The latter being only marginally consistent with the observed extragalactic gamma rays background.

[1]Numerical values quoted here should be intended as reference values being background photons distributed over energy and not monochromatic.

4 Conclusions

The most important physical task in the physics of UHECR is certainly a clear identification of the sources. As discussed above, a key observable is the mass composition as it fixes few fundamental characteristics of the sources (γ_g and E_{max}) and it regulates the production of secondary cosmogenic neutrinos and gamma rays. A pure proton composition is, theoretically, a natural possibility. Proton is the most abundant element in the universe and several different astrophysical objects, at present and past cosmological epochs, could provide efficient acceleration even if it requires very high luminosities and maximum acceleration energies. The complexity of the scenario based on a composition with heavy nuclei disfavours astrophysical sources placed at high redshift because of the lacking of heavy elements. To reproduce observations, mixed composition requires sources with flat injection spectra for heavy nuclei, or needs particular dynamics in the source environment, such as photo-disintegration on strong local photon fields. A remarkable feature of the mixed composition scenario is the relative low maximum energy required at the source.

We conclude highlighting the two principal avenues on which the study of UHECR should develop in the near future. From one hand, as discussed above, a firm experimental determination of the mass composition is an unavoidable step forward in this field of research. On the other hand, the highest energy regime, typically energies $E \gtrsim 5 \times 10^{19}$ eV, still remains less probed with not enough statistics to firmly detect possible anisotropies in the arrival directions and the exact shape of the suppression. Current technologies, such as Auger and (extended) TA, can reach, in several years, one order of magnitude more in the number of observed events at the highest energies, which seems not enough to firmly detect anisotropies, to probe new physics or to detect neutrinos. New technologies are needed and future space observatories, with improved photon detection techniques, like the proposal POEMMA (Probe of Extreme Multi-Messenger Astrophysics) [18], promise a new era in the physics of UHECR, allowing the needed statistics to detect anisotropies, cosmogenic neutrinos and to probe new physics, [1, 2, 18].

References

[1] R. Aloisio, PTEP **2017**, 12A102 (2017), 1707.08471
[2] R. Aloisio, P. Blasi, I. De Mitri, S. Petrera, Springer Book (2018), 1707.06147
[3] A. Aab et al. (Pierre Auger), Nucl. Instrum. Meth. **A798**, 172 (2015), 1502.01323
[4] P. Tinyakov (Telescope Array), Nucl. Instrum. Meth. **A742**, 29 (2014)
[5] A. Aab et al. (Pierre Auger), Science **357**, 1266 (2017), 1709.07321
[6] A. Aab et al. (Pierre Auger), Phys. Rev. **D90**, 122005 (2014), 1409.4809
[7] R.U. Abbasi et al., Astropart. Phys. **64**, 49 (2015), 1408.1726
[8] *Auger and TA: Joint Contributions to the ICRC 2017*, 1801.01018
[9] R. Aloisio, S. Matarrese, A.V. Olinto, JCAP **1508**, 024 (2015), 1504.01319
[10] R. Aloisio et al, JCAP **1510**, 006 (2015), 1505.04020
[11] P. Abreu et al. (Pierre Auger), JCAP **1302**, 026 (2013), 1301.6637
[12] R. Aloisio, V. Berezinsky, P. Blasi, JCAP **1410**, 020 (2014), 1312.7459
[13] P. Abreu et al. (Pierre Auger), Astrophys. J. **762**, L13 (2012), 1212.3083
[14] R. Aloisio, V. Berezinsky, A. Gazizov, Astropart. Phys. **39-40**, 129 (2012), 1211.0494
[15] M.G. Aartsen et al. (IceCube) (2015), 1510.05225
[16] P. Abreu et al. (Pierre Auger), Adv.High Energy Phys. **2013**, 708680 (2013), 1304.1630
[17] M. Ackermann et al. (Fermi-LAT), Astrophys. J. **799**, 86 (2015), 1410.3696
[18] A.V. Olinto et al., PoS **ICRC2017**, 542 (2018), 1708.07599

Directional dark matter search with the NEWSdm experiment

Giovanni De Lellis[1,2,a]

[1] *Dipartimento di Fisica "E. Pancini", Università "Federico II" di Napoli, Naples, Italy*
[2] *Istituto Nazionale di Fisica Nucleare, Naples, Italy*

Abstract. The nature of Dark Matter is one of the fundamental questions to be answered. Direct Dark Matter searches are focussed on the development, construction, and operation of detectors looking for the scattering of Weakly Interactive Massive Particles (WIMPs) with target nuclei. The measurement of the direction of WIMP-induced nuclear recoils is a challenging strategy to extend dark matter searches beyond the neutrino floor and provide an unambiguous signature of the detection of Galactic dark matter. Current directional experiments are based on the use of gas TPC whose sensitivity is strongly limited by the small achievable detector mass. NEWSdm is an innovative directional experiment proposal based on the use of a solid target made by newly developed nuclear emulsion films and read-out systems achieving a position accuracy of 10 nm.

1 Introduction

Current experimental efforts for direct Dark Matter searches are devoted to the search for rare interactions between galactic halo WIMPs and nuclei with detectors located in underground laboratories. Since dark matter detectors are rapidly improving their sensitivity, in the next decade they will encounter the neutrino background where Solar, atmospheric, and diffuse supernova neutrinos mimic the dark matter signal. Neutrinos are the ultimate background for these searches as they produce recoils with similar rates and energy spectra. Moreover, the controversy in the low-mass (order of 10 GeV/c^2) WIMP region where some dark matter hints were shown, highlights the need for additional discrimination power between WIMP events and backgrounds to unambiguously prove WIMP signals. New generation detectors capable of measuring the direction of nuclear recoil tracks resulting from the WIMP elastic scattering off target nuclei would provide the discrimination of neutrino background and the unambiguous identification of galatic WIMPs as dark matter candidates.
Several approaches have been proposed [1]. The use of a solid target is the only approach to achieve large masses and therefore high sensitivity in the low cross-section region. Nevertheless, in a solid medium, the track length of a nuclear recoil would be of the order of a few hundred nanometers, thus requiring a detector with ultra-high tracking resolution. This is the challenge addressed by the NEWSdm proposal [2].

The NEWSdm apparatus consists of a nuclear emulsion detector acting both as target and as nanometric tracking device. Nuclear emulsions are made of silver halide crystals embedded in a gelatine matrix. When ionizing particles pass through it, some of the halide crystals are modified

[a]e-mail: giovanni.de.lellis@cern.ch

in such a way that they are turned into grains of silver after the *developing* process. The three-dimensional trajectory can be reconstructed with an optical microscope by connecting all the silver grains. NEWSdm employs a novel emulsion technology called Nano Imaging Trackers (NIT) [3, 4] with AgBr crystals of nanometric size, one order of magnitude smaller than those used in the OPERA experiment [5]. The detector is a bulk of NIT surrounded by a shield to reduce the external background. It is placed on an equatorial telescope to absorb the Earth rotation, thus keeping its orientation fixed towards the apparent WIMP flux. The angular distribution of the WIMP-scattered nuclei is therefore expected to be strongly anisotropic.

2 Optical microscope read-out system

NIT have a linear density of about 11 crystals/μm [3], thus making the reconstruction of trajectories with path lengths shorter than 100 nm possible, if analyzed by means of microscopes with enough resolution. The search for signal candidates requires the scanning of the whole emulsion volume. The read-out system has therefore to fulfill two main requirements: a fast, completely automated, scanning system is needed to analyse the target volume over a time scale comparable with the exposure; the spatial resolution has to reach the challenging value of a few tens of nanometers, well beyond the diffraction limit.
The NIT analysis is performed with a two-step approach: a fast scanning with a state-of-the-art resolution for the signal preselection followed by a pin-point check of preselected candidates with unprecedented nanometric resolution. A fast scanning is performed at first with an improved version of the optical microscope used for OPERA films [6]. An R&D program has achieved a speed of about 200 cm^2/h [7, 8]. Given the intrinsic resolution of the optical microscope ($\sim$ 200 nm), the sequence of several grains making a track of a few hundred nanometers may appear as a single cluster. Nevertheless, a cluster made of several grains tends to have an elliptical shape with the major axis along the direction of the trajectory, while a cluster produced by a single grain tends to have a spherical shape. The shape analysis with an elliptical fit is indeed the first approach to select signal.
In order to reproduce the effect of a WIMP-induced nuclear recoil and to measure the efficiency and the resolution of the new optical prototype, a test beam with low velocity ions was performed. Kr ion beam with energies of 200 and 400 keV [9] and a C ion beam with energies of 60, 80 and 100 keV were used. Silver grains belonging to the tracks appear as a single cluster. An elliptical fit of the cluster shape allows a clear separation between fog grains and signal tracks [10].

The second step of the analysis exploits the plasmon resonance effect occurring when nanometric silver grains are dispersed in a dielectric medium [11]. The polarization dependence of the resonance reflects the shape anisotropy and can be used to infer the presence of non-spherical nanometric silver grains within a cluster made of several grains. This effect is used to retrieve track information beyond the diffraction limit. Images of the same cluster taken with different polarization angles show a displacement of the position of its barycenter. The analysis of this displacement allows to distinguish clusters made of a single grain from those made of two or more grains. An unprecedented accuracy better than 10 nm has been achieved in both coordinates with this method as shown in Fig. 1. The clusters showing a barycenter displacement larger than 3σ are defined as "moving grains" and those below are defined as "static grains". Moreover, when more than one grain is reconstructed, the corresponding track is called micro-track. The plot in Fig. 2 shows the angular distribution for these three different components produced by the implantation of 100 keV Carbon ions: static grains are uniformly distributed as expected since they are mostly due to isolated grains produced by thermal excitation, while a clear angular peak along the direction of the implantation is shown for moving grains. Micro-tracks are also produced by implanted carbon ions when their track lengths exceeds about 200 nm and they also show an angular peak. This plot demonstrates the directionality of the emulsion technique.

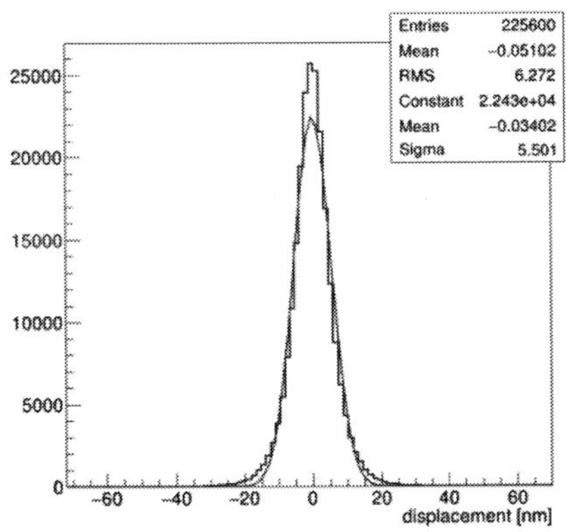

Figure 1. Achieved position accuracy better than 10 nm.

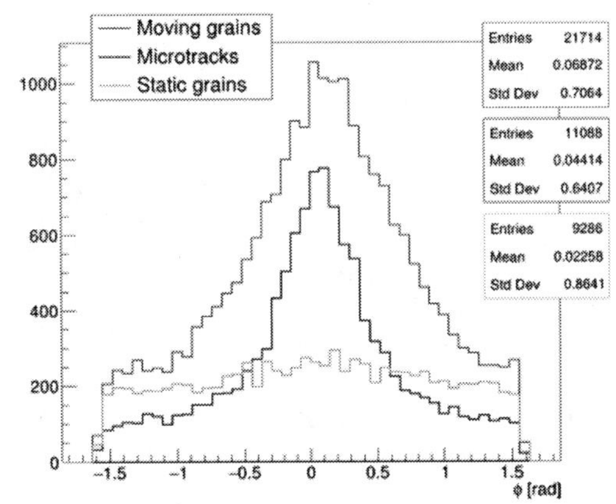

Figure 2. Angular distribution of single grains and micro-tracks.

3 Experimental setup

NEWSdm has setup at Gran Sasso a facility for the emulsion handling and film production. A dedicated structure was constructed to shield a detector of 10g mass against the environmental background sources over an exposure time of about 1 month. The aim is to measure the detectable background and validate estimates from simulations. This test will pave the way for a pilot experiment with an exposure of about 10 kg year. The experimental setup is located in the Hall B of the Gran Sasso Laboratory and it consists of a shield from environmental backgrounds and a cooling system to ensure the required temperature for the NIT detector. A thickness of 40 cm for the polyethylene blocks and of 10 cm for the lead plates was chosen to get a negligible contribution from environmental background. The whole structure installed in Hall B is shown in Fig. 3.

4 Background and sensitivity

Background sources for dark matter searches are α and β particles, γ-rays and neutron induced recoils, while NIT are essentially insensitive to minimum ionizing particles (MIP). The main sources of α-particles are U and Th radioactive chains and Radon. These α-particles have MeV energies and their range in emulsion is a few tens of microns, by far longer than WIMP-induced nuclear recoils. α-particles can therefore be identified and discarded in the emulsions by an upper cut on the track length. The β-rays produced in ^{14}C decay can be discarded by using a cryogenic environment [12].
Neutron induced recoils are not distinguishable from the expected WIMP signal, except for the isotropic angular distribution and for the typical track length, largely exceeding the range expected for WIMP-induced recoils. While the external neutron flux can be reduced to a reasonable level with an appropriate shield, the intrinsic emulsion radioactivity would be responsible of an irreducible neutron yield through (α, n) and ^{238} U spontaneous fission reaction. The detectable neutron-induced background has been measured to be about 0.06 per year per kilogram [13], thus allowing to design a detector for an exposure of 10 kg year without any further purification.

Figure 3. Picture of the shielded structure installed in the underground Hall B at Gran Sasso.

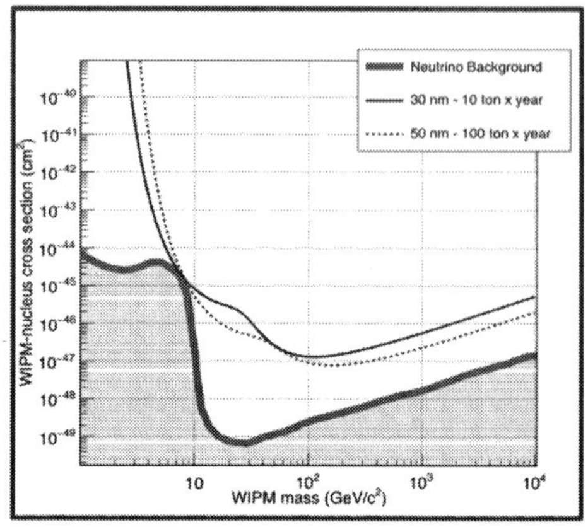

Figure 4. Discovery potential of an emulsion detector with different track length thresholds and exposures.

Fig. 4 shows the discovery potential of the NEWSdm experiment [2]. Different track length thresholds and detector exposures are assumed: the dotted red line represents a 50 nm threshold with 100 ton year exposure while the solid blue line assumes a 30 nm threshold with an exposure of 10 ton year. Both curves assume a negligible background. The presence in the emulsions of lighter nuclei such as carbon in addition to the heavier nuclei of silver and bromine results in a good sensitivity to WIMPs in the mass range between 10 to 100 GeV/c^2. We have also compared the exclusion limits with the thick orange curve representing the so-called neutrino floor for a Xe/Ge detector [14]. Given the nanometric accuracy achieved, it is realistic to assume a position resolution of several tens of nanometers provided that the grain density is sufficiently high. A mass of about 10 ton would provide sensitivity to explore the region beyond the neutrino floor.

5 Summary and plans

NEWSdm aims at providing an unambiguous signature of the Galactic dark matter and at extending dark matter searches beyond the neutrino floor. The directionality has been proven with Carbon ions of the energies relevant for dark matter searches. The current intrinsic radioactive level allows to design an experiment with a 10 kg year exposure. The selection of the emulsion components could further increase the radiopurity, thus allowing larger exposures. We plan a pilot experiment with ~ 10 kg year exposure and explore the region indicated by the DAMA experiment with a powerful and complementary approach, acting also as a demonstrator of this new directional technique.

References

[1] Battat J B R et al, *Physics Reports* **662** (2016) 1
[2] Agafonova N et al, NEWSdm Collaboration, *Eur. Phys. J.* **C78** (2018) no.7, 578
[3] Natsume N et al, *Nucl. Instr. Meth.* A **575** (2007) 439
[4] Naka T et al, *Nucl. Instrum. Meth.* A **718** (2013) 519
[5] Agafonova A et al, OPERA Collaboration, *JINST* **4** (2009) P04018.
[6] Alexandrov A et al, *JINST* **10** (2015) P11006

[7] Alexandrov A et al, *JINST* **11** (2016) P06002
[8] Alexandrov A et al, *Nature Scientific Reports* **7** (2017) 7310
[9] Naka T et al, *EAS Publ. Ser.* **53** (2012) 51
[10] Kimura M and Naka T, *Nucl. Instrum. Meth.* A **680** (2012) 12
[11] Tamaru H et al, *Applied Phys. Lett.* **80** (2002) 1826
[12] Kimura M et al, *Nucl. Instrum. Meth.* **A845** (2017) 373-377.
[13] Alexandrov A et al, *Astropart. Phys.* **80** (2016) 16
[14] Billard J, Strigari L and Figueroa-Feliciano E, *Phys. Rev.* D **89** (2014) 2

Observation of the Energy Dependence of Primary and Secondary Cosmic Rays with the AMS Detector on the International Space Station

V. Formato[1,a]
on behalf of the *AMS-02* collaboration

[1] *INFN - Sezione di Perugia, Via A. Pascoli s.n.c., 06123 Perugia (PG)*

Abstract. Precision study of cosmic nuclei provides detailed knowledge on the origin and propagation of cosmic rays. AMS is a multi-purpose high energy particle detector designed to measure and identify cosmic ray nuclei with unprecedented precision. It is able to provide precision studies of nuclei simultaneously to multi-TeV energies. In 7 years on the Space Station, AMS has collected more than 120 billion both primary and secondary cosmic rays. Primary cosmic rays, such as p, He, C and O, are believed to be mainly produced and accelerated in supernova remnants, while secondary cosmic rays, such as Li, Be and B are thought to be produced by collisions of heavier nuclei with interstellar matter. Primary cosmic rays such as He, C, and O are found to have identical rigidity dependence, similarly to secondary cosmic rays (such as Li, Be and B) which share the same the same spectral shape. The peculiar case of Nitrogen being a mixture of a primary and secondary component will also be shown.

1 Introduction

Cosmic rays (CRs) are charged particles coming isotropically from outer space to the Earth atmosphere with an energy range spanning over several decades. It is still unclear where and how CRs are accelerated in the Galaxy, as well as how they propagate through the interstellar medium (ISM), and both these aspects are subjects of current research. Another fundamental aspect in CR physics is the search for any signal of new physics in the antimatter component of CRs that could be linked to the decay of dark matter (DM) particles or to the existence of primordial antimatter. The bulk of cosmic rays is made of protons ($\sim 88\%$) and helium nuclei ($\sim 9\%$), while the remaining fraction is composed mostly of electrons ($\sim 2\%$) and heavier nuclei ($\sim 1\%$) with small traces of antimatter (positrons and antiprotons). Depending on wheter a cosmic ray reaches the solar system after being accelerated at its source or after being created in a anelastic interaction of other CRs with the ISM, it is labelled a *primary* or a *secondary* cosmic ray. Examples of primary species are protons, ^{4}He nuclei, carbon and oxygen nuclei, and electrons, while typical examples of secondary species are ^{2}H, ^{3}He, lithium, beryllium and boron nuclei.

Being created directly during the propagation of primary CRs in the Galaxy, secondary CRs carry important information on how the transport process modifies the original CR energy spectrum after the

[a] e-mail: valerio.formato@pg.infn.it

acceleration at the source. This is important for the theoretical prediction of the amount of secondary antimatter produced by CR propagation in the Galaxy, since this represents the natural background in the search for new physics in the antimatter channel. Thus the hadronic component of CRs holds valuable information on how CRs are accelerated and on how their energy spectrum is affected by the propagation in the galactic magnetic field. By measuring also secondary species, important informations can be obtained about CR propagation, especially when the flux of a secondary species is divided by the flux of a primary species, preferably one of its progenitors.

Not counting protons, the most abundant species in the hadronic component of CRs is helium, followed by carbon and oxygen wich make up the bulk of the primary component. The secondary component is made up by boron, lithium and beryllium nuclei. Nitrogen defies a clear classification as primary or secondary, as it is supposed to be an admixture of both a primary component and a secondary component.

Light nuclei in CRs with charge $Z \leq 8$ have been recently measured by the AMS-02 collaboration in Refs. [1–3] and the results will be summarized in section 3.

2 The AMS-02 experiment

AMS-02 is a general-purpose high-energy physics particle detector operating onboard the International Space Station (ISS) since May 2011. The instrument will be active for the entire ISS lifetime which is currently forseen to last until 2024. With such a long exposure time and its large geometrical acceptance (~ 0.5 m^2 sr), AMS-02 is capable to provide high-quality data on CR fluxes at the TeV energy scale with unprecedent precision and sensitivity.

The AMS-02 detector is described in details in ref. [4]. It is composed by several sub-detector systems that allow precise particle identification as well as redundant measurements of the main characteristics of CR particles, such as arrival direction, charge and magnetic rigidity ($R = pc/Ze$). The particle trajectory in the AMS-02 magnetic field is reconstructed from the position measurements along the 9 silicon layers of the tracking system, which provides a spatial resolution of 10μm in the y (bending) view and 30μm in the x (non-bending) view. From the reconstructed trajectory the particle direction and rigidity can be measured. The velocity $\beta = v/c$ can be determined from the transit time between the upper and lower TOF scintillator planes (with a resolution, for $Z = 1$, $\Delta\beta/\beta^2 \sim 2\%$), or more precisely using the RICH sub-detector (with a resolution, for $Z = 1$, $\Delta\beta/\beta \sim 10^{-3}$). The central part of AMS-02 is surrounded by an anti-coincidence system (ACC). The detector is completed with a Transition Radiation Detector (TRD), which is located between the first layer of the silicon tracker and the upper TOF, and a 18-layer electromagnetic calorimeter (ECAL), which is placed at the bottom.

3 Data analysis

The key elements used in this measurement are the permanent magnet, the silicon tracker, and the four planes of time of flight (TOF) scintillation counters. The dataset used corresponds to the first 60 months of AMS data taking, from May 19th 2011 to May 26th 2016, during which AMS recorded 8.5×10^{10} cosmic ray triggers. Events of a given species were required to be downward going and to have a reconstructed track in the inner tracker which passes through the first tracker layer (L1). In the highest rigidity region, $R \leq 1.13$ TV, the track is also required to pass through the last tracker layer (L9). Track fitting quality criteria such as a χ^2/d.o.f. < 10 in the bending coordinate are applied, similar to [5–7].

Charge measurements on L1, the inner tracker, the upper TOF, the lower TOF, and, for $R >$ 1.13 TV, L9 are required to be compatible with charge $Z = 2$ to 8 for helium to oxygen events

	N. of events	Residual background at 2 GV / 3.3 TV
Helium	90×10^6	$< 0.1\%$
Lithium	1.9×10^6	5% / 2%
Beryllium	0.9×10^6	8% / 13%
Boron	2.6×10^6	5% / 8%
Carbon	8.6×10^6	$< 0.5\%$
Nitrogen	2.2×10^6	$< 3\%$ / 6%
Oxygen	7.0×10^6	$< 0.5\%$

Table 1. Number of events for each species after background subtraction, the amount of residual background subtracted.

respectively. The number of events obtained after background subtraction [1–3] is reported in table 1. With this selection, the charge confusion from non-interacting nuclei is negligible over the whole rigidity range. The residual background comes from heavier nuclei which interact above the tracker L2. The background resulting from interactions in the material between L1 and L2 (TRD and upper TOF) is evaluated by fitting the charge distribution of the tracker L1 charge with charge distribution templates of Li, Be, B, C, N and O. Furthermore an additional cut on the L1 charge is applied as explained in [1–3].

The background from interactions on materials above the tracker L1 (a thin support structures made by carbon fiber and aluminum honeycomb) has been estimated from simulation using MC samples generated according to the AMS flux measurements as detailed in [7]. The estimated residual background levels at 2 GV and 3.3 TV are shown in table 1. The bin-to-bin migration of events was corrected using the unfolding procedure described in Ref. [6]. A detailed description of the systematic uncertainties can be found in Refs. [1–3].

4 Conclusions

The fluxes obtained are shown in Fig. 1 as a function of rigidity with the total errors, the sum in quadrature of statistical and systematic errors. As seen, cosmic ray species can be divided into two distinct families, primary CRs exhibiting a $R^{-2.7}$ power-law behaviour below 300 GV, and secondary CRs with a $R^{-3.1}$ power-law behaviour below 300 GV. Besides a different normalisation, all primaries fluxes seem to have the same rigidity shape in common, and all the secondaries flux seem to share the same rigidity dependence although different from the primary shape. Remarkably all the species show a deviation from the single power-law shape at ~ 200 GV. This can be also seen examining the evolution of the spectral index as a function of rigidity, shown in Fig. 2. The nitrogen flux is the only one not following either of the two behaviors but, as shown in Ref. [3], it can be well described by a mixture of a primary component and a secondary component both following the spectral shape of the primary and secondary elements respectively.

References

[1] Aguilar M. *et al.*, Phys. Rev. Lett. 119 251101 (2017)
[2] Aguilar M. *et al.*, Phys. Rev. Lett. 120 021101 (2018)
[3] Aguilar M. *et al.*, Phys. Rev. Lett. 121 051103 (2018)

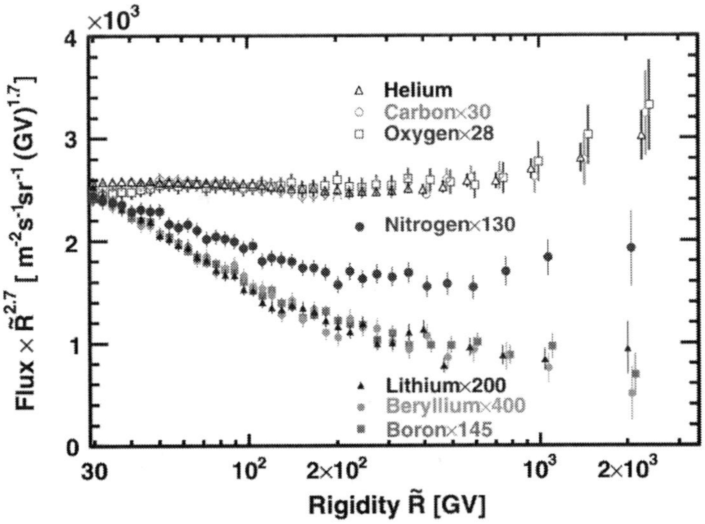

Figure 1. Fluxes of elements from helium to oxygen as measured by the AMS-02 experiment, multiplied by $R^{2.7}$. Each flux is scaled to better show the similar shape of primary and secondary species.

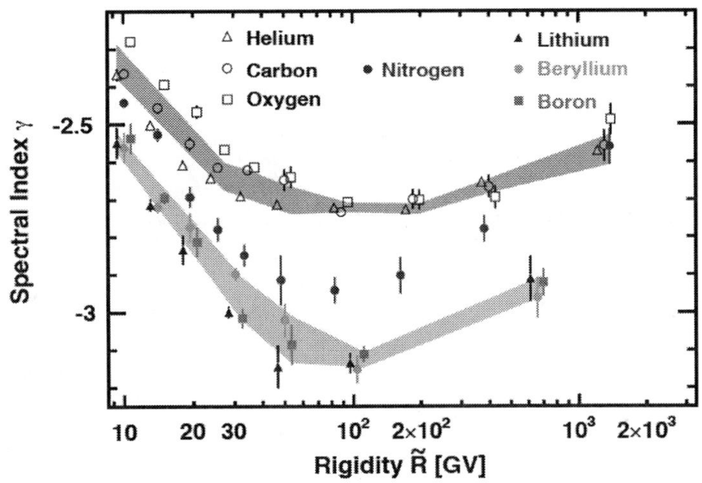

Figure 2. Spectral indices of elements from helium to oxygen as measured by the AMS-02 experiment.

[4] S. Ting, Nucl. Phys. B, Proc. Suppl. 243–244, 12 (2013)
[5] Aguilar M. *et al.*, Phys. Rev. Lett. 115 211101 (2015)
[6] Aguilar M. *et al.*, Phys. Rev. Lett. 114 171103 (2015)
[7] Aguilar M. *et al.*, Phys. Rev. Lett. 117 231102 (2016)

The Large Magellanic Cloud with the Cherenkov Telescope Array

María Isabel Bernardos[1,a], María Benito[2], Fabio Iocco[2], Salvatore Mangano[1], Olga Sergijenko[4], Ekaterina Karukes[2], and Lili Yang[3]for the CTA Consortium

[1] *Centro de Investigaciones Energéticas, Medioambientales y Tecnológicas*
[2] *ICTP-South American Institute for Fundamental Research. Instituto de Física Teórica - UNESP.*
[3] *University of Nova Gorica*
[4] *National Academy of Sciences of Ukraine, Kiev, Ukraine*

Abstract. The Large Magellanic Cloud (LMC) is a spiral galaxy, satellite of the Milky way with a high star formation activity. It represents a unique laboratory for studying an extended and spatially resolved star-forming galaxy through gamma-ray observatories. Therefore, the LMC survey is one of the key science projects for the Cherenkov Telescope Array (CTA), the next-generation ground-based gamma-ray observatory. In this document we present the work performed over the last year by the CTA working group dedicated to the LMC, in order to offer a first characterization of the LMC at TeV energies. We have performed detectability forecasts based on the expected CTA performance for all sources in the region of interest of the LMC with known emission at GeV energies and above. Based on previous observations made by Fermi-LAT and H.E.S.S. we have characterized all point sources, extended sources and diffuse emission produced by cosmic-ray propagation, extrapolating their spectra to CTA energies. Finally, we have characterized the signal expected by different annihilation mechanisms of dark matter (DM) particles within the LMC, computing the detection sensitivity curve for this target in the cross-section-to-mass plane.

1 Introduction

The LMC is a satellite galaxy of the Milky Way with unique characteristics that make it a valuable target for gamma-ray observations. It is located in the Southern hemisphere at a galactic latitude of -32.9° off the galactic plane, with a disk shape almost faced on (with a small inclination angle of 30-40°). The LMC is one of the nearest active star-forming galaxies, at a distance of about 50 kpc (with an uncertainty of 2% [1]. In this work, we study the forecasts on detecting the already known gamma-ray emission coming from sources in LMC with the Southern Array of CTA, the next-generation of ground-based gamma-ray observatories which is already in construction. We will cast simulations of a survey of the LMC, based on an emission model built from catalogs of present gamma-ray telescopes. In section 2 We talk about the previous gamma-ray observations of the LMC performed by Fermi-LAT and H.E.S.S. In section 3 the analysis procedure is described. A study on the sensitivity of CTA South for the detection of a DM annihilation gamma-ray signal coming from the LMC in presented in section 4.

[a] e-mail: mariaisabel.bernardos@ciemat.es

2 Previous observations of the LMC in gamma-ray

The LMC has been already observed in the gamma ray by Fermi-LAT and H.E.S.S., with the identification of several gamma-ray sources of different nature: Supernova Remnant 132D, the Pulsar Wind Nebula 157B, the very interesting superbubble 30 Doradus C [2] and the gamma-ray binary J0534.1-6732/HESS J0536-675, whose period was recently measured [3]. Also, H.E.S.S. acquired more than 200 h of data pointing at the position where the Supernova (SN) 1987A exploded in the last century, but they did not detect any gamma-ray emission [4]. However, they calculated upper limits on the flux of SN1987A which fall far above the sensitivity of CTA South (for 50 h of observations), making it a key target for CTA.

3 Forecasts on detection of known gamma-ray sources

We aimed to test the capability of CTA to detect and study the already known gamma-ray emission from the LMC. To do so, we have retrieved information of the gamma-ray sources detected in the LMC by other experiments to build an emission model. We have used one of the proposed analysis softwares for CTA, the *ctools* [5] combined with a custom pipeline to do a likelihood fit over a simulated survey of the LMC, and we have calculated the significances of the individual sources as seen with CTA.

3.1 Emission model of the LMC

We have retrieved a list of gamma-ray sources detected in the surroundings of the LMC, a Region of Interest (ROI) of 10° radius using the 3FGL [6] and the 3FHL [7] catalogs from Fermi-LAT, the *Fermi* collaboration paper [8] and H.E.S.S. papers [2],[3]. We have combined the data of the two experiments and have extrapolated the spectra of the sources to CTA energies (up to 100 TeV). Our LMC emission model has three main components: 10 point sources, 3 extended sources and a diffuse emission produced by cosmic-ray interaction with the interstellar medium. For the latter we have used models for leptonic and hadronic emission components (courtesy of Pierrick Martin [9]). A skymap of the full emission model is shown in Figure 1.

3.2 Simulations and analysis: Likelihood fitting

We have simulated 340 h of observation of the LMC divided in 6 pointings (56.6 h each) surrounding the LMC center(see Figure 1). Then we have performed a binned analysis with a spatial resolution of 0.1° per pixel, and 20 energy bins from 100 GeV to 100 TeV. We have used the Prod3b calibration function `South_z40_average_50h`. With the tool *ctmodel* we have built a model of predicted number counts of the observations (to be used as mock data) and for each indiviual source in the emission model (to build a model from the combination of all sources). We performed a binned likelihood analysis using a simple Poisson likelihood function:

$$L = \prod_{i}^{N} \frac{m_i^{n_i}}{n_i!} e^{-m_i} \tag{1}$$

In Equation 1, N is the total number of bins (energy and spatial bins), n_i is the number of data counts (photons) in the *i-th* bin, and m_i is the estimated number of counts in the *i-th* bin by the model. In the fit, the free parameters are the normalizations of the spectral models for the different sources. We defined the significance of a source in terms of the test statistics (TS)(see Equation 2), where L is the maximum likelihood, and L_{null} is the Likelihood of the null hypothesis for that specific source.

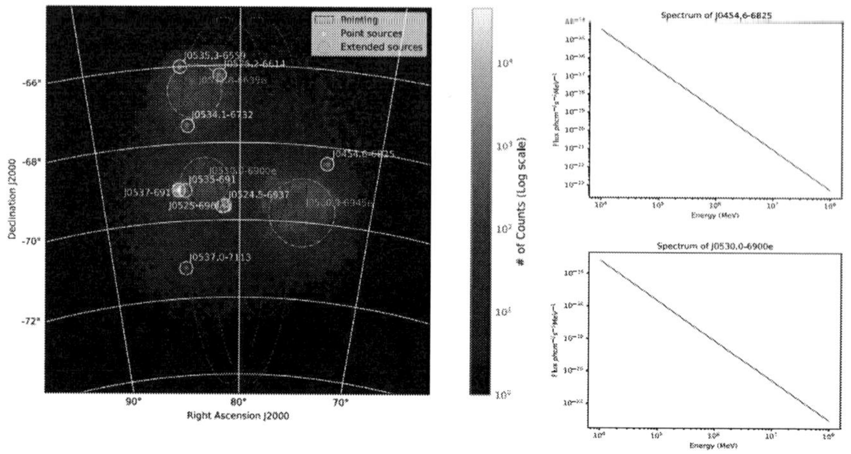

Figure 1: *Left*:Skymap of the LMC simulated in this work obtained with *ctobssim* and *ctskymap* tools. Green dashed lines represent the 6 pointings of the LMC survey. White circles mark the position of point sources and cyan point-dashed circles mark the position of the three extended sources. The atmospheric cosmic-ray background has been substracted in this image for a better appreciation of the structure details. *Top right*: Spectrum of one of the point sources, J0454-6825. *Bottom right*: Integrated spectrum of the diffuse source J0530-6900e

$$TS = 2\log \frac{L}{L_{null}} \tag{2}$$

3.3 Significances and Correlations

The results of the likelihood fit are listed in Table 1 with the significances for the sources. We also have studied the correlations between extended sources, which can affect their significances as photons from one source can be assigned to another one with which the former has a high correlation. In particular, we have noticed a considerable anti-correlation (correlation coefficient $r \approx -0.5$) between the leptonic emission model and the extended source J0530.0-6900e. We can do a temptative explanation of this correlation knowing that the source is identified as the west region of the star-forming region 30 Doradus, whose activity can be enhancing the leptonic diffuse emission produced by electrons inverse-compton scattering the radiation field, that experience very strong losses and don't travel very far form their sources [10].

4 Indirect detection of Dark Matter annihilation in the LMC with CTA

The LMC can be an interesting source for indirect DM searches. Having an astrophysical factor for DM annihilation $J\ 10^{20}GeV^2cm^-5$ [11] and being an extended object, we can use the spatial structure of its baryonic emission to unveil a possible diffuse DM signal coming from its halo. We have studied the sensitivity of CTA to detect a signal from different DM models depending on DM particle mass, annihilation channel and density profile. We have calculated 95% CL limits on the DM annihilation cross section through a maximum-likelihood analysis over 300 realizations of the LMC data. As a result, the sensitivity curve with 95% and %84 containment bands (same bands used by [11]) expected

Table 1: Significance of the astrophysical sources considered in the ROI of the LMC for this work

Source name	Significance(σ)
J0500.9-6945e	88.41
J0530.0-6900e	19.98
J0531.8-6639e	111.97
J0537-691	264.43
J0524.5-6937	116.62
J0534.1-6732	44.58
J0525.2-6614	83.40
J0535.3-6559	64.72
J0454.6-6825	47.81
J0537.0-7113	10.66
J0535-691	21.05
J0525-696	40.58
J0509.9-6418	0.50

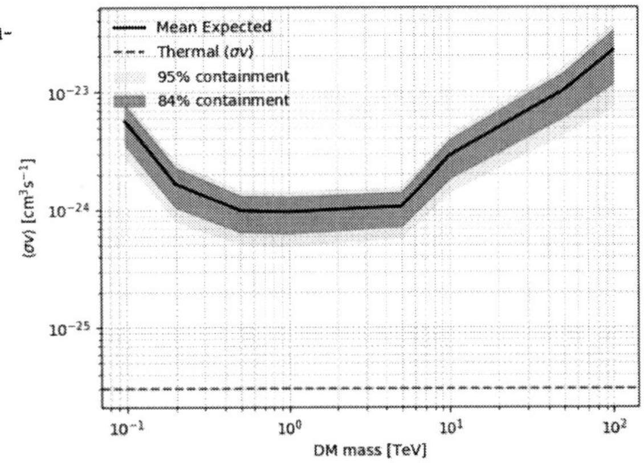

Figure 2: Sensitivity curve for W^+W^- annihilation channel, using a NFW density profile for the LMC DM halo.

for the W^+W^- annihilation channel with a Navarro-Frenk-White (NFW) density profile is shown in Figure 2. A visual inspection reveals that the sensitivity curve is several orders of magnitude above the thermal cross section, compatible with the results for DM searches in the LMC from [11]. This preliminary result confirms that the LMC is not a prime target for searches of signal of annihilating Weakly Interacting Massive Particles (WIMPs) DM.

Acknowledgements

We gratefully acknowledge financial support from the agencies and organizations listed here: http://www.cta-observatory.org/consortium_acknowledgments. This research has made use of the CTA instrument response functions provided by the CTA Consortium and Observatory, see https://www.cta-observatory.org/science/cta-performance/ (version prod3b-v1) for more details.

References

[1] G.e.a. Pietrzyński, **495**, 76 (2013), 1303.2063
[2] A. Abramowski et al. (H.E.S.S.), Science **347**, 406 (2015), 1501.06578
[3] N. Komin, M. Haupt (HESS) (2017), 1708.03171
[4] N.e.a.f. Komin, ArXiv e-prints (2012), 1201.0639
[5] J.e.a. Knödlseder, "Astron. Astrophys." **593**, A1 (2016), 1606.00393
[6] F. Acero et al. (Fermi-LAT), Astrophys. J. Suppl. **218**, 23 (2015), 1501.02003
[7] M. Ajello et al. (Fermi-LAT) (2017), 1702.00664
[8] M. Ackermann et al. (Fermi-LAT), Astron. Astrophys. **586**, A71 (2016), 1509.06903
[9] Cherenkov Telescope Array Consortium, ArXiv e-prints (2017), 1709.07997
[10] P. Martin, Astron. Astrophys. **564**, A61 (2014), 1402.0383
[11] M.R.e.a. Buckley, Phys. Rev. **D91**, 102001 (2015), 1502.01020

Latest results of the ANTARES neutrino telescope

Juan-de-Dios Zornoza[1a], on behalf of the ANTARES Collaboration
[1]*Instituto de Física Corpuscular, IFIC (UV-CSIC), c/ Catedrático José Beltrán 2, 46980 Paterna, Valencia (Spain)*

Abstract. Neutrino astronomy is in an exciting moment. The discovery of a cosmic flux of high energy neutrinos by IceCube heralds a new era in which neutrinos have finally joined the multi-messenger study of the Universe. This new important window complements more "traditional" probes (as cosmic rays or photons), given the particular combination of characteristics of neutrinos (neutral, stable and weakly interacting). The ANTARES detector, built in the Mediterranean Sea, has succeeded in two key points. First, it has shown the feasibility of the technique of underwater neutrino telescopes, which offers important advantages in terms of performance (better angular resolution, better visibility of the Galaxy if built in the Northern Hemisphere). This has paved the way for the next step, KM3NeT, already in construction. Second, the physics harvest of ANTARES is very rich, including many results that show the particular advantages of being in the Mediterranean, as mentioned above. The analyses performed include the search for point-like sources, diffuse fluxes, transient phenomena, dark matter, etc. In this talk we will review this long list of achievements.

1 Introduction

Neutrino astronomy is part of the new era in our way to observe de Universe. During many years, photons and cosmic rays have been our only probes for that task, but this has changed. The first milestones in this change were the discovery of high energy cosmic neutrinos by IceCube [1, 2] and the discovery of gravitational waves by LIGO experiment [3]. More recently, multi-messenger astronomy really started with the observation of a correlation of gravitational waves and electromagnetic signals [4] and of high energy cosmic neutrinos and electromagnetic signals [5]. The importance of these discoveries comes from the fact that gamma rays and cosmic rays have intrinsic limitations that constrain our capability to observe the sky with them. Gamma rays and cosmic rays interact with radiation and matter. Moreover, cosmic rays, being charged particles, are bent by magnetic fields in the Universe, so directional information is smeared. Neutrinos, however, travel unaffected by magnetic fields and only interact weakly, so they are very useful messengers to study regions very far or dense regions of the Universe.

Cosmic neutrinos are expected to be produced in the interaction of cosmic protons and neutrons at their sources. The measurement of the flux very high energy cosmic rays allowed to set the scale of the required neutrino telescopes. This size is of the order of about one cubic kilometre, which is the size of IceCube, in the South Pole.

[a] Corresponding author: zornoza@ific.uv.es

In the quest to access the neutrino window, the two neutrino telescope projects in the Mediterranean Sea, ANTARES and KM3NeT, are also key players, as it will be discussed in the following. Here we will focus on the results already obtained by ANTARES. An overview of the status and prospects of its successor, KM3NeT can be found in the corresponding proceedings of this conference.

2 The ANTARES detector

The ANTARES (Astronomy with a Neutrino Telescope and Abyss environmental Research) telescope [6] is installed off the French coast, at about 40 km from Toulon, at a depth of 2500 meters. It was completed in 2008 (by 2007 five lines were already in operation and taking data). It consists of 12 lines anchored at the sea bed, with a length of 450 m and separated by about 70 m (see Fig. 1). These lines hold a three-dimensional array of 885 photomultipliers (PMTs) to detect the Cherenkov light induced by relativistic leptons produced in interactions of high energy neutrinos with nucleons in the surrounding water and rock. The PMTs and their electronics are hold in pressure resistant spheres called Optical Modules (OMs). These OMs are grouped in triplets for a better rejection of the optical background and point 45 degrees downwards. Additional electronics are housed in titanium cylinders called Local Control Modules (LCMs). The total instrumented volume is about 0.05 km^3. In addition to the OMs, the lines are also equipped with optical beacons, tiltmeters, compasses and hydrophones for calibration and positioning purposes. These devices provide a precision of ~1 ns in time [7] and ~10 cm in position [8].

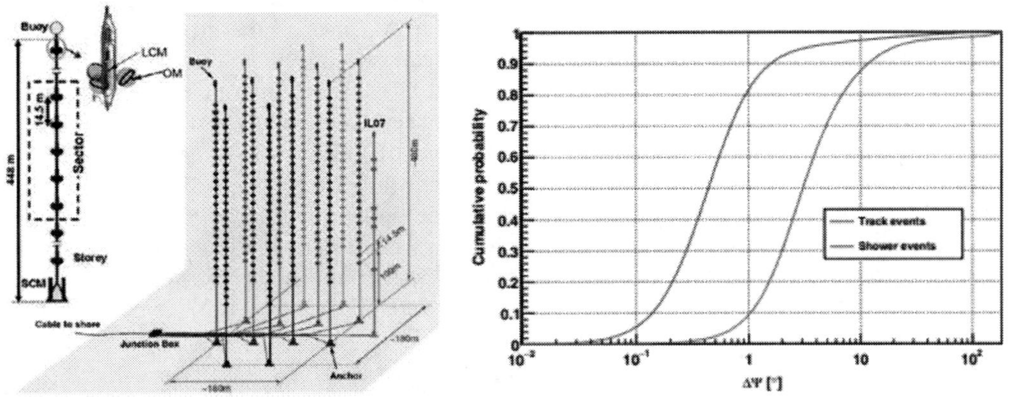

Figure 1. Left : Layout of the ANTARES detector. Right : Angular resolution for the sample of events for point-like source searches.

There are two main event topologies in neutrino telescopes. Muons, produced in muon-neutrino charged-current interactions, are very penetrating, so they produce long (hundreds of meters) tracks of Cherenkov light. Given the long lever-arm of these events, their direction can be reconstructed with very good accuracy (about 0.4° of median) (see Fig. 1). On the other hand, neutral-current interactions of any flavour and charged-current interactions of electron and tau neutrinos are seen as bright spheres of light (cascade events), for which it is harder to reconstruct well the direction. Still, given the excellent optical properties of deep sea water, the angular resolution is as good as 4 degrees.

The physical background in neutrino telescopes when searching for astrophysical sources comes from atmospheric muons and atmospheric neutrinos, both produced by cosmic rays interacting in the atmosphere (in a very similar way as the neutrino signal is produced at the sources). In order to reject the (dominant) component of atmospheric muons, the usual strategy is to select only events which are reconstructed upwards, since atmospheric muons cannot traverse the Earth. However, given the large rate of atmospheric muons, it is also necessary to set selection cuts to ensure to avoid that misreconstructed events from downgoing atmospheric muons contaminate the neutrino sample. Concerning the background from atmospheric neutrinos, different strategies can be followed, depending on the analysis. Among them, we can mention the fact that the spectral index for this background is softer that what is expected for cosmic signals, so a cut in the energy improves the purity of the sample.

2 Astronomy

Best candidates to produce high energy neutrinos detectable by neutrino telescopes are some of the most violent process in the Universe, where nucleons are accelerated at energies and produce charged pions after their interactions with ambient radiation or matter. These pions will decay and produce neutrinos and muons, which in turn also decay and produce more neutrinos. In addition, gamma rays are also expected to be produced in these interactions, coming from the production of neutral pions. Astrophysical sources where these processes should happen include supernova remnants, micro-quasars and soft-gamma repeaters (among the Galactic sources) and gamma-ray bursts, active galactic nuclei and star burst galaxies (among the extragalactic sources).

2.1 Diffuse fluxes

The signal of cosmic neutrinos detected by IceCube seems to correspond to a diffuse, extra-galactic flux. A (subdominant) Galactic component is also possible, as maybe hinted by the differences in the best fits for the different searches performed by IceCube.

Searches for diffuse fluxes take advantage of a larger signal, since all the flux in the visible sky is integrated. The price to pay is a larger background, which can be reduced by a cut in the energy of the events. In a recent analysis [9], ANTARES has searched in a data sample of nine years, including tracks and showers and found a small (non-significant) excess compatible with the IceCube signal. This analysis excludes the absence of a cosmic flux at a level of 1.6σ.

Another example of search for a diffuse flux is the analysis of the region of the so-called Fermi Bubbles [10] (see Fig. 2). The gamma rays detected by the Fermi experiment could be explained by hadronic mechanisms which would also produce high energy neutrinos.

Finally, a search in the Galactic plane has also been recently published [11]. No significant excess has been found, but constraints to close to the predictions of the KRA model [12] have been set.

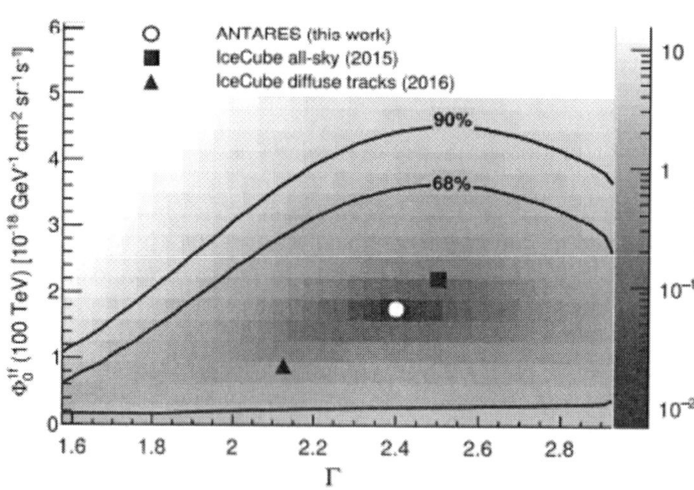

Figure 2. 2D log-likelihood scan of the diffuse cosmic flux normalization and spectral index. Black lines indicate 90% and 68% confidence levels.

2.2 Point-like sources

The search for point-like sources is one of the key analyses in ANTARES, since it has a clear competitive advantage to observe the Southern sky, where most of the Galaxy, including the Galactic Centre, is best observed by neutrino telescopes, since the Earth can be used as an efficient filter for atmospheric muon rejection as previously explained. In a recent analysis [13], ANTARES has carried out several types of searches for point-like sources using data from 2007 to 2015, including 1) full-sky, 2) search in a list of locations of sources candidates for emitting high-energy neutrinos, 3) a region of about 20° around the Galactic Centre, and 4) a search around Sagittarius A*. Figure 3 shows the sensitivities and limits obtained in the list of candidates sources. These recent searches included, for the first time, the contribution from cascade events. Although the angular resolution of these events is not as good as for tracks, in water it is still good enough to be used for these kind of searches. No excess has been found in these searches (see Fig. 3).

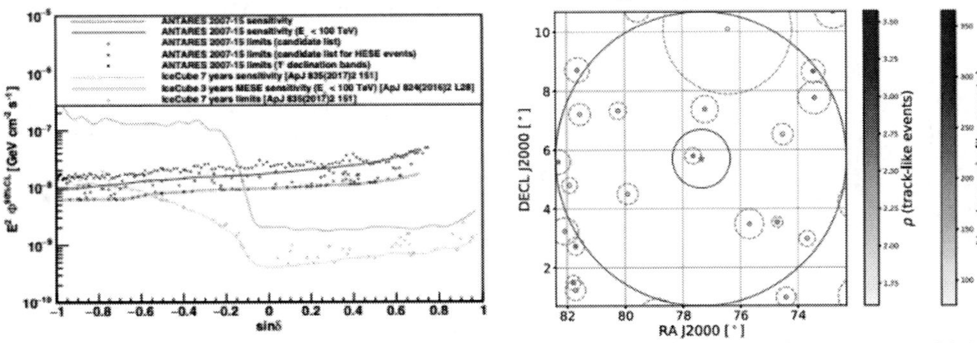

Figure 3. Left: Limits in the cosmic neutrino flux as a function of the declination of the source. Sensitivities are also shown assuming a E^{-2} spectrum with no cut off (solid line) and a cut off in the neutrino energy at 100 TeV. Recents reults from IceCube are also shown. Right: Events around the most significant cluster in the all-sky search.

2.3 Multi-messenger

As mentioned before, the recent years have witnessed the birth of a multi-messenger astronomy, in which all the cosmic messengers are participating. ANTARES has participated in several of these searches, looking for correlations of its data with gravitational wave events, high energy neutrinos and electromagnetic signals. As an example of this effort, Figure 4 shows the results of the search for a correlation of events in the ANTARES data with the gravitational wave event GW170104. No neutrino candidates were found in a window of ±500 s around the event and limits on the neutrino fluence were set [14].

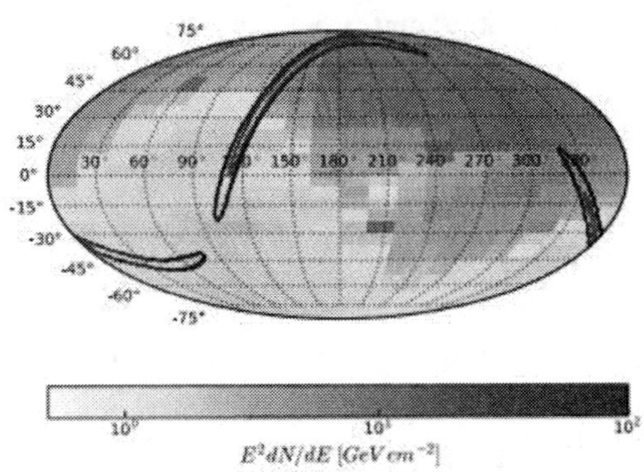

Figure 4. All-sky upper limits on the neutrino fluence from GW170104 assuming a E^{-2} spectrum emitted at the source. Countrous for 68% and 90% CL are also shown. The skymap is shown in equatorial coordinates.

3 Particle physics

Particle physics is also a main domain among the scientific goals of ANTARES. This include dark matter, neutrino properties and magnetic monopoles.

3.1 Dark matter

Dark matter is one of the most pressing question in fundamental physics today. Since we do not know what dark matter is made of, it is impossible to pin-point the best strategy to discover it. Each method has its own advantages. Neutrino telescopes offer a complementary way to explore this unknown side of Nature. Among several interesting sources, the Sun and the Galactic Centre are the best candidates to look for. In the case of the Sun, we expect that dark matter particles that encountered by the Sun as it travels around the centre of the Galaxy would scatter off the nucleons in the Sun, lose energy and eventually become gravitationally trapped. These dark matter particles are expected to self-annihilate to several products which would include high energy neutrinos as secondaries. A potential signal of high energy neutrinos in the direction of the Sun would be a smoking gun of dark matter, since, contrary to most of other indirect searches, the astrophysical backgrounds are known and low. If we assume that, as it is expected, there is an equilibrium between capture and self-annihilation, we can set limits on the scattering cross section of DM-nucleon [15].
For the Galactic Centre, limits are set on the thermally averaged annihilation cross-section of dark matter particles. For masses above 30 TeV, ANTARES has set the best limits worldwide [16] (see Fig. 5).

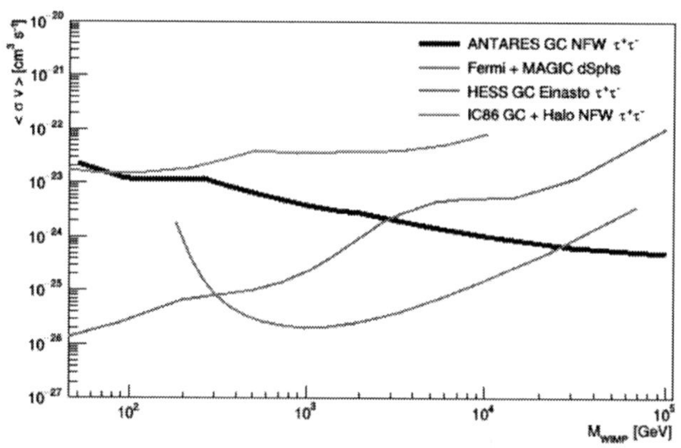

Figure 5. ANTARES limits on the thermally averaged annihilation cross-section as a function of the dark matter particle mass. Results of other experiments are also shown for comparison.

3.2 Oscillations

ANTARES has also shown capability to observe neutrino oscillations. As a matter of fact, the first hints of neutrino oscillations set by neutrino telescopes were obtained by this experiment [17]. In a more recent analysis, using ten years of data, a measurement of θ_{23} and Δm_{32}^2 has been made, consistent with the world best fit values. Moreover, these data have also been used to constrain 3+1 models of sterile neutrinos. These results are shown in Fig. 6 [18].

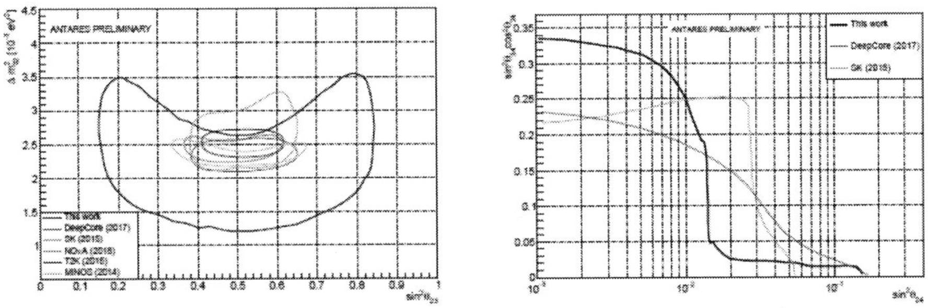

Figure 6. Left: Contour at 90%CL in the plane of θ_{23} and Δm_{32}^2 using ANTARES data from 2007 to 2016. Right: Constraints on 3+1 models of sterile neutrinos in the parameter space of $|U_{\mu4}|^2 = \sin^2\theta_{24}$ and $|U_{\tau4}|^2 = \sin^2\theta_{34}\cos^2\theta_{24}$.

4 Conclusion

ANTARES started taking data in 2007. Since then, a very rich harvest of physics analyses has been produced, both for astronomy and fundamental particle physics. These results include searches for diffuse fluxes, point-like sources, multi-messenger, dark matter and neutrino properties. This effort has shown that underwater neutrino telescopes are feasible and that they have the capability to profit from the advantages of neutrino astronomy in the Mediterranean: great angular resolution and very good visibility of our Galaxy.

The successful experience of ANTARES is a solid base for the next great leap, the KM3NeT detector, the construction of which has already started.

References

1. IceCube Collaboration. Science. 342 (2013) 1242856.
2. M.G. Aartsen, M. Ackermann, J. Adams, J.A. Aguilar, M. Ahlers, M. Ahrens, D. Altmann, T. Anderson, C. Arguelles, T.C. Arlen. Phys Rev Lett. 113 (2014) 101101.
3. B. P. Abbott et al. (LIGO Scientific Collaboration and Virgo Collaboration), Phys. Rev. Lett. 116, 061102
4. B. P. Abbott et al., ApJL 848 L12 (2017)
5. IceCube and other colls., Science 12 Jul 2018
6. M. Ageron et al., Nucl. Instrum. Meth. A 656 (2011) 11-38
7. J.A. Aguilar et al., Astropart. Phys. 34 (2011) 539-549,
8. S. Adrian-Martinez et al., JINST 7 (2012) T08002
9. A. Albert et al., ApJL 853, L7 (2018)
10. S. Adrián-Martínez et al., Eur. Phys. J. C (2014) 74:2701
11. Adrián-Martínez et al., Physics Letters B 760 (2016) 143
12. D. Gaggero, D. Grasso, A. Marinelli, A. Urbano, M., Valli, ApJL, 815:L25 (2015)
13. A. Albert et al., Phys. Rev. D 96, 082001 (2017)
14. A. Albert et al., ApJL 850 L35 (2017)
15. Adrián-Martínez et al., Physics Letters B, Volume 759, 10
16. A. Albert et al., Phys. Let. B 769 (2017) 249,
17. S. Adrian-Martinez et al., Phys. Lett. B 714 (2012) 224-230
18. Salvadori, I. (2018), Zenodo. http://doi.org/10.5281/zenodo.1300869

Latest results on dark matter searches with H.E.S.S

Lucia Rinchiuso[1,a] on behalf of the H.E.S.S. Collaboration

[1] *IRFU, CEA, Université Paris-Saclay, F-91191 Gif-sur-Yvette, France*

Abstract. The nature of Dark Matter (DM) is one of the most debated questions of contemporary physics. Ground-based arrays of Cherenkov telescopes such as the High Energy Spectroscopic System (H.E.S.S.) search for DM signatures through the detection of Very-High-Energy (VHE, $E > 100$ GeV) gamma-rays. DM particles could self-annihilate in dense environments producing VHE γ-rays in the final states that could be eventually detected by H.E.S.S.. The H.E.S.S. observation strategy for DM search focuses towards the Galactic Centre (GC) region and nearby dwarf galaxy satellites of the Milky Way. The GC dataset provides the most stringent constraints on the DM annihilation cross section in the mass range 300 GeV - 70 TeV. Searches have been carried out towards classical and ultra-faint dwarf galaxies to test specific heavy DM models. The latest results towards the GC and dwarf galaxies are shown.

1 The H.E.S.S. experiment

The High Energy Spectroscopic System (H.E.S.S.) is an array of Imaging Atmospheric Cherenkov Telescopes (IACTs) located in Namibia. During the first phase (H.E.S.S. I) the experiment consisted of four telescopes of 12 m diameter. The phase 2 (H.E.S.S. II) has started with the addition of a 28 m diameter telescope in the middle of the array to lower the energy threshold down to a few tens GeV. H.E.S.S. can detect Very-High-Energy (VHE, $\gtrsim 100$ GeV) γ-rays, that interact in the atmosphere creating electromagnetic showers. The showers then produce Cherenkov light that is detected by the telescopes situated in the light cone. The position and energy of the primary γ-ray can be reconstructed from the shower images recorded in the camera. The γ-rays are separated from the hadronic background based on image shape parameters of the shower. The H.E.S.S. array has a Field of View (FoV) of 5°, angular resolution per γ better than 0.1° and energy resolution $10\%E$. For observations at zenith towards a point-like source, it reaches a flux sensitivity 1% of the Crab flux in 25 hours.

2 Indirect dark matter search

2.1 Dark matter annihilation

IACTs can probe Dark Matter (DM) indirectly, *i.e.* through the detection of VHE γ-rays produced in the final state of self-annihilation of DM. The expected γ-ray flux from DM annihilation writes:

$$\frac{d\phi_\gamma}{dE} = \frac{1}{4\pi}\frac{\langle\sigma v\rangle}{2m_{\rm DM}}\sum_{\rm i} Br_{\rm i}\frac{dN_{\rm i}}{dE} \times J(\Delta\Omega)\,,$$

[a] e-mail: lucia.rinchiuso@cea.fr

where m_{DM} is the DM mass, $\langle\sigma v\rangle$ the thermally-averaged velocity-weighted annihilation cross section, $Br_i dN_i/dE$ the spectrum of the annihilation channel i weighted by its branching ratio Br_i. The J-factor J describes the density distribution of DM in the target. We refer to the prompt photon annihilation channel as *γ-line*, and the annihilation channels with photons due to decay or hadronization of vector bosons, quarks and leptons in the final state as the *continuum*. The expected DM signal is large in environments with large J. The signal region, referred to as the *ON region*, is defined as a disk around the target, while the background is measured in signal-free or lower-signal regions in the same FoV (*OFF region*).

2.2 Likelihood analysis technique

The γ-ray events are measured in the *ON* and *OFF* regions. If no significant excess is observed between the *ON* and *OFF* regions a Log-Likelihood Ratio Test Statistic (LLR TS) is performed to set 95% confidence level (C.L.) limits on $\langle\sigma v\rangle$ for a fixed DM density profile and a specific annihilation channel. The likelihood function used in H.E.S.S. in the latest analyses is binned in two dimensions (2D): energy and space. This choice is made to take advantage of the peculiar spectral shape and spatial distribution of DM signal compared to the power-law-like and spatially-isotropic background. The likelihood function contains two Poisson terms for the *ON* and for the *OFF regions*, respectively:

$$\mathcal{L}_{ij} = Poiss(N_{ON,ij}, N_{S,ij} + N_{B,ij}) \times Poiss(N_{OFF,ij}, N'_{S,ij} + \alpha_i N_{B,ij}).$$

N_{ON} and N_{OFF} are the measured events, N_S and N'_S the expected signal in the *ON* and *OFF regions*, respectively, N_B is the expected background in the *ON region* and α is the ratio between the size of the *OFF* and *ON regions*.

3 The Galactic Center region

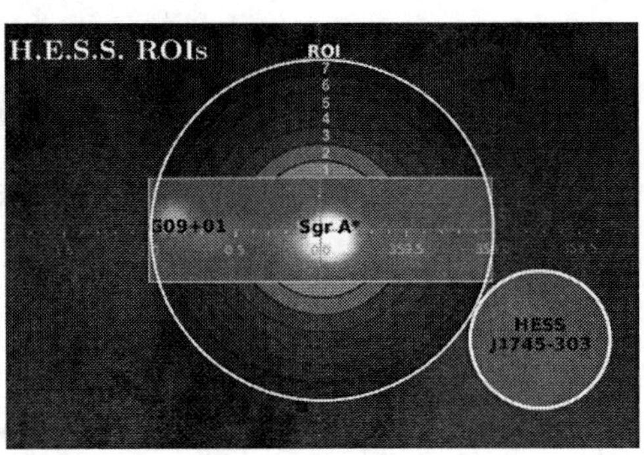

Figure 1. Galactic Center sky map in Galactic longitude and latitude. The RoIs are shown in red and the excluded regions as shaded areas. The standard VHE γ-ray sources Sgr A*, G09+01 and HESS J1745-303 are drawn.

The region around the Galactic Center (GC) is expected to harbor a large amount of DM. For this reason and due to its proximity it is an optimal target for DM detection. In addition H.E.S.S. is in a preferential position to observe it being situated in the Southern hemisphere. H.E.S.S. I has collected about 250 hours of observations towards the GC in ten years. Due to the large dataset and the telescope pointing strategy an *ON* region up to one degree in radius can be defined around the target. It is then split into seven concentric subregions (RoI) with inner radii from 0.3° to 0.9°, that make up the spatial bins for the likelihood function.

The *OFF region* is defined for each RoI and each observation as the region symmetric to the *ON* one with respect to the pointing position of the observation. Thus, the *ON* and *OFF regions* have the same solid angle size

($\alpha = 1$) and are in the same FoV: they are observed under the same conditions, no acceptance correction is needed due to azimuthal symmetry. However, the GC region harbors several VHE γ-ray sources that must be rejected for the DM search analysis. Fig. 1 shows a GC sky map with overlaid the *ON region* split in seven RoIs. The VHE γ-ray sources HESS J1745-290 coincident with the supermassive black hole Sagittarius A* [1], G09+01 [2] and HESS J1745-303 [3] are shown in the map. In the DM search analysis masks (shaded areas in Fig. 1) are used over these sources and the Galactic plane diffuse emission to avoid contamination from γ-rays produced in standard astrophysical processes. These areas are excluded both in the *ON* and *OFF regions* in order to maintain the same solid angle size.
No significant excess is observed in the *ON region* versus the OFF region. So the 2D-likelihood analysis is performed in the *continuum* [4] and γ *line* [5] channels, assuming an Einasto DM density profile. The LLR TS is applied to test the hypothesis of presence of DM vs the no-DM hypothesis and to obtain the 95% C.L. upper limits on the annihilation cross section. The results for the γ-*lines* search are shown in Fig. 2 left (red dots) together with the expected limits (black solid line), their containment bands at 1σ and 2σ (green and yellow boxes, respectively) and the previous H.E.S.S. limits (blue squares) [6]. Thanks to the novel 2D-likelihood technique, the more performing raw data analysis [7] and the ten-year data set, the constraints are improved of a factor about 6 at 1 TeV, reaching $\langle\sigma v\rangle \simeq 4\times10^{-28}$ cm^3 s^{-1}. The H.E.S.S. limits become competitive with Fermi-LAT [8] at a few hundred GeV. The two experiments together put the strongest constraints so far on $\langle\sigma v\rangle$ in the GeV-TeV DM mass range.

4 The dwarf galaxies satellites of the Milky Way

Satellite dwarf Spheroidal galaxies (dSph) of the Milky Way are among the most DM dominated objects in the Universe. In addition, they do not show VHE γ-ray emission from standard sources, e.g. there is no hints for recent star formation. Thus, they are expected to have a large signal-to-noise ratio and are good targets that could provide an unambiguous detection of a DM signal. H.E.S.S. I has accumulated a total of about 130 hours of observations towards Carina (23 h), Coma Berenices (11 h), Fornax (6 h), Sculptor (12 h) and Sagittarius (85 h). On these targets the *ON region* measures 0.2° (0.3° for Coma Berenice) accordingly to the DM density profile and is split in RoIs of width 0.1°. The *OFF regions* are defined for each RoI and each observation with the *multiple OFF* technique ($\alpha > 1$). No significant excess is observed in the *ON region* with respect to the *OFF*, in any of the dSphs. The 2D LLR TS is applied to compute 95% C.L. upper limits on the different targets for the *gamma-line* signal. The best combined constraint of 5×10^{-25} cm^3 s^{-1} is obtained at 1 TeV. A further analysis has been conducted on specific pure WIMP (Weakly Interacting Massive Particle) models. Limits have been computed for the Minimal DM fermionic triplet (3plet) and quintuplet (5plet) [9] using recent calculations [10]. The full WIMP spectrum is used in the analysis, including W^+W^- tree-level and ZZ, $Z\gamma$ and $\gamma\gamma$ one-loop contributions. Fig. 2 right [10] shows the observed limits on the 5plet for single dSph observation and the combination of the five dSph observation (black solid line). The predicted cross section (gray dashed line) and thermal $m_{\rm DM}$ (blue band) are shown. A significant $m_{\rm DM}$ range is excluded for the 5plet (red boxes).

5 Prospects

H.E.S.S. is pursuing a DM program focused on the search for DM annihilating towards the GC and dSph. A new strategy has been developed in the GC region for further DM studies. A survey at higher latitudes from the GC is ongoing with the full telescopes array. These new observations are

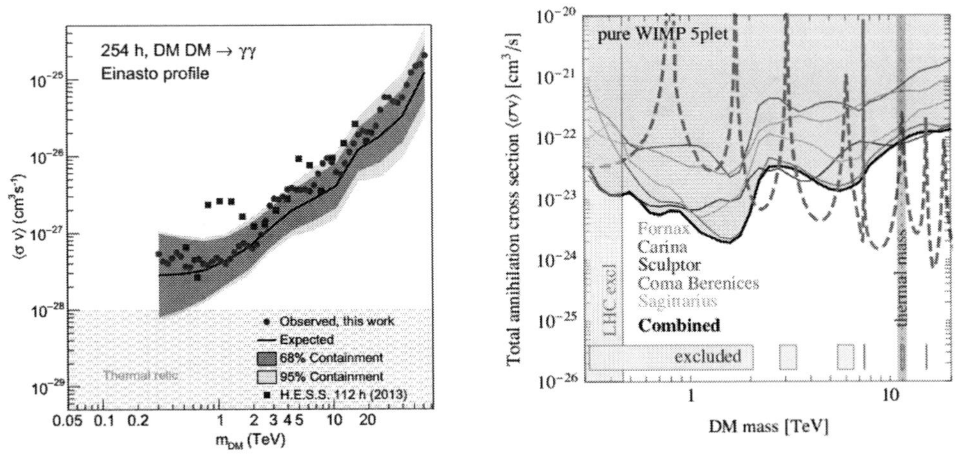

Figure 2. 95% C. L. upper limits on $\langle\sigma v\rangle$ as a function of m_{DM} . *Left:* Limits for the *gamma-line* derived from H.E.S.S. observations taken over ten years (254 h live time) of the inner 300 pc of the GC region. Observed limits (red dots) and mean expected limit (black solid line) are shown together with the 1σ (green band) and 2σ (yellow band) containment bands. *Right:* Limits for the 5plet towards dwarf galaxies are shown for single galaxy observation and for their combination (black solid line). The predicted cross section (gray dashed line), thermal mass (blue band) and excluded masses (red boxes) are represented.

expected to provide more than doubled statistics and give the possibility to increase the size of the *ON region*, pushing the limits to smaller $\langle\sigma v\rangle$ values. Two projects have also been started for the dSph. An observation campaign of some of the recently-discovered DES dSph [11] is ongoing. A project has begun among the H.E.S.S., MAGIC and VERITAS collaborations for a common effort to jointly search for DM towards dSph exploiting the complementarity of the different instruments and available datasets for more powerful analyses.

References

[1] F. Aharonian et al. (H.E.S.S. Collaboration), Astron. Astrophys. 503, 817 (2009)
[2] F. Aharonian et al. (H.E.S.S. Collaboration), Astron. Astrophys. 432, L25 (2005)
[3] F. Aharonian (H.E.S.S. Collaboration), Astron. Astrophys. 483, 509 (2008)
[4] H. Abdalla et al. (H.E.S.S. Collaboration), Phys. Rev. Lett. 117, 111301 (2016)
[5] H. Abdalla et al. (H.E.S.S. Collaboration), Phys. Rev. Lett. 120, 201101 (2018)
[6] A. Abramowski et al. (H.E.S.S. Collaboration), Phys. Rev. Lett. 110, 041301 (2013)
[7] M. De Naurois and L. Rolland, Astropart. Phys. 32, 231 (2009)
[8] M. Ackermann et al. (Fermi-LAT Collaboration), Phys. Rev. D91, 122002 (2015)
[9] M. Cirelli et al., Nucl.Phys.B787:152-175 (2007)
[10] H. Abdalla et al. (H.E.S.S. Collaboration), arxiv:1810.00995 (2018)
[11] A. Drlica-Wagner et al. (DES Collaboration), ApJ 813 109 (2015)

The quest for dark matter in dwarf spheroidal galaxies with the Cherenkov Telescope Array

Francesco G. Saturni[1,2,a], Gonzalo Rodríguez-Fernández[3], and Aldo Morselli[3] for the CTA Consortium

[1] *INAF – Osservatorio Astronomico di Roma, Via Frascati 33, I-00040 Monte Porzio Catone (RM), Italy*
[2] *ASI – Space Science Data Center, Via del Politecnico snc, I-00133 Roma, Italy*
[3] *INFN – Sezione di Roma Tor Vergata, Via della Ricerca Scientifica 1, I-00133 Roma, Italy*

Abstract. Dwarf spheroidal galaxies are among the best environments that can be studied with Cherenkov telescopes for indirect searches of γ-ray signals coming from dark matter self-interaction (annihilation or decay), due to their proximity and negligible background emission. We present new determinations of the dark-matter amount – i.e. the astrophysical factors J and D – in dwarf-galaxy halos obtained through the MCMC Jeans analysis of their brightness and kinematic data. Such factors are of great importance to test the performances of the next-generation γ-ray instruments such as the Cherenkov Telescope Array in detecting dark-matter signals from astronomical environments, or constraining the limits to dark-matter physics parameters (particle mass and lifetime, annihilation cross section).

1 Introduction

The problem of establishing the nature of Dark Matter (DM; [1]) is one of the major open challenges in modern astrophysics. Several efforts, both on the side of elementary particles (e.g. Weakly Interacting Massive Particles, WIMPs; [2]) and macroscopic objects (Massive Compact Halo Objects, MACHOs; [3]), have been made to identify plausible DM candidates. However, the parameter space covered by such candidates ranges over several orders of magnitude in masses and cross sections [4]. The current framework for the astronomical searches for DM signals is based on the possibility that DM particles self-interact via annihilation or decay to produce final-state γ-ray photons [5] whose detection is potentially at reach of next-generation Cherenkov telescopes such as the Cherenkov Telescope Array (CTA; [6]). Here we present new estimates of the DM content in the dwarf spheroidal galaxies (dSphs; [7]), which appear to be the best observational targets for indirect DM searches in terms of proximity and lack of background emission [8, 9]. In particular, we focus our analysis on the classical dSph Draco I (Dra I) and the ultra-faint Coma Berenices (CBe), Canes Venatici I (CVn I), Segue 1 (Seg 1), Reticulum II (Ret II) and Triangulum II (Tri II). With respect to previous works, all of the results shown here are obtained with a common methodology for all targets, and will be included, along with other promising dSphs, in a future comprehensive publication on this research topic.

[a] e-mail: francesco.saturni@inaf.it

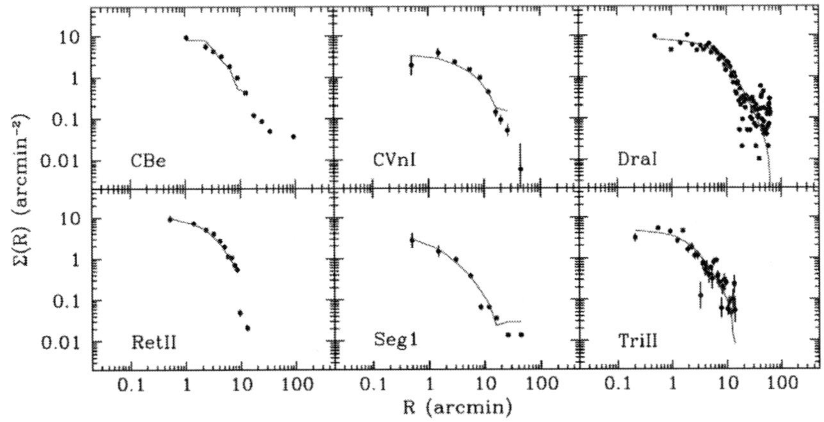

Figure 1: Best-fit brightness profiles $\Sigma(R)$ of the analyzed dSphs as a function of the object's projected (2D) radial coordinate R from the dSph centroid. In each panel, the projected 3D profile resulting from the fit (*red line*) is shown superimposed to the corresponding data set (*black dots*; [10–14]).

2 Indirect DM detection in the γ-ray domain

The γ-ray flux expected from DM self-interaction depends on the type of DM reaction, the mass of the DM particle and the amount of DM along the line of sight. In case of annihilation of two DM particles, the latter quantity is summarized into the astrophysical factors for DM annihilation $J(\Delta\Omega)$ or decay $D(\Delta\Omega)$:

$$J(\Delta\Omega) = \int_{\Delta\Omega} d\Omega \int_{\mathrm{l.o.s.}} \rho_{\mathrm{DM}}^2(\ell, \Omega) d\ell \tag{1}$$

$$D(\Delta\Omega) = \int_{\Delta\Omega} d\Omega \int_{\mathrm{l.o.s.}} \rho_{\mathrm{DM}}(\ell, \Omega) d\ell \tag{2}$$

Several attempts to get such values with increasing accuracy exist in the literature [15–19]. Here we present new determinations of the astrophysical factors for six of the best dSphs known to date in the field of DM indirect searches, based on the procedure described in [16] which exploits the capabilities of the Markov-Chain Monte Carlo (MCMC) Jeans analysis performed by the CLUMPY code [20]. At variance with them, we adopt a common framework for the treatment of surface-brightness and kinematic data sets of the analyzed dSphs, including the most promising newly-discovered targets.

2.1 Input priors and data sets for the CLUMPY set-up

The CLUMPY software allows to perform the dynamical analysis of the DM halos around dSphs assuming that such galaxies can be considered steady-state collisionless systems in spherical symmetry and with negligible rotation, in which the contribution of the stellar component to the total mass can be neglected. In these conditions, their second-order Jeans equation [24] can be solved to obtain the DM density profile $\rho_{\mathrm{DM}}(r)$ along the object's 3D radial coordinate r, once parametric forms of the stellar number density, radial velocity dispersion and velocity anisotropy are given. We follow the prescription by [25], fitting the brightness density $n^*(r)$ of each galaxy with a 3D Navarro-Frenk-White (NFW) profile [26] projected onto the corresponding surface brightness (see Fig. 1 and Tab. 1) and adopting an Einasto profile [27] for ρ_{DM}, so as to be directly comparable with the majority of the values of J and D available in the literature. For the dSph stellar kinematics, we use the same data sets of [16], [22] and [23], also adopting the definition of stellar membership probabilities P by [16].

Table 1: *Top:* best-fit parameters – heliocentric distance $d_\odot$ [21]; 3D brightness scale density ρ_s^*, radius r_s^* and profile indexes; dSph virial radius R_{vir}; 3D DM density scale ρ_s, radius r_s and index α [16] – for the six analyzed dSphs, classified for type ('cls': classical, 'uft': ultra-faint). *Bottom:* logarithmic astrophysical factors log J and log D for DM self-interaction, computed at their optimal angles ($\alpha_J = 2r_s^*/d_\odot$ or $\alpha_D = r_s^*/d_\odot$ respectively) and at 0.5 deg for a direct comparison with the literature [16, 22, 23].

Name	Type	$d_\odot$ (kpc)	ρ_s^* (10^5 L$_\odot$ kpc^{-3})	r_s^* (kpc)	α^*	β^*	γ^*	R_{vir} (kpc)	ρ_s (10^6 M$_\odot$ kpc^{-3})	r_s (kpc)	α
CBe	uft	44 ± 4	3.7 ± 1.7	0.092 ± 0.008	3.8	3.8	1.4	$4.2^{+14.5}_{-2.7}$	37^{+54}_{-28}	$1.0^{+4.6}_{-0.7}$	$0.64^{+0.25}_{-0.29}$
CVn I	uft	218 ± 10	2.15 ± 0.50	0.568 ± 0.033	3.2	3.8	0.2	$12.2^{+3.2}_{-4.0}$	$2.5^{+6.7}_{-1.4}$	$1.9^{+2.8}_{-1.1}$	$0.32^{+0.32}_{-0.45}$
Dra I	cls	76 ± 6	470 ± 140	0.170 ± 0.016	2.8	3.8	0.2	$2.68^{+0.45}_{-0.32}$	85^{+68}_{-29}	$0.28^{+0.33}_{-0.20}$	$0.41^{+0.23}_{-0.21}$
Ret II	uft	30 ± 3	157 ± 24	0.023 ± 0.003	3.2	3.8	0.4	$1.7^{+5.9}_{-1.1}$	65^{+220}_{-50}	$0.41^{+2.36}_{-0.34}$	$0.56^{+0.29}_{-0.27}$
Seg 1	uft	23 ± 2	13 ± 11	0.034 ± 0.007	3.6	3.8	0.8	$0.28^{+1.94}_{-0.23}$	$1.6^{+22.7}_{-1.3}$	$0.32^{+4.02}_{-0.25}$	$0.53^{+0.33}_{-0.26}$
Tri II	uft	30 ± 2	45 ± 24	0.020 ± 0.002	3.6	3.2	0.2	$6.2^{+15.0}_{-4.7}$	800^{+2070}_{-660}	$0.64^{+1.76}_{-0.55}$	$0.52^{+0.32}_{-0.27}$

	α_J (deg)	log $J(\alpha_J)$ (GeV^2 cm^{-5})	log J(0.5 deg) (GeV^2 cm^{-5})	α_D (deg)	log $D(\alpha_D)$ (GeV cm^{-2})	log D(0.5 deg) (GeV cm^{-2})
CBe	0.24	$19.1^{+0.7}_{-0.6}$	$19.4^{+0.9}_{-0.7}$	0.12	$17.7^{+0.6}_{-0.4}$	$18.7^{+0.7}_{-0.6}$
CVn I	0.30	$17.1^{+0.2}_{-0.2}$	$17.2^{+0.2}_{-0.2}$	0.15	$17.0^{+0.1}_{-0.1}$	$17.6^{+0.2}_{-0.2}$
Dra I	0.26	$18.6^{+0.1}_{-0.2}$	$18.6^{+0.1}_{-0.1}$	0.13	$17.4^{+0.1}_{-0.1}$	$18.0^{+0.1}_{-0.1}$
Ret II	0.09	$18.6^{+0.5}_{-0.5}$	$19.3^{+0.9}_{-0.7}$	0.05	$16.9^{+0.4}_{-0.3}$	$18.5^{+0.7}_{-1.0}$
Seg 1	0.17	$16.3^{+2.1}_{-2.1}$	$16.7^{+2.3}_{-2.5}$	0.09	$16.0^{+1.2}_{-1.2}$	$17.0^{+1.4}_{-2.2}$
Tri II	0.08	$20.8^{+0.6}_{-0.5}$	$21.6^{+1.0}_{-0.9}$	0.04	$17.8^{+0.5}_{-0.4}$	$19.6^{+0.8}_{-1.0}$

2.2 Astrophysical factors from the MCMC spherical Jeans analysis of the dSph halos

We run 10 MC chains with 10^4 realizations each for every target, choosing the unbinned likelihood analysis [25]. We then derive the distribution of virial radii R_{vir} for each galaxy from the output profiles as made by [16]. Finally, we compute the profiles of $J(\alpha_{int})$ and $D(\alpha_{int})$ as functions of the integration angle α_{int} by running the CLUMPY executable over the posterior distributions of the DM profile parameters. In Tab. 1, we report the values of the astrophysical factors obtained in this way for different integration angles, in order to favor a direct comparison with the results already available in the literature. In Fig. 2, we also show the profiles of J and D for the dSph with the highest expected DM amount (Tri II) and that with the most robust DM halo parameters (Dra I).

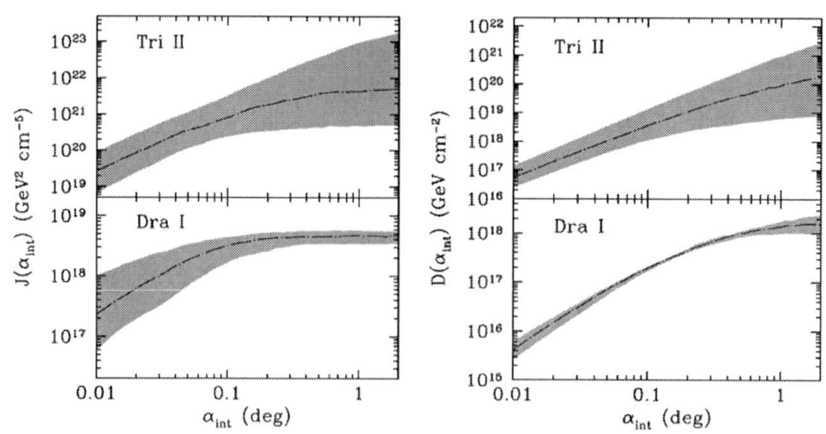

Figure 2: *Left panels:* DM-annihilation astrophysical factor profiles $J(\alpha_{int})$ as functions of the integration angle α_{int} for the most massive halo (Tri II) and the halo with the most robust parameter determination (Dra I). *Right panels:* equivalent profiles $D(\alpha_{int})$ for DM decay.

3 Discussion

Our MCMC examination of the first six DM halos in a completely uniform framework of input data allows us to produce J and D values for dSphs in agreement with the literature. The most discrepant result is $J(0.5\text{ deg})$ for Dra I, which is offset by a factor of ~4 from e.g. the value given by [16]. Nevertheless, CLUMPY is only able to perform a *spherical* Jeans analysis of the DM halos, thus not taking into account additional uncertainties arising from a possible halo triaxiality (~0.4 dex). We also note that, within the analyzed data sets, the dSph Tri II is potentially observable with CTA as a single targets in ~100 h integration. However, the limited statistics available on its stellar kinematics (which is also possibly contaminated by binary stars [28]) prevents to draw any firm conclusion, again emphasizing the importance of collecting good and clean stellar data for such targets in order to evaluate them as viable candidates for DM searches with Cherenkov telescopes.

References

[1] F. Zwicky, Ac. H. Ph. **6**, 110 (1933)
[2] E. Aprile, J. Aalbers, F. Agostini, et al., Ph. Rv. L. **121**, 111302 (2018)
[3] P. Tisserand, L. Le Guillou, C. Afonso, et al., A&A **469**, 387 (2007)
[4] L. Roszkowski, Pramana **62**, 389 (2004)
[5] M. Cirelli, G. Corcella, A. Hektor, et al., JCAP **3**, 051 (2011)
[6] CTA Consortium, *Science with the Cherenkov Telescope Array* (World Scientific Pub., 2019)
[7] L.E. Strigari, J.S. Bullock, M. Kaplinghat, et al., Nature **454**, 1096 (2008)
[8] N.W. Evans, F. Ferrer, S. Sarkar, Ph. Rv. D **69**, 123501 (2004)
[9] M. Doro, J. Conrad, D. Emmanoulopoulos, et al., A. Ph. **43**, 189 (2013)
[10] M. Irwin, D. Hatzidimitriou, MNRAS **277**, 1354 (1995)
[11] N.F. Martin, J.T.A. de Jong, H.W. Rix, ApJ **684**, 1075 (2008)
[12] R.R. Muñoz, M. Geha, B. Willman, AJ **140**, 138 (2010)
[13] K. Bechtol, A. Drlica-Wagner, E. Balbinot, et al., ApJ **807**, 50 (2015)
[14] B.P.M. Laevens, N.F. Martin, R.A. Ibata, et al., ApJL **802**, L18 (2015)
[15] V.A. Acciari, T. Arlen, T. Aune, et al., ApJ **720**, 1174 (2010)
[16] V. Bonnivard, C. Combet, M. Daniel, et al., MNRAS **453**, 849 (2015)
[17] A. Geringer-Sameth, S.M. Koushiappas, M. Walker, ApJ **801**, 74 (2015)
[18] G.D. Martinez, MNRAS **451**, 2524 (2015)
[19] K. Hayashi, K. Ichikawa, S. Matsumoto, et al., MNRAS **461**, 2914 (2016)
[20] M. Hütten, C. Combet, D. Maurin, Co. Ph. C. **235**, 336 (2019)
[21] A.W. McConnachie, AJ **144**, 4 (2012)
[22] V. Bonnivard, C. Combet, D. Maurin, et al., ApJL **808**, L36 (2015)
[23] A. Genina, M. Fairbairn, MNRAS **463**, 3630 (2016)
[24] J. Binney, S. Tremaine, *Galactic Dynamics: Second Edition* (Princeton University Press, 2008)
[25] V. Bonnivard, C. Combet, D. Maurin, et al., MNRAS **446**, 3002 (2015)
[26] J.F. Navarro, C.S. Frenk, S.D.M. White, ApJ **490**, 493 (1997)
[27] J. Einasto, Tr. Alm. **5**, 87 (1965)
[28] E.N. Kirby, J.G. Cohen, J.D. Simon, et al., ApJ **838**, 83 (2017)

Highlights from H.E.S.S.

Mathieu de Naurois[1,a] for the H.E.S.S. collaboration
[1] *LLR - IN2P3/CNRS - Ecole Polytechnique*

Abstract. In the 15 years since its construction, the H.E.S.S. gamma-ray observatory has allowed the study of the Very High Energy gamma-ray sky at resolutions and sensitivities which were never before possible. During this period H.E.S.S. has discovered a rich zoo of both galactic and extra galactic source classes, made measurements of the galactic cosmic ray spectrum and placed limits on fundamental physical processes.
A summary of the latest H.E.S.S. results for these source classes has been presented at the conference, describing the most interesting new observations and their physical interpretation. Additionally the latest upgrades and improvements to the H.E.S.S. hardware and data analyses, and the future science prospects for the experiment have been described.

1 Introduction

The H.E.S.S. array, inaugurated 15 years ago, revolutionized our view on the Very High Energy (VHE) gamma-ray sky with more than 100 sources discovered. A second phase of the experiment started in 2012, with the installation of the fifth, largest telescope (28 m diameter), allowing the threshold to be lowered from $\sim$ 100 GeV down to $\sim$ 20 GeV and bridge the gap with space instruments such as Fermi-LAT. To date, H.E.S.S. is the only *hybrid*[1] in the field, thus paving the way for the next generation Cherenkov Telescope Array. In the last years, tremendous efforts went into the design and optimization of analysis software able to exploit simultaneously monoscopic & stereoscopic events in a consistent way. At the same time, the field of VHE γ-ray astronomy is experiencing a kind of *phase-transition*, where single discoveries tend to have less importance and where population studies and detailed analysis of sources of specific interest are opening new fields. In this context, a new paradigm of simulation (Section 2.3) is now emerging and opening new science capabilities, with unprecedented resolution and understanding of the instrument.

The last years have seen the long awaited emergence of multi-wavelength, multi-messenger astronomy, advocated for since decades, with the discovery of gravitational waves from a binary neutron star merger [1, 2] and more recently, with the association of a very high energy neutrino observed by IceCube with an active galactic nuclei [3]. H.E.S.S., thanks to its very fast reaction time and slewing speed, is actively involved in this new *time-domain* astronomy.

After 15 years of operations, the H.E.S.S.-I cameras have been fully refurbished in 2016 with new electronics, inspired from the design of CTA, to optimize the overlap with CT5 (reduced threshold and dead-time) and ease maintenance. Ongoing negotiations to operate the array beyond 2019 are progressing well, and a wealth of new results are expected in the coming years.

[a] e-mail: denauroi@in2p3.fr
[1] i.e. array consisting of telescopes of different size operated together

2 The H.E.S.S. view of the Milky Way

The H.E.S.S. collaboration recently released a bundle of 14 publications in a special edition of *Astronomy & Astrophysics* [4], focusing on the H.E.S.S.-I analysis of the Galactic Plane. This major release contains in particular:

- the H.E.S.S. Galactic Plane survey [5], cornerstone of the publication and the major project of the collaboration over the last 10 years;
- the first population studies, on Pulsar Wind Nebula (PWN) [6] and Supernova Remnants (SNR) [7];
- detailed analysis on several sources, and in particular first evidence of emission beyond the shell of a SNR [8];
- a comprehensive view of the Galactic Center region, including diffuse emission [9];
- upper limits on several specific objects and/or potential new source classes.

2.1 H.E.S.S. Galactic Plane Survey

The H.E.S.S. Galactic plane survey (HGPS) [5] is a major long-term project, corresponding to ~ 2700 hours of high-quality observations acquired with the H.E.S.S.-I telescope array from 2004 to 2013. Results have already been published on a small (~ 10%) fraction of the current data set [10, 11].

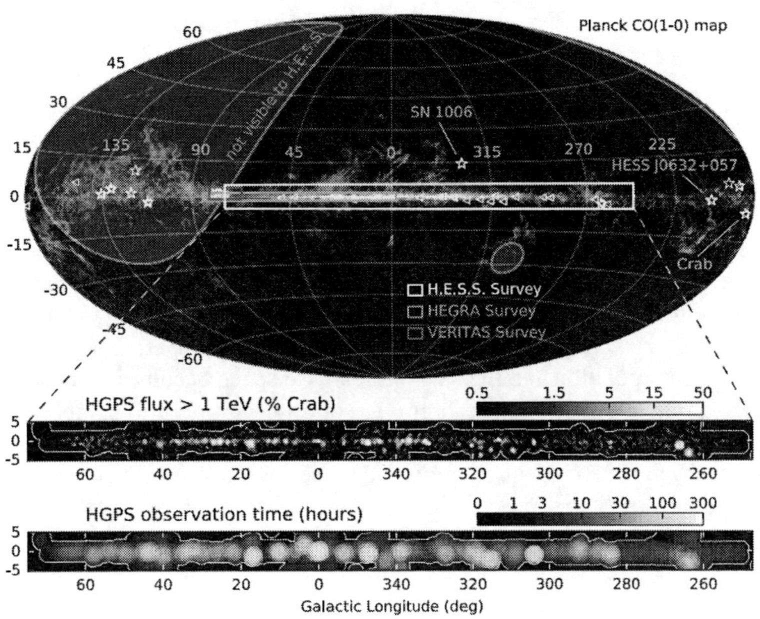

Figure 1. H.E.S.S. Galactic Plane Survey region, flux map and exposure map (from top to bottom). The all-sky image on the top panel shows a Planck CO Map with *Fermi*-LAT identified Galactic 1FHL sources (triangles) and the 15 known Galactic TeV sources (white stars) outside the HGPS region. The HEGRA Galactic Plane Survey [12] and the VERITAS Cygnus survey [13, 14] regions are illustrated in blue and green, respectively. From [5].

The region of the Milky Way covered by the HGPS (Galactic longitude between -110 and 65 degrees and Galactic latitude $|b| < 3.5$ degrees) is depicted as a white rectangle in Fig. 1 and compared to the HEGRA (in blue) and VERITAS Cygnus (in green) surveys.

A catalogue of 77 cosmic accelerators was derived from the HGPS using a semi-automatic pipeline, out of which 6 are new sources that were previously unknown or unpublished. Less than half of the sources (31) are firmly identified, the largest population consisting of pulsar wind nebulæ followed by supernova remnants and binary systems. Most of the remaining sources are confused (having several possible counterparts) while a significant fraction (12 sources) have no known counterparts at other wavelengths.

Maps and tables from the HPGS were released, for the first time, as FITS file to allow better exploitation by the community[2].

2.2 Population Studies

For the first time, the existence of a large scale catalogue of TeV sources made population studies on the dominant source classes, PWN [6] and SNRs [7] possible. One major result is that most young and powerful pulsars have a PWN detected in VHE. H.E.S.S. data also exhibit a correlation of TeV surface brightness with pulsar spin-down power $\dot{E}$, likely caused by an increase of extension with decreasing $\dot{E}$, and hence with time, and also by a mild decrease of TeV gamma-ray luminosity with decreasing $\dot{E}$. This allows a PWN evolution model to be constructed for the first time, from which it is possible to estimate how many of these objects will be visible by CTA. Regarding SNRs, typical ambient density values around shell-type SNRs can be constrained to $n \leq 7\ \mathrm{cm}^{-3}$ and electron-to-proton energy fractions above 10 TeV to $\epsilon_{\mathrm{ep}} \leq 5 \times 10^{-3}$. Comparisons of VHE with radio luminosities also provide support for the theory of magnetic field amplification at shell-type SNRs.

2.3 Resolving sources down to arcmin scale

The resolution power of current Imaging Atmospheric Cherenkov Telescopes is presently restricted to scales of a few arc-minutes, and is limited first by statistics and second by the uncertainties in the characterisation of the instrument Point Spread Function (PSF). A new paradigm for the simulation of IACTs has been developed during the last years [15], aiming at overcoming the limitations of current approaches, and to more efficiently cover the simulation phase space. Indeed with the advent of hybrid systems consisting of a large number of IACTs with different sizes, types, and camera configurations, the complexity of the interpolation and the size of the phase space becomes increasingly prohibitive.

In the *Run-Wise* simulation (RWS) paradigm, in contrast with traditional approach using a fixed grid of simulation parameters, the actual observation conditions as well as the individual telescope configurations of each observation run of a given data set, are taken into account. These run-wise simulations exhibit considerably reduced systematic uncertainties compared to the existing approach. They are also more computationally efficient and simple and therefore opens new physics cases. In particular, the PSF is much better understood and reproduced with the RWS simulation, allowing for the first time the study of the extension of sources down to arc-minute scale, well below the actual PSF size.

Using the RWS simulation scheme,the extension of the Crab nebula at TeV gamma-ray energies was resolved for the first time [16], with a width of $52''$ assuming a Gaussian source shape (Fig. 2). The H.E.S.S. extension covers a good fraction of the optical nebula and is significantly larger than the one measured with Chandra. This result can be explained by the radiative cooling of electrons, the

[2]These files can be downloaded from https://www.mpi-hd.mpg.de/hfm/HESS/hgps/

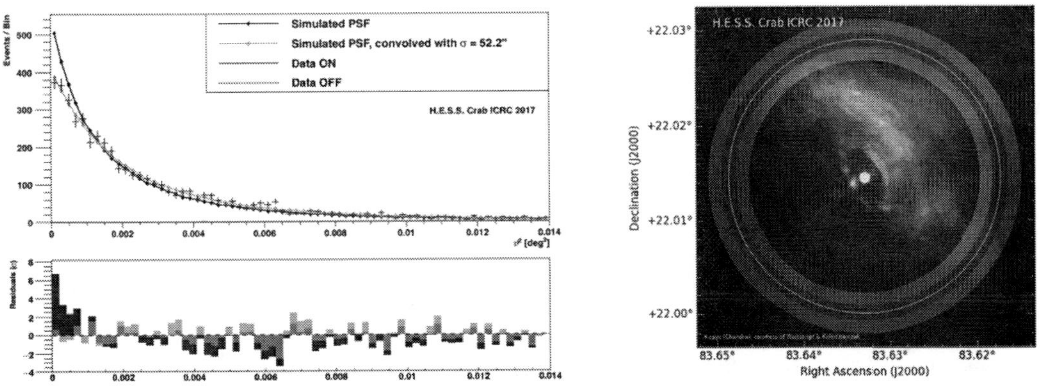

Figure 2. Left: θ^2 histogram of γ-like events from the Crab nebula (ON) and control regions (OFF). Simulated PSF and the PSF convoluted with the best-fit Gaussian are shown as well. **Right:** Chandra X-ray image of the Crab nebula. The H.E.S.S. extension is shown as solid white circle overlaid on top of shaded annulus indicating the statistical and systematic uncertainties of our measurement. From [16].

size of the nebula decreasing with increasing energies, as measured in other PWNs [e.g. 17], coupled with the fact that the energies of electrons producing the TeV gamma rays are well below those of electrons emitting the hard X-rays measured by Chandra [e.g. 18].

3 Extragalactic Observations

3.1 Rise of FSRQs

The advent of the fifth H.E.S.S. telescope, with a threshold of ~ 20 GeV at zenith, opened up a whole new lower-energy window from which extragalactic observations benefited the most. In particular, several bright FSRQ AGN have been detected recently during outburst episodes: PKS 0736+017 ($z \approx 0.19$) in February 2015, PKS 1510-089 ($z \approx 0.36$) in May 2016, and 3C 279 ($z \approx 0.54$) in July 2015 [19].

The detection of VHE emission from FSRQs during their outbursting episodes was unexpected due to the presence of the broad line region (BLR), an intense radiation field in the vicinity of the AGN, which acts as a barrier to the escape of VHE emission from within it. The detection of VHE emission from these sources can therefore be used to place strong constraints on the position of the emission site with respect to the BLR location. With gamma-ray emission beyond 200 GeV detected from each of these systems during their outbursts, the emission site is found to be constrained to sit at a distance beyond $r_{\rm BLR} \approx 10^{17}\,{\rm cm}\left(L_{\rm disk}/10^{45}\,{\rm ergs}^{-1}\right)^{1/2}$, where = $L_{\rm disk}$ is the thermal luminosity of the accretion disk [20].

3.2 Multi-Wavelength, Multi-Messenger astronomy

Due to its very large effective area and low threshold, H.E.S.S.-II has a competitive advantage over space instruments such as Fermi-LAT for transient phenomena in the overlapping energy range. As speed of reaction is critical for GRBs and other transient phenomena, a rapid repointing system for the telescopes has been implemented [21], which has been designed to respond as fast as possible, without human intervention, to targets of opportunity (ToOs). The average overall response time has been reduced to a timescale of ~few minutes, half of the sky being reachable in less than 60 s.

An example of the recent improvement in response time to neutrino alerts is the follow-up observations of the ANTARES neutrino event on the 30th January 2017. These observations took placed only 32 seconds after the reconstructed neutrino event occurred. Although searches for a gamma-ray counterpart within this data set were unsuccessful, the success of the automatic response system has already proved itself through this considerable reduction in the follow-up delay time to neutrino event alerts. The same applies to gravitational waves, where automatic scheduling is done for events occurring during the night. In the case of the neutron star merger GW170817, the H.E.S.S. telescopes were on target 5.3 h after the neutron star merger [1, 2], but less than 5 min after the revised LIGO-VIRGO alert was distributed and a handful of seconds after the start of the astronomical night. These 5 min latency was used to implement an observation strategy based on the probability density region obtained by weighting the three-dimensional GW map with galaxies from the GLADE catalog.

4 Electron spectrum

The spectrum of high energy electrons, up to ~ 1 TeV is traditionally measured using space based multi-purpose instruments such as AMS [22] or Fermi-LAT [23], due to much better hadronic rejection than IACTs. Dedicated space based instruments such as CALET or DAMPE are planning to extend the measurement of the electron spectrum up to ~ 10 TeV.

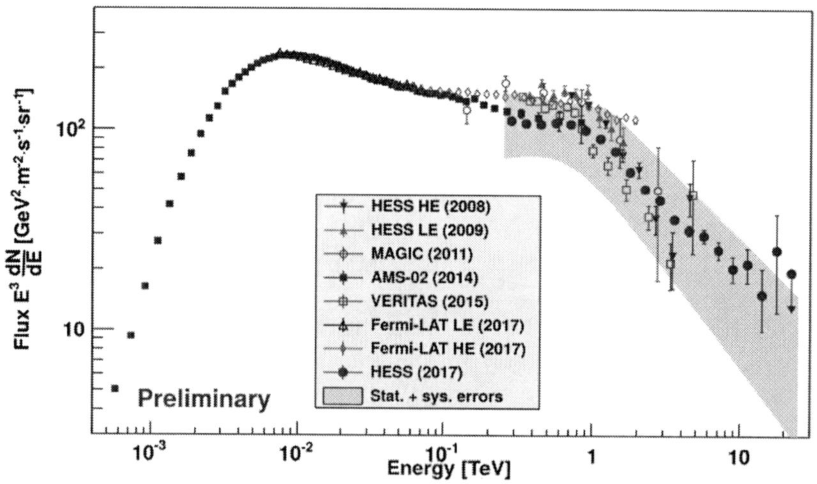

Figure 3. Cosmic-ray electrons energy spectrum measured with H.E.S.S. in 2017 (red dots) compared to previous measurements from various experiments.

In contrast, ground-based Cherenkov telescopes, thanks to their very large effective areas, are able to access the highest energies where the flux becomes very low. More than 9 years after the first electron spectrum measurement with H.E.S.S.[24] up to 5 TeV, subsequent observations have increased fourfold the amount of available data. In addition, analysis techniques have improved significantly, leading to a much better suppression of the background of cosmic-ray nuclei. These improvements allow a measurement of cosmic-ray electrons up to energies of $\sim$ 20 TeV, presented for the first time at the 35th ICRC, Busan, South Korea (Fig. 3). The measured electron spectrum can be described by two different regimes: up to $\sim$ 1 TeV, the spectrum appears quite regular with a constant slope, and becomes steeper above this energy.

This break in the spectral slope most probably signals the transition between a regime where a large number of sources contribute to the spectrum, to a regime where only the few, closest sources to the Earth are able to contribute. The very high energies reached in this measurement allow models of nearby sources of cosmic-ray electrons in which one source is very prominent to be tested. Such models are often invoked as a possible explanation for the excess of positrons measured by some experiments such as Pamela and AMS.

5 Conclusions & Perspectives

The results presented here represents only a small fraction of the recent highlights from H.E.S.S.. Other results were presented at the conference, including in particular – but not only – the discovery of new binary system in the Large Magellanic Cloud, first indications of γ-ray emission extending beyond the shell of a SNR, a detailed view of the Galactic Center, and evidence for VHE emission from pulsars. After more than 15 years of operation, the H.E.S.S. array is fully refurbished and ready to exploit the opening window of Multi-wavelength, Multi-messenger astronomy before the CTA era.

References

[1] B.P. Abbott, et al. (LIGO-VIRGO Scientific Collaboration), ApJL **848**, L12 (2017)
[2] H. Abdalla, et al. (H.E.S.S. Collaboration), ApJL **850**, L22 (2017)
[3] M.G. Aartsen, et al. (IceCube Collaboration), Science **361**, eaat1378 (2018)
[4] H. Abdalla, et al. (H.E.S.S. Collaboration), A&A **612**, E1 (2018)
[5] H. Abdalla, et al. (H.E.S.S. Collaboration), A&A **612**, A1 (2018)
[6] H. Abdalla, et al. (H.E.S.S. Collaboration), A&A **612**, A2 (2018)
[7] H. Abdalla, et al. (H.E.S.S. Collaboration), A&A **612**, A3 (2018)
[8] H. Abdalla, et al. (H.E.S.S. Collaboration), A&A **612**, A6 (2018)
[9] H. Abdalla, et al. (H.E.S.S. Collaboration), A&A **612**, A9 (2018)
[10] F. Aharonian, et al (H.E.S.S. Collaboration), Science **307**, 1938 (2005)
[11] F. Aharonian, et al (H.E.S.S. Collaboration), ApJ **636**, 777 (2006)
[12] F.A. Aharonian, et al., A&A **395**, 803 (2002)
[13] A. Weinstein, et al. (VERITAS Collaboration), ArXiv e-prints (2009), 0912.4492
[14] A. Popkow (VERITAS Collaboration), PoS(ICRC) p. 750 (2015)
[15] M. Holler, et al., PoS(ICRC2017) **35**, 755 (2017)
[16] M. Holler, et al. (H.E.S.S. Collaboration), PoS(ICRC2017) (2017)
[17] F. Aharonian, et al. (H.E.S.S. Collaboration), A&A **460**, 365 (2006)
[18] A.M. Atoyan, F.A. Aharonian, MNRAS**278**, 525 (1996)
[19] A.M. Taylor, et al. (H.E.S.S. Collaboration), PoS(ICRC2017) (2017)
[20] G. Ghisellini, F. Tavecchio, MNRAS**397**, 985 (2009), 0902.0793
[21] C. Hoischen, et al. (H.E.S.S. Collaboration), PoS(ICRC2017) **35**, 636 (2017)
[22] M. Aguilar, et al. (AMS Collaboration), Phys. Rev. Lett. **113**, 221102 (2014)
[23] S. Abdollahi, et al. (The Fermi-LAT Collaboration), Phys. Rev. D **95**, 082007 (2017)
[24] F. Aharonian, et al. (H.E.S.S. Collaboration), Phys. Rev. Lett. **101**, 261104 (2008)

AGILE γ-ray sources coincident with cosmic neutrino events

Fabrizio Lucarelli[1,2,*] and *Marco* Tavani[3,4,**] on behalf of the AGILE Team.

[1]INAF–OAR, via Frascati 33, I-00078 Monte Porzio Catone (Rome), Italy
[2]ASI Space Science Data Center (SSDC), Via del Politecnico snc, I-00133 Rome, Italy
[3]INAF/IAPS--Roma, Via del Fosso del Cavaliere 100, I-00133 Roma, Italy
[4]Univ. "Tor Vergata", Via della Ricerca Scientifica 1, I-00133 Roma, Italy

Abstract. Using data obtained by the γ-ray imager on board of the AGILE satellite, we systematically searched for transient γ-ray sources above 100 MeV in temporal and spatial coincidence with high-energy neutrino IceCube events. We found some significant γ-ray transients possibly associated with HESE neutrinos recently detected by the IceCube experiment. In this talk, we review the AGILE results and we will show that the probability of this association to be by chance is very low. One of the sources detected by AGILE is related to the blazar AGN TXS 0506+056. For the other sources, there are no obvious known counterparts.

1 Introduction

The clear observation of an isotropic flux of very high energy (VHE) neutrinos by the IceCube experiment [1] has opened the quest for the identification of their possibile electro-magnetic (EM) counterparts. Up to now, only two extra-terrestrial sources of neutrinos have been clearly identified: the Sun and SN1987a.

Theoretically, it is expected that VHE neutrinos with energies above 10 TeV are produced in cosmic beam dumps where cosmic rays (protons and heavy ions), accelerated via Fermi mechanisms in regions very near to compact objects or at the shock fronts of expanding supernova shells, collide with matter or radiation fields surrounding the central engine or filling an ejected jet of plasma (see [2] for a review). High-energy γ-ray photons above MeV–GeV are expected from neutral pions' decays produced in these proton-proton ($p - p$) or proton-photon ($p - \gamma$) collisions, with flux intensities that may vary depending on the source characteristics and its environment [3].

Active Galactic Nuclei (AGNs) of the blazar category are considered as the main cosmic neutrino source candidates [4]. Contribution from other types of active galaxies (starburst galaxies, radio-galaxies) [5, 6], AGN winds [7] as well as Galactic sources (supernovae remnants expanding in dense molecular clouds, microquasars, hidden compact objects) should also be also taken into account [8].

Observation of transient EM emission, spatially and temporally compatible with the IceCube neutrinos, is crucial to identify their possible counterpart. Since April 2016, the

*e-mail: fabrizio.lucarelli@inaf.it
**e-mail: marco.tavani@iaps.inaf.it

IceCube Collaboration alerts the astronomical community whenever a high-energy single neutrino-induced event (HESE/EHE)[1] is detected [9].

Thanks to the IceCube real time system and the fast space- and ground-based observations which followed the alert, a neutrino event recorded by IceCube on Sept. 22nd, 2017, has been significantly associated for the first time to a quasi-simultaneous γ-ray flare from the AGN blazar TXS 0506+056 [10]. Using IceCube archival data, a second strong neutrino flare was observed from the same direction between 2014 and 2015 [11]: the BL Lac TXS 0506+056 has thus been suggested as the first most-likely extra-galactic neutrino source ever observed.

The AGILE γ-ray mission [12] is also contributing to the new-born multi-messenger neutrino astronomy. Thanks to the very large field of view of its main detector, the γ-ray imager GRID, and the spinning observation mode, AGILE is a very suitable instrument for performing searches for short transient γ-ray counterparts to multi-messenger events like the neutrino observed on Sept. 2017.

In what follows, we report about the results of the AGILE search for transient γ-ray sources consistent *in time* and *space* with publicly announced IceCube HESE starting events.

2 AGILE search for γ-ray transient emission consistent with recent public IceCube events

The *AGILE* space mission is devoted to high-energy γ-ray astrophysics in the 30 MeV—30 GeV energy range. After an initial period of pointed observations, since 2009 the satellite has been put in spinning observation mode. In this way, AGILE is capable to image about 70-80% of the whole sky every day, with good sensitivity and angular resolution to γ-rays above 100 MeV[2].

One of the AGILE main astrophysical goal is the rapid identification of transient γ-ray sources over time-scales of a few days. To accomplish this issue, the AGILE Team has developed a very fast ground segment alert system and a robust Quick Look (QL) scientific analysis, which allows to automatically detect significant γ-ray excesses above 100 MeV over time-scales of 1-2 days [14]. This automatic procedure routinely runs daily over predefined AGILE count maps covering 48-hrs of data.

To find γ-ray signals consistent with a sample of 10 public IceCube events published since Apr. 2016[3], we have explored the database of automatic AGILE QL detections searching for transient γ-ray emission with the following characteristics: 1) positionally compatible, within the AGILE angular resolution, with the reconstructed arrival directions of the IceCube HESE/EHE neutrinos; 2) temporally occurring within a fixed search time window of ±4 days around the neutrino event time T_0.

From the DB search[4], we found three significant AGILE detections which satisfy the temporal and spatial association criteria defined above in correspondence of the following three IceCube events: IC-160731, IC-170321, and IC-170922.

Table 1 reports the three AGILE detections and the corresponding identification IceCube parameters (ID and occurence date). For each of them, we estimated the *post-trial probability*

[1]The events are announced through the GCN/AMON network and published at https://gcn.gsfc.nasa.gov/amon_hese_events.html (HESE events) or at gcn.gsfc.nasa.gov/amon_ehe_events.html for the highest energy events (EHE events).

[2]Sensitivity (at 5σ detection level) to γ-ray fluxes above 100 MeV: $(2 \div 3) \times 10^{-6}$ ph cm^{-2} s^{-1} over 2-day integration time intervals. Mean angular resolution (FWHM): 2.5° in the 100—400 MeV energy range; 1.2° in the 400 MeV—1 GeV band [13].

[3]The IceCube neutrinos considered to be most likely background events have been discarded from the search.

[4]Based on the typical AGILE angular resolution, we used the values of 1.0, 1.5, and 2.0 degrees for the database search radius around the initial IceCube input sky positions.

Table 1. The three AGILE QL detections close in time and space to IceCube HESE/EHE neutrinos. Column 1 displays the AGILE name assigned to the transient. Columns 2 and 3 show the main parameters of the corresponding IceCube event (event ID, neutrino event time T_0). Columns 4 to 6 show, respectively, the AGILE γ-ray flux (above 100 MeV) estimated over the QL 2-day integration time bin, the distance in time Δt from the QL detection centroid and T_0, the false alarm rate (FAR) expected for each detection, and the corresponding *post-trial* false alarm probability P_i.

AGILE source	IceCube event	T_0 (MJD)	$F_\gamma(E > 100\ MeV)$ $\times 10^{-6}$ (ph cm^{-2} s^{-1})	Δt (days)	FAR	P_i *post-trial*
A (AGL J1418+0008)	IC–160731	57600.079	(1.8 ± 0.7)	-2.0	5.9×10^{-4}	2.0×10^{-3}
B (AGL J0634–1504)	IC–170321	57833.314	(1.5 ± 0.6)	-2.2	1.5×10^{-3}	5.7×10^{-3}
C (AGL J0508+0459)	IC–170922	58018.871	(1.7 ± 0.7)	-2.8	1.0×10^{-3}	5.0×10^{-3}

that the proximity *in time* and *space* of the AGILE and IceCube events is *by chance*, taking into account the distance in time between the two detections and the AGILE occurence of false detections due to background noise (*false alarm rate* - FAR) [15], directly estimated from ten years of data [16].

Each detection results to have a post-trial false alarm probability P_i of the order of 3 Gaussian equivalent standard deviations. If we calculate the *joint* post-trial probability for the three detection as $P_{joint}(\mathrm{post-trial}) = 1 - (1 - \mathrm{P_A} * \mathrm{P_B} * \mathrm{P_C})^N$ [5], we find that the chance probability to observe an AGILE/IceCube correlation for 3 out of 10 events is around 1.7×10^{-6} ($\sim$ 4.7 Gaussian equivalent one-sided standard deviations).

3 Summary

In this contribution, we presented the results of the AGILE observations of the uncertainty regions of a selected sample of IceCube HESE neutrinos announced since April 2016 through the GCN/AMON system. Searching the database of automated AGILE QL detections, we found three significant γ-ray detections above 100 MeV within 1.5 degrees from the IceCube best-fit centroids, occurring within two days from the neutrino event time T_0. We showed that the global post-trial false alarm probability to find 3 out of 10 AGILE/IceCube correlations is quite low ($\sim 10^{-6}$), and significantly hints towards an astrophysical connection between the γ-ray and neutrino events.

In case of IC-170922 (AGILE Source C/AGL J0508+0459), we observed a significant temporal correlation between the neutrino event and the almost simultaneous γ-ray activity in HE and VHE bands observed from the BL Lac type blazar TXS 0506+056 [10]. This is suggestive of this AGN sub-class of blazars being one of the main VHE neutrino emitters from hadronic processes. For the other two cases of possible γ/neutrino correlation, IC-160731/AGL J1418+0008 (AGILE Source A) and IC-170321/AGL J0634–1504 (AGILE Source B), there was not any clear evidence of flaring activity from any known EM source inside the AGILE/IceCube common position uncertainty regions. Searching for possible EM counterparts from existing radio and X-ray source catalogs using the cross-search SSDC tools[6], we identified a new possible HBL candidate[7], within the IC-160731/AGL J1418+0008 error circle: the elliptical galaxy SDSS J141457.72-002058.6 (recently classified as the extreme blazar 3HSP J062753.3-151957 [21]). For IC-170321/AGL J0634–1504, several HBL candidates appear within the IceCube error box, among them the weak γ-ray source 3FGL J0627.9-1517 (also classified as the extreme blazar 3HSP J062753.3-151957 [21]).

[5] N is the number of global trials given by the product of two contributions: the total number of IceCube events considered (10), and the number (3) of optimizations of the AGILE detections DB search radius.

[6] https://tools.ssdc.asi.it

[7] A sub-class of blazars considered as one of the most promising neutrino–emitter candidates [19, 20].

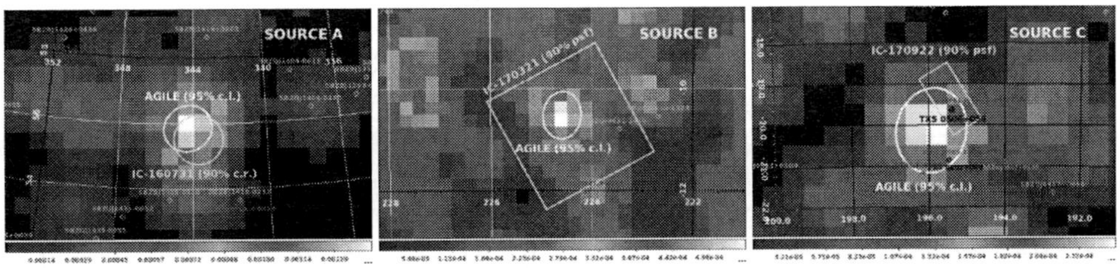

Figure 1. AGILE intensity maps above 100 MeV, in (ph cm^{-2} s^{-1} sr^{-1}) and galactic coordinates, centered at the position of the IceCube events IC-160731 (*left panel*), IC-170321 (*middle panel*), and IC-170922 (*right panel*), for the three AGILE detections A, B and C near the IceCube T_0's. The AGILE 95% confidence level location contours, obtained with the maximum likelihood AGILE analysis [17], are shown in white; the IceCube position uncertainty boxes/circles in yellow. The positions of the classified AGNs from the BZCAT Catalog [22] and the 3FGL γ-ray sources [23] are shown in cyan and magenta colors, respectively.

Given the lack of clear blazar counterparts, the IceCube/AGILE detections for Sources A and B might be explained by different classes of bursting extra-galactic sources (radio-galaxies, starburst galaxies, eruptions from AGN cores, etc.). Moreover, the IceCube neutrino fluxes can be produced by γ-ray hidden sources, where the targets for $p-p$ or $p-\gamma$ absorption processes are highly enhanced due to the high matter/radiation density surrounding the central engine. In such cases of enhanced target density, we might expect then to observe a soft γ-ray component peaking at MeV/sub-GeV due to reprocessing of VHE photons emitted by pions decay. The AGILE detections of γ-ray sources near IC-160731 and IC-170321 indicate the possibility that, from time to time, under particularly favorable conditions, the neutrino source may become transparent to these MeV/GeV γ-rays. AGILE, with its optimized sensitivity to soft γ-ray emission in the 100—400 MeV energy band and fast response to transient events, is thus particularly adapted to observe such softer spectrum sources.

To conclude, this is the first time that evidence of multiple γ-ray sources in close spatial and temporal coincidence with cosmic neutrinos is presented. More simultaneous neutrino and γ-ray events are in any cases needed to strengthen the correlation indicated in the current AGILE data analysis. Continuous blazar monitoring is also needed to confirm the association of BL Lac-type flaring sources with neutrino events. Given the low Galactic latitude of the IC-170321 event ($b = -10.75$ deg), in principle, also galactic sources should be taken into account and included in future searches of EM counterparts to neutrino-emitter.

Acknowledgements: AGILE is an ASI space mission developed with scientific and programmatic support from INAF and INFN. Research partially supported through the ASI grant no. I/028/12/2. Part of this work is based on archival data, software and online services provided by the ASI - Space Science Data Center (SSDC).

References

[1] Aartsen, M. G., et al., Evidence for high-energy extraterrestrial neutrinos at the IceCube detector. Science 342, 1242856 (2013).

[2] Halzen, F., High-energy neutrino astrophysics. Nature Physics 13, 232-238 (2017).

[3] Meszaros, P., Astrophysical sources of high-energy neutrinos in the IceCube era. Annual Rev. of Nuclear and Particle Science 67, 45-67 (2017).

[4] Mannheim, K., High-energy neutrinos from extragalactic jets. Astrop. Physics 3, 295 (1995).

[5] Loeb, A., and Waxman, E., The cumulative background of high energy neutrinos from starburst galaxies. Journal of Cosmology and Astroparticle Physics, Issue 05, id. 003 (2006).

[6] Becker Tjus, J., et al., High-energy neutrinos from radio galaxies. Phys. Rev. D, Vol. 89, Issue 12, id.123005 (2014).

[7] Lamastra, A., et al., Extragalactic γ-ray background from AGN winds and star-forming galaxies in cosmological galaxy-formation models. A&A 607, A18 (2017).

[8] Anchordoqui, L. A., et al., Cosmic neutrino pevatrons: a brand new pathway to astronomy, astrophysics, and particle physics. Journal of High Energy Astrophysics 1-2, 1-30 (2014).

[9] Aartsen, M. G., et al., The IceCube realtime alert system. Astroparticle Physics 92, 30-41 (2017).

[10] Aartsen, M. G., et al., Multimessenger observations of a flaring blazar coincident with high-energy neutrino IceCube-170922A. Science 361, eaat1378 (2018).

[11] Aartsen, M. G., et al., Neutrino emission from the direction of the blazar TXS 0506+056 prior to the IceCube-170922A alert. Science 361, 147-151 (2018).

[12] Tavani, M., Barbiellini, G., Argan, A., et al., The AGILE Mission, A&A 502, 995 (2009).

[13] Sabatini, S., Donnarumma, I., Tavani, M., et al., On the angular resolution of the AGILE γ-ray imaging detector, ApJ 809, 60 (2015).

[14] Bulgarelli, A., Trifoglio, M., Gianotti, F., et al., THE AGILE alert system for γ-ray transients, ApJ 781, 19 (2014).

[15] Connaughton, V., Burns, E., Goldstein, A., et al., FERMI GBM observations of LIGO gravitational-wave event GW150914, ApJL 826, L6 (2016).

[16] Lucarelli, F., et al., AGILE detection of γ-ray sources coincident with cosmic neutrino events, submitted for publication in ApJ (2019).

[17] Bulgarelli, A., Chen, A. W., Tavani, M., et al., Evaluating the maximum likelihood method for detecting short-term variability of AGILE γ-ray sources, A&A, 540, A79 (2012).

[18] Lucarelli, F., et al., AGILE Detection of a Candidate Γ-Ray Precursor to the ICECUBE-160731 Neutrino Event, ApJ 846, 121 (2017).

[19] Padovani, P., Resconi, E., Giommi, P., Arsioli, B., Chang, Y. L., Extreme blazars as counterparts of IceCube astrophysical neutrinos, MNRAS 457, 3582 (2016).

[20] Resconi, E., Coenders, S., Padovani, P., Giommi, P., Caccianiga, L., Connecting blazars with ultrahigh-energy cosmic rays and astrophysical neutrinos, MNRAS 468, 597 (2017).

[21] Chang, Y.L., Arsioli, B., Giommi, P., Padovani, P., Brandt, C., The 3HSP catalog of extreme and High Synchrotron Peaked blazars, submitted for publication in A&A (2019).

[22] Massaro, E., Maselli, A., Leto, C., et al., The 5th edition of the Roma-BZCAT. A short presentation, Astrophysics and Space Science, 357, 75 (2015).

[23] Acero, F., Ackermann, M., Ajello, M., et al., FERMI-LAT third source catalog, Ap. J. Supp., 218, 23 (2015).

Temporal studies on GRB light curves for neutrino flux predictions in multi-collision zone models

Angela Zegarelli[1,2,a]

[1]*Dipartimento di Fisica, Università La Sapienza, P.le Aldo Moro 2, 00185 Roma, Italy*
[2]*INFN - Sezione di Roma, P.le Aldo Moro 2, 00185 Roma, Italy*

Abstract. For the analysis here presented, the most luminous GRB detected by Konus-*WIND* until now was selected, namely GRB 110918A. Through the NeuCosmA code, simulating multi-zone collisions during the GRB prompt emission, a synthetic light curve representative of the observed one has been obtained, in order to study how characteristic parameters of the inner engine activity influence the light curve. This is achieved by varying the input parameters values in the simulation. In agreement with other previous studies, the inner engine is found to be more variable than the observed light curve. Furthermore, the Empirical Mode Decomposition method is applied and its stability is analyzed. Finally, the first multi-collision neutrino flux estimate for a real GRB is presented. Assuming the flux evaluated through the simulation reproducing the GRB gamma-ray light curve and taking into account the data taking conditions and the detector response during the GRB episode, the number of expected signal in the ANTARES neutrino telescope from GRB 110918A was calculated to be $\mu_s = 6.1 \times 10^{-4}$, which is found to be a factor 20 lower than previous estimations with one-zone models.

1 Introduction

Gamma-Ray Bursts are the most powerful explosions in the Universe. They constitute highly beamed sources of gamma rays and possibly, in case of hadronic acceleration, also of high-energy neutrinos. In the internal shock scenario, blobs of plasma emitted from a central engine are believed to collide within a relativistic jet and form shocks, responsible for the particle acceleration. Searching for a temporal and spatial coincidence among the GRB prompt emission and high-energy neutrinos is crucial to provide information about the acceleration mechanism in GRBs. The detection of neutrino events would allow to identify these sources as hadronic accelerators and as candidates for Ultra High Energy Cosmic Rays (UHECRs). However, an open issue regarding GRBs is represented by the way in which the central engine operates. In order to shed light on this topic, the information on the temporal behavior of individual GRBs, as observed in gamma rays, is very precious.
Hence, after a brief description in Section 2 of the GRB considered in this work, the simulation model used to reproduce its photon light curve is explained in Section 3. In Section 4, a novel approach for GRB timing studies, which adopts the Hilbert-Huang transform of the light curve for the identification of the characteristic signal modes, is analyzed. However, this method is not found to be stable with

[a]e-mail: Angela.Zegarelli@roma1.infn.it

respect to Poissonian fluctuations of the signal. In order to clarify this point, still some investigation is required. Finally, the neutrino flux prediction for this GRB and the number of expected muon neutrino signals in the ANTARES neutrino telescope is presented in Section 5.

2 Selected candidate: GRB 110918A

GRB 110918A [1] is the brightest long GRB ever detected by Konus-*WIND* (KW) and the most luminous one observed until now. At its redshift ($z = 0.984 \pm 0.001$), the high energy fluence measured by KW, $F_\gamma = 7.5 \times 10^{-4}$ erg cm^{-2}, implies a huge isotropic equivalent energy released in the source frame, $E_{\rm iso} = (2.1 \pm 0.1) \times 10^{54}$ erg, and an isotropic-equivalent peak luminosity $L_{\rm iso} \sim 4.4 \times 10^{54}$ erg s^{-1}. The GRB light curve (Figure 1) starts with an extremely bright, hard and short pulse followed by three weaker, softer and partly overlapping pulses in the next $\sim$ 25 s, showing a long quiescent period ($\sim$ 10 s) between the second and the third peak. The duration of the burst is $T_{90} = 19.6 \pm 0.1$ s. Furthermore, assuming that the shortest timescale in GRB prompt emission (variability time t_v) coincides with the shortest pulse width, the value $t_v = 0.25$ s was estimated [1].

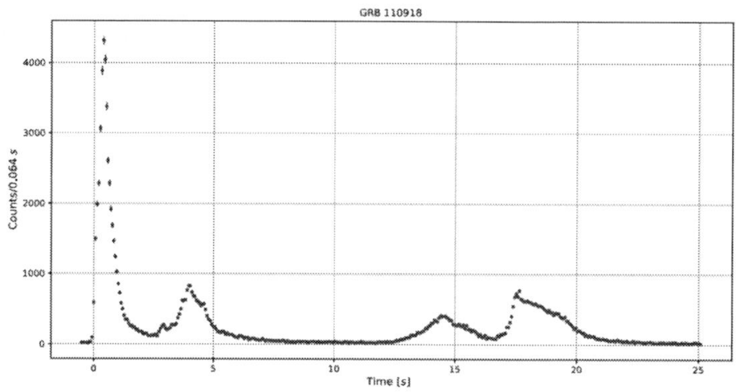

Figure 1: Konus-*WIND* photon light curve observed from GRB 110918A.

This GRB has been selected because it is ultraluminous, it has peculiar light curve characteristics, and the behavior of its emission is highly non linear, as derived from a temporal study of its Probability Density Function [2]. This latter point allows to use the Hilbert-Huang transformation, as shown in Section 4.

3 GRB light curve study with multi-collision approach

In order to simulate a GRB light curve similar to the one of GRB 110918A, the NeuCosmA code [3] has been used. A *multi-collision approach* [4] allows to simulate collisions at different radii in a GRB jet considering a system of plasma shells with a distribution of Lorentz factors, masses and widths. In this way, an entire light curve is derived and precise secondary flux predictions are obtained, including neutrinos. Therefore, a synthetic GRB light curve reproducing that of GRB 110918A has been constructed. After choosing several input parameters (e.g. radius and Lorentz factor distributions of the shells), 500 simulations have been performed and the averaged light curve has been computed

(Figure 2). The variability time $t_v = 0.220 \pm 0.008$ s, the burst duration $T_{90} = 17.136 \pm 0.003$ s and the number of collisions $N_{coll} = 78 \pm 3$ for the averaged simulated light curve are consistent with the quantities observed for GRB 110918A: $t_v = 0.25$ s, $T_{90} = 19.6 \pm 0.1$ s and $N_{coll} \simeq 79$. As the number of observed pulses ($N_{pulses} = 5$) is much smaller than the number of ejected shells ($N_{shells} = 110$), the inner engine results more variable than the observed light curve [5].

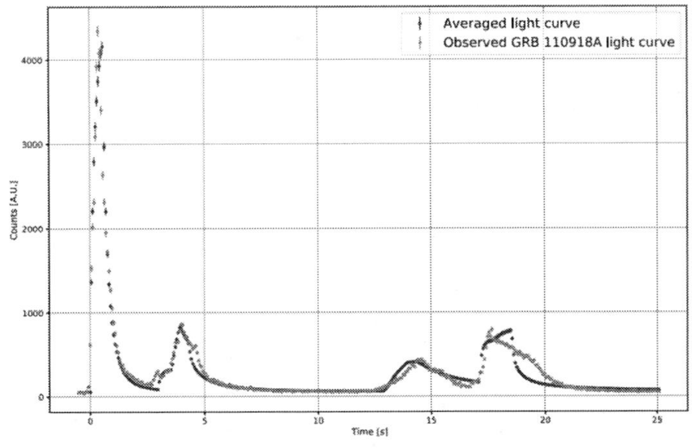

Figure 2: GRB 110918A Konus-*WIND* observed photon light curve (in orange, same as in Figure 1) and photon light curve simulated with the multi-collision NeuCosmA code (in blue).

4 Hilbert-Huang transform: a novel approach for GRB timing studies

Non-linear and non-stationary time series allow for an analysis through the Hilbert-Huang transformation (HHT) [6]. Its use for the GRB timing studies, a powerful tool to understand the physics of the GRB prompt emission and its variability, is a recent idea, given the lack of direct observations of the inner engine. Such an analysis consists of two steps:

1. Empirical Mode Decomposition (EMD): decomposition of the signal into individual modes embedded in the data (Intrinsic Mode Functions, IMFs), that relies on the assumption that data consist of different simple intrinsic modes of oscillation.

2. Hilbert Spectral Analysis: the Hilbert transform is applied to each IMF in order to construct the energy-frequency-time distribution, called *Hilbert spectrum*, in which the time localities of events is preserved.

The stability of this method has been investigated: while translation and addition of a constant factor to the light curve do not influence results in individual IMFs, the addition of random fluctuations produces relevant modification to the EMD. For this reason, it was not possible to compare the results of HHT obtained with the simulated and with the observed light curve.

5 Neutrino flux prediction

In this work the first multi-collision neutrino flux estimate for a real GRB is presented (Figure 3). Assuming the neutrino flux evaluated through the simulation reproducing the GRB light curve and taking into account the data taking conditions and the detector response during the GRB episode [7], the number of expected signal in the ANTARES neutrino telescope from GRB 110918A was calculated: $\mu_s = 6.1 \times 10^{-4}$. This value is smaller than the one previously calculated by the ANTARES Collaboration according to the one-zone approach, namely $\mu_s = 1.3 \times 10^{-2}$ [8]. This is connected with the fact that the novel modeling presented here accounts for the whole inner engine behavior, as dictated by the emitted light curve, while past calculations simplified such a description within a unique collision able to reproduce the gamma-ray spectrum. The limited expected event rate justifies the lack of spatial and temporal coincidence in ANTARES between high-energy neutrinos and GRB events until now [9][10].

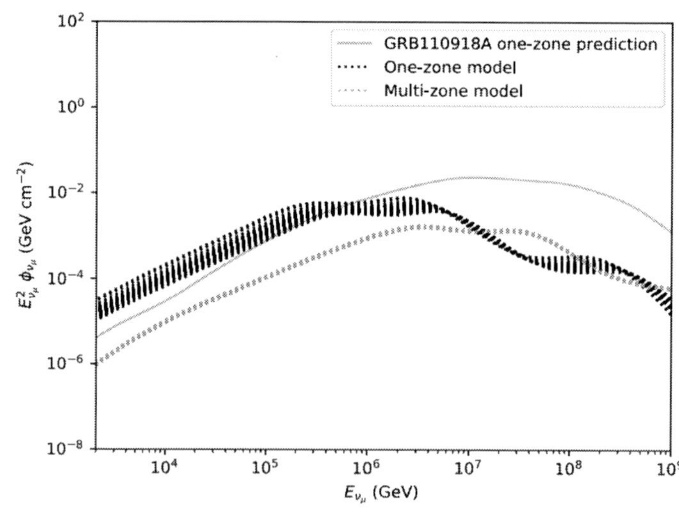

Figure 3: Expected $\nu_\mu + \bar{\nu}_\mu$ fluence for the GRB 110918A. Green solid line: former one-zone prediction, extracted from the gamma-ray spectrum [8]. Dashed black lines: one-zone prediction obtained by averaging over 500 complete light curve simulations performed using the NeuCosmA code (this work). Dashed orange lines: multi-zone prediction obtained by averaging over 500 complete light curve simulations performed using the NeuCosmA code (this work).

References

[1] D. D. Frederiks, et al., ApJ, 779, 151, 2013
[2] N. Bhatt & S. Bhattacharyya, MNRAS, 420, 1706, 2012
[3] S. Hümmer, et al., ApJ, 721, 630, 2010
[4] M. Bustamante, et al., Astrophys. J., 837, 33, 2017
[5] E. Nakar & T. Piran, MNRAS, 331, 40, 2002

[6] N. E. Huang, et al., Proc. R. Soc. Lond. A, 454, 903, 1998
[7] S. Celli, Proceeding of RICAP16, EPJ Web of Conferences, 136 , 04006, 2017
[8] A. Albert, et al., MNRAS, 469, 906, 2017
[9] S. Adrián-Martínez et al., ANTARES Collaboration, A&A 559, A9, 2013
[10] S. Celli, Proceeding of ICRC17, Proceedings of Science, 301, 998, 2017

VERITAS: Status and Recent Results

John Quinn[1,a] for the VERITAS Collaboration[2]

[1] *School of Physics, University College Dublin, Belfield, Dublin 4, Ireland*
[2] *http://veritas.sao.arizona.edu*

Abstract. VERITAS has just completed its 11th year of full four-telescope scientific operations and continues to function with excellent efficiency. Its science program, encompassing galactic, extragalactic, and fundamental physics, entails dedicated observations of specific targets as well as multi-messenger target-of-opportunity observations. The current operational status of VERITAS is presented, as well as recent science highlights, and the future plans for the observatory are discussed.

1 Introduction

The Very Energetic Radiation Imaging Telescope Array System (VERITAS) [1] is an array of four imaging atmospheric Cherenkov telescopes, located at the Fred Lawrence Whipple Observatory in southern Arizona, U.S.A., which has been in full scientific operation since 2007. The VERITAS Collaboration has 25 member institutions from Canada, Germany, Ireland and the U.S.A., in addition to approximately 20 active associate members who contribute through multi-wavelength observations and theoretical work. VERITAS conducts a broad observation program focussed on galactic, extragalactic and fundamental science, which incorporates significant flexibility through dedicated target-of-opportunity (ToO) and director's discretionary time (DDT) to respond to multi-messenger alerts. Self-triggering from real-time analysis of VERITAS's own observations has also proved invaluable. In this proceeding the status and future prospects of the VERITAS observatory will be presented along with select recent science highlights from its scientific program.

2 VERITAS Technical and Operational Status

VERITAS consists of four 12 m diameter optical reflectors, each equipped with 499 pixel photomultiplier-tube (PMT) cameras, that stereoscopically image the Cherenkov light generated by extensive air showers that develop when gamma and cosmic rays interact with the atmosphere. VERITAS entered full four-telescope scientific operation in 2007 and has undergone several significant upgrades, including the move of Telescope 1 to improve the geometric layout of the array in 2009 [2] and the upgrade of the cameras with higher quantum-efficiency PMTs in 2012 [3], to significantly improve its performance compared to the original configuration. A prototype Schwarzschild-Couder telescope (pSCT) for the Cherenkov Telescope Array (CTA) has been constructed on the original site

[a] e-mail: John.Quinn@ucd.ie

of VERITAS Telescope 1, which will allow simultaneous operation with VERITAS to evaluate its performance and to conduct joint science studies. VERITAS operates in the energy range 85 GeV to 30 TeV, has an angular resolution of 0.08° at 1 TeV, a peak effective area $\sim 10^5$ m^2, and can detect a 1% Crab source in less than 25 hrs. A detailed overview of VERITAS performance can be found in [4]. New analyses methods with improved sensitivity are currently under development and undergoing validation.

VERITAS typically obtains 950 hours of dark-sky observations per year and observations have been extended into moderate moonlight conditions to obtain approximately 200 additional hours per year. The excellent on-site technical support results in typically >95% four-telescope operation with 97% achieved in the most recent observing season. VERITAS is fully funded for operations through 2019 and in late 2018 will apply for funds for continued operations until at least 2022.

3 Recent Science Highlights

At the time of writing, the VERITAS catalogue consists of 63 objects, with 23 galactic and 40 extragalactic sources. The extragalactic sources are dominated by Active Galactic Nuclei (AGN), which constitute 39 of the detected sources, with starburst galaxy M82 being the one non-AGN extragalactic source detected. In addition to source detections, VERITAS observations have been used to derive dark matter limits, constrain the extragalactic background light intensity and intergalactic magnetic field strength, and to study the cosmic ray iron and electron spectra. A selection of recent science highlights are presented below.

3.1 Galactic Science Highlights

3.1.1 PSR J2032+4127 / MT91 213

TeV J2032+4130 was the first unidentified TeV gamma-ray source, discovered by the HEGRA observatory [5] and was later found by VERITAS to be extended [6]. A Fermi-LAT pulsar, PSR J2032+4127, is located in the southeast corner of TeV J2032+4130, which in 2015 was identified as being in a binary system with the 15 $M_\odot$ star MT91 213, with an orbit of eccentricity ~0.95 and period ~50 years, and periastron due to occur on 13 November 2017. Extensive multiwavelength observations were conducted during the passage through periastron resulting in clear detections ($> 20\sigma$) of a variable point-source by both VERITAS (181 hrs of observations) and MAGIC (88 hrs), and variable X-ray flux by Swift XRT (135 hrs) [7]. The VHE spectrum changes shape during the periastron passage and the flux and spectral variability as a function of orbital phase is more complicated than predicted.

3.1.2 Supernova Remnants: Tycho's SNR, Cassiopeia A and IC 443

Supernova remnants (SNR) are among the leading candidates to be the sources of galactic cosmic rays with energies up to the 'knee' ($\sim$ 1 PeV - 10^{15} eV, "PeVatron"), and young SNR are the best candidates to provide observational evidence as the highest energy cosmic rays, which are believed to be produced early on, will not yet have escaped from the site of production. The sites of cosmic ray acceleration should produce corresponding gamma-ray emission through the interaction of the accelerated cosmic rays with interstellar material.

VERITAS conducted a deep multi-year exposure on Tycho's SNR, a well studied young (~450 yr) Type Ia in a clean environment, yielding 147 hours of data, which were combined with 84 months of *Fermi*-LAT observations. The centroids of both the VERITAS and *Fermi*-LAT observations are consistent with each other and with the centre of the remnant while the measured power-law spectral index at energies > 400 GeV of $2.92 \pm 0.42_{stat} \pm 0.20_{sys}$ compares to a power-law of index $2.14 \pm 0.09_{stat} \pm 0.02_{sys}$ measured in the *Fermi*-LAT band [8]. The VERITAS results point to a likely lower maximum particle energy than anticipated if such SNR are the PeVatrons.

VERITAS also conducted a deep, 65 hr, exposure on Cassiopeia A, a young (~350 year old) SNR of Type IIb. In the energy range from 300 GeV to 7 TeV the spectrum is best fit with a power-law of index $2.75 \pm 0.10_{stat} \pm 0.20_{sys}$ [9]. When compared to *Fermi*-LAT observations, the spectrum softens at energies >300 GeV, again indicting a cut-off in the maximum particle energy in the remnant. The VERITAS results on both Tycho's SNR and Cassiopeia A indicate that more sources in addition to young SNR may be required to explain the origins of galactic Cosmic Rays.

IC 443 is a resolved middle-aged (~3,000-30,000 yrs) mixed-morphology SNR that VERITAS has observed for over 150 hrs of data since 2007. The overall nebula, and the nebula sub-divided into four regions, has been analysed and compared to an analysis of the same regions in *Fermi*-LAT. The morphology seen with both instruments is highly consistent, suggesting a single population of cosmic rays across a wide range of energies. The spectral shape is best fit with a broken power-law, with a break at ~40 GeV when jointly fitting the 1 GeV to 10 TeV data set.

3.1.3 VERITAS follow-up of HAWC sources

The HAWC observatory's 2HWC catalog [10], resulting from a survey incorporating 507 days of observations, contains 39 sources, with 20 being associated with previously-detected TeV objects. A VERITAS study entailing a total of 218 hrs archival and dedicated follow-up observations of 14 of the objects without know counterparts was conducted and combined with the analysis 8.5 years for *Fermi*-LAT observations [11]. The VERITAS observations resulted in the detection of just one object, DA 495, a pulsar wind nebula coinciding with 2HWCJ1953+294. The *Fermi*-LAT analysis detected GeV emission from a known TeV pulsar wind nebula, SNRG54.1+0.3 (VERJ1930+188), and a 2HWC source, 2HWCJ1930+188. The lack of VERITAS detections of the other HAWC sources is possibly due to source extensions as the VERITAS analyses were limited to moderately extended sources with extensions $< 0.3°$.

3.1.4 VERITAS observations of the Galactic Centre region

The Galactic Centre region is a high-priority target for VERITAS, which observes it at large zenith angles (typically $\gtrsim 60°$), resulting in an energy threshold of ~2 TeV and an increased effective area of $\sim 5 \times 10^5$ m^2 at energies above a few TeV. Observations are conducted offset 0.7° from the source ("wobble" mode) with dedicated off source observations $\sim 5°$ away. The VERITAS observations have resulted in a detection of Sgr A* at $> 35\sigma$, consistent with a point source, with a spectrum best fit with a power law with exponential cut off at $\sim$ 11 TeV. An analysis of the diffuse extended emission, using the regions as described in HESS [12] reveal a hard spectrum consistent with a power law up to the highest observed energy of $\sim$ 40 TeV, supporting the hypothesis that the Galactic Centre region contains a PeVatron. A VERITAS publication detailing the latest results is in preparation.

3.2 Extragalactic Science Highlights

3.2.1 3C 264

3C 264 is an FR-I-type radio galaxy at redshift of $z = 0.0216$. It has a rapidly evolving knot structure as seen from radio and optical observations and is detected in MeV to GeV gamma-rays by *Fermi*-LAT with an extrapolated flux at energies >200 GeV of 1.6% Crab. Ten hours of VERITAS observations in 2017 revealed a hint of a signal, prompting follow-up observations in 2018. These observations resulted in the detection of the 3C 264 by VERITAS initially at the 5.4σ level in 12 hours of data [13]. This detection prompted additional VERITAS and multi-wavelength observations, with VERITAS detecting it at the 8σ level. The observations revealed it have a flux of ~0.5% that of the Crab Nebula and to be weakly variable.

3.2.2 Ton 599

VERITAS observations of Ton 599, an FSRQ at redshift $z = 0.72$, were triggered by an exceptional period of outburst seen by the *Fermi*-LAT instrument in early November 2017 where its flux reached 20 times its 3FGL value [14]. While Ton 599 was initially at too low an elevation to be observed by VERITAS, the outburst seen at MeV-GeV energies persisted and went through a second exceptional enhancement in December 2017, prompting follow-up observations that resulted in the VERITAS detection of it at $\sim 8\sigma$ in two hours of observations, with a flux level of ~12% that of the Crab Nebula [15]. Ton 599 is the third-most-distant VHE gamma-ray source.

3.2.3 OJ 287

OJ 287 is a BL Lac object at redshift $z = 0.3056$ with quasi-periodic pairs of optical outbursts which have been interpreted as the accretion-disk crossings of a 150 million $M_\odot$ black hole in a 12-year precessing orbit about an 18 billion $M_\odot$ black hole [16]. In late 2016 the *Swift-XRT Monitoring of Fermi-LAT Sources of Interest* web site [17] revealed that OJ 287 was undergoing a period of enhanced activity. VERITAS conducted observations and detected OJ 287 at a level of ~1.3% Crab [18]. Copious multi-wavelength observations were taken, including radio observations that indicate the enhanced emission may coincide with emission from a new feature in the jet.

3.2.4 BL Lacerate

A very rare rapid flare from a low-peaked BL Lac (LBL) object was observed by VERITAS from BL Lacertae, with a rise time of ~2.3 hr. and a decay time of ~ 36 min. on 2016 October 5 [19]. The peak flux in 4-min. bins reached 180% that of the Crab Nebula. Contemporaneous multi-wavelength variability was also detected, with a possible superluminal feature passing the core of the jet identified in the VLBA observations at 43 GHz.

3.2.5 TXS 0506+056

The IceCube neutrino observatory detected a neutrino event of potential astrophysical origin directionally consistent with being from the blazar TXS 0506+056 on 2017 September 22, directly linking the acceleration of hadronic particles to the jets of active galaxies. Within 12 hours of the ICECUBE event VERITAS conducted one hour of observations under non-optimal weather conditions and no gamma-ray source detected. VERITAS conducted a further five hours of follow-up observations in the period 2017 September 28 to September 30 in response to the announcement that *Fermi-LAT* had detected TXS 0506+056 in enhanced gamma-ray state. While the VERITAS observations did not reveal a detection, MAGIC detected the source between 2017 September 28 and October 3. The results of the initial observations from IceCube, VERITAS and several other observatories were reported in [20]. VERITAS continued observations of the object over the next 5 months, accumulating 35 hours of data, and achieved a detection at the 5.8σ significance level [21]. The source had an average integral flux above 110 GeV of 1.6% that of the Crab Nebula, consistent with being steady over the period of observation, with a spectrum fit by a power-law with index of 4.8 ± 1.3. The broadband spectrum has a softening between the *Fermi*-LAT and VERITAS energy that cannot be explained by EBL absorption alone, which has implications for the distribution of the parent population of particles producing the electromagnetic emission [21].

3.3 Dark Matter and Cosmic Ray Highlights

3.3.1 Dark Matter Limits

Recent VERITAS observations and analyses of ~230 hours of data on five dwarf spheroidal galaxies (Boötes I, Draco, Segue I, Ursa Minor and Willman I) have been published in [22]. No evidence for gamma-ray emission was found from either galaxy individually, not from a joint analysis (Willman I excluded), and upper limits on the dark matter annihilation cross section are derived at 1 TeV for the bottom quark ($b\bar{b}$), tau lepton ($\tau^+\tau^-$) and gauge boson ($\gamma\gamma$) final states respectively of 1.35×10^{-23} cm^3 s^{-1}, 2.85×10^{-24} cm^3 s^{-1} and 1.32×10^{-25} cm^3 s^{-1}. Work is in progress to analyse extensive data sets on Ursa Major II and the galactic centre region in the context of dark matter limits.

3.3.2 Cosmic Ray Electron Spectrum

Cosmic ray electrons at TeV energies detectable at the Earth should have been produced within one kpc since they lose energy rapidly. A recent analysis of ~ 300 hours of VERITAS data for energies between 300 GeV and 5 TeV determined that the spectrum cannot be fit with a single power law but requires broken power law with break energy at $710 \pm 40_{stat} \pm 140_{sys}$ GeV [23].

3.3.3 Cosmic Ray Iron Spectrum

A novel template-based analysis of 71 hours of VERITAS data utilising the direct-Cherenkov method has been used to derive an energy spectrum for cosmic ray iron nuclei in the energy range 20 TeV to 500 TeV [24]. The spectrum was well fit with power law of index $2.83 \pm 0.3_{stat}$, in agreement with partially overlapping measurements at lower energies, and extends the spectrum higher energies.

4 Summary and Conclusions

VERITAS is operating extremely well with excellent reliability, and the 2017-2018 observing season was its most efficient yet. The collaboration has a multi-faceted science program covering galactic, extragalactic and fundamental physics. A selection of recent science highlights have been presented here, and many more studies using already-obtained data are in progress. VERITAS is fully funded for operations through end of 2019 and in late 2018 will apply for operations through 2022.

5 Acknowledgements

This research is supported by grants from the U.S. Department of Energy Office of Science, the U.S. National Science Foundation and the Smithsonian Institution, and by NSERC in Canada. We acknowledge the excellent work of the technical support staff at the Fred Lawrence Whipple Observatory and at the collaborating institutions in the construction and operation of the instrument.

References

[1] T.C. Weekes et al., Astroparticle Physics **17**, 221 (2002), 0108478
[2] J.S. Perkins, G. Maier (for the VERITAS Collaboration), Fermi Symposium (2009), 0912.3841
[3] N. Otte (for the VERITAS Collaboration), Proc. 32nd ICRC **9**, 247 (2011), 1110.4702
[4] N. Park (for the VERITAS Collaboration), Proc. 34th ICRC **34**, 771 (2015), 1508.07070
[5] F. Aharonian et al., A&A **393**, L37 (2002), 0207528
[6] E. Aliu et al., ApJ **783**, 16 (2014), 1401.2828
[7] A.U. Abeysekara, et al., ApJL **867**, L19 (2018), 1810.05271
[8] S. Archambault et al., ApJ **836**, 23 (2017), 1701.06740
[9] S. Kumar (for the VERITAS Collaboration), Proc. 34th ICRC (2015), 1508.07453
[10] A.U. Abeysekara et al., ApJ **843**, 40 (2017), 1702.02992
[11] A.U. Abeysekara et al., ApJ **866**, 24 (2018), 1808.10423
[12] HESS Collaboration, Nature **531**, 476 (2016), 1603.07730
[13] R. Mukherjee (for the VERITAS Collaboration), The Astronomer's Telegram **11436** (2018)
[14] C.C. Cheung, D. Gasparrini, S. Buson, The Astronomer's Telegram **10931** (2017)
[15] Mukherjee (for the VERITAS Collaboration), The Astronomer's Telegram **11075** (2017)
[16] M.J. Valtonen et al., ApJ **819**, L37 (2016), 1603.04171
[17] M.C. Stroh, A.D. Falcone, ApJS **207**, 28 (2013), 1305.4949
[18] S. O'Brien (for the VERITAS Collaboration), Proc. 35th ICRC (2017), 1708.02160
[19] A.U. Abeysekara et al., ApJ **856**, 95 (2018), 1802.10113
[20] IceCube Collaboration (plus Fermi-LAT, MAGIC, AGILE, ASAS-SN, HAWC, H.E.S.S., INTEGRAL, Kanata, Kiso, Kapteyn, Liverpool Telescope, Subaru, Swift/NuSTAR, VERITAS and VLA/17B-403 teams), Science **361**, eaat1378 (2018), 1807.08816
[21] A.U. Abeysekara et al., ApJL **861**, L20 (2018), 1807.04607
[22] S. Archambault et al., PhysRevD **95**, 082001 (2017), 1703.04937
[23] A. Archer et al., PhysRevD **98**, 062004 (2018), 1808.10028
[24] A. Archer et al., PhysRevD **98**, 022009 (2018), 1807.08010

Measurement of the cosmic ray spectrum with the Pierre Auger Observatory

Daniela Mockler[1,a] for the Pierre Auger Observatory[2,b]

[1] *Karlsruhe Institute of Technology, Institute of Experimental Particle Physics (ETP), Karlsruhe, Germany*
[2] *Full author list and acknowledgments at: http://www.auger.org/archive/authors_2018_09.html*

Abstract. The flux of ultra-high energy cosmic rays above 3×10^{17} eV has been measured with unprecedented precision at the Pierre Auger Observatory. The flux of the cosmic rays is determined by four different measurements. The surface detector array provides three data sets, two formed by dividing the data into two zenith angle ranges, and one obtained from a nested, denser detector array. The fourth measurement is obtained with the fluorescence detector. By combing all four data sets, the all-sky flux of cosmic rays is determined. The spectral features are discussed in detail and systematic uncertainties are addressed.

1 Introduction

The measurement of the energy spectrum of Ultra-High Energy Cosmic Rays (UHECRs) is key to answer the fundamental questions of the origin and propagation of these particles. The data taking of the Pierre Auger Observatory [1] started in 2004 and the high-quality data taken since then led to an unprecedented measurement of the UHECR spectrum. Two spectral features have been established undoubtedly: a flattening of the spectrum at an energy around 5×10^{18} eV (the *ankle*), and a steep flux suppression at the highest energies. The precise measurement of the flux, in combination with the mass composition and the distribution of the arrival directions of the particles over the sky, are important ingredients for astrophysical models describing the origin and the propagation of UHECRs.

The Pierre Auger Observatory exploits a hybrid detection technique to measure the extensive air showers induced by UHECRs impinging the Earth's atmosphere. The ground-based Surface Detector array (SD) records the lateral distribution of the secondary particles, while the Fluorescence Detector (FD) measures the calorimetric energy stored in the atmosphere. The SD is composed of 1600 water-Cherenkov detectors, which are arranged in a hexagonal grid. With a spacing of 1500 m, the SD covers an area of 3000 km^2. The array is fully efficient for energies above 3×10^{18} eV [2]. A nested array with a reduced detector spacing of 750 m was built to increase the SD sensitivity towards lower energies. This array consists of 120 stations, of which 49 stations have been deployed in addition, and the remaining 71 stations are shared with the standard SD. The spanned area amounts to 27 km^2 and the energy of full efficiency is lowered to 3×10^{17} eV [3]. The SD is sensitive to the electromagnetic and muonic shower component and the duty cycle is almost 100% [4, 5]. The atmosphere above the

[a] e-mail: daniela.mockler@kit.edu
[b] e-mail: auger_spokespersons@fnal.gov

SD is overlooked by 27 optical telescopes distributed into five buildings along the periphery of the array. The FD records the fluorescence light emitted by the nitrogen molecules that are excited by the secondary particles traversing through the atmosphere as well as the Cherenkov light induced by the secondaries. Therefore, the FD measures the longitudinal development of the air shower. The FD operation is only possible in clear and moonless nights, resulting in a duty cycle of about 13% [6, 7].

We present the measurements of the flux of cosmic rays obtained with the four data sets and their combination to the all-sky flux spanning an energy range of almost three decades.

2 Flux measurements with the SD and FD

The reconstruction of the arrival direction and the core position with the SD array is based on the spatial and temporal combination of the individual station trigger times. The recorded signals are calibrated to the signal produced by a Vertical Equivalent Muon (VEM) [5, 8]. The attenuation in the atmosphere differs for the electromagnetic and muonic shower components, which necessitates different reconstruction methods for different zenith angle ranges. The SD data are divided into *vertical* events ($\theta < 60°$) and *inclined* events ($60° \leq \theta < 80°$). The reconstruction of the vertical events is based on describing the recorded signals in the stations with a lateral distribution function. The expected signal $S(r_{\mathrm{opt}})$ at an optimal distance r_{opt} from the shower axis gives the energy estimator for the SD measurements. The optimal distance depends on the grid spacing, resulting in the energy estimators $S(450)$ and $S(1000)$ for the 750 m and 1500 m arrays respectively. These energy estimators are zenith angle dependent. For a fixed energy, the observed signals at ground depend on the amount of traversed atmosphere due to the attenuation of the shower particles. This attenuation is corrected for by using a Constant Intensity Cut method [9]. Taking the median zenith angle of 38° (35°) for the 1500 m (750 m) array, the expected signal $S(r_{\mathrm{opt}})$ is corrected to a signal a shower would have produced had it arrived at 38° (35°). For the 750 m array, only events with zenith angles below 55° are accepted, which results in a median angle of 35°. The reconstruction of inclined events differs due to the dominant muonic shower component. Due to the increased amount of traversed atmosphere by the shower, the electromagnetic component is largely absorbed [10]. The muons are deflected by the geomagnetic field and the cylindrical symmetry at ground, which is seen for vertical showers, is lost. The observed signal distributions at ground are fitted to muon density distributions predicted from simulated air showers. The energy estimator of inclined events is based on determining the relative muon content N_{19} in the showers. The reference muon density corresponds to the one produced by a simulated proton shower with an energy of 10^{19} eV [11]. The energy of the primary particle is inferred from N_{19}. For a high-quality SD event, the station with the highest signal has to be surrounded by a hexagon of six active stations. The calculation of the SD exposure is based on the determination of the geometrical aperture of the array for the respective zenith angle range and the observation time [2]. The exposure values up to 31 Dec 2016 for the different arrays are given in fig. 1a.

The energy calibration of the SD events is based on a data-driven approach by selecting a subset of high-quality events (called golden hybrid events) which have been measured by both the SD and FD [12]. The relation between the energy estimators $\hat{S}$ of the SD data sets, i.e. S_{35}, S_{38}, and N_{19}, and the energy E_{FD} measured by the telescopes is well-described by a power law $E_{\mathrm{FD}} = A \cdot \hat{S}^{B}$. The golden hybrid events are depicted in fig. 1b for the three different SD data sets. Superimposed to the individual events are the calibration functions resulting from maximum-likelihood fits. The respective numbers of the golden hybrid events are given in addition. Due to the finite detector resolution, migration effects occur. Due to the steeply falling flux, more events migrate from lower energy bins towards higher energy bins. Therefore, the measured spectra are smeared with respect to the true one. To obtain the true spectra, a deconvolution is performed and the spectra are corrected by less than 10% in the energy range of interest.

The four independent energy spectra derived from the data measured by Auger are shown in fig. 2a. As the SD spectra are calibrated with the energies measured by the FD, the SD energies share the uncertainty of the FD energy scale of 14%. The hybrid sample is built from showers measured by the FD in coincidence with at least one triggered station of the SD. Combining the measurement of the FD with the timing information of the SD station allows for a very accurate determination of the shower geometry in the hybrid approach. The detector exposure is influenced by several factors concerning the atmospheric and data taking conditions. To mimic these conditions as closely as possible, the hybrid exposure is determined with the help of extensive Monte Carlo simulations that take into account the exact conditions of the observatory [7]. The derived energy spectrum starts at an energy of 10^{18} eV and overlaps at lower energies with the spectrum obtained SD-750 m data and subsequently with the spectrum deduced from the SD-1500 m data.

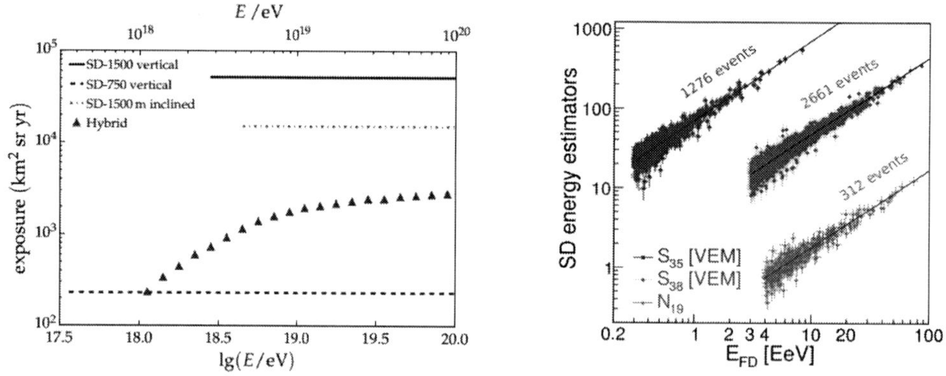

(a) (b)

Figure 1: (a) Integrated exposure of the different detectors as a function of the energy. The SD exposure is flat above an energy corresponding to the full trigger efficiency. (b) Correlation between the SD energy estimates S_{35}, S_{38}, N_{19} and the energy determined by the FD.

3 Combined energy spectrum

The combination is performed by means of a global maximum-likelihood fit. The flux normalizations are used as additional constraints to obtained the flux scaling factors that match the spectra: $(-0.8 \pm 0.2)\%$ for the SD-1500 vertical, $(-1 \pm 4)\%$ for the SD-750, $(5.4 \pm 0.7)\%$ for the SD-1500 horizontal and $(-6 \pm 2)\%$ for the hybrid. The spectral features are then derived by fitting the combined spectrum with the following function:

$$J_{\text{unf}}(E) = \begin{cases} J_0 \left(\frac{E}{E_{\text{ankle}}}\right)^{-\gamma_1} & E < E_{\text{ankle}} \\ J_0 \left(\frac{E}{E_{\text{ankle}}}\right)^{-\gamma_2} \left[1 + \left(\frac{E_{\text{ankle}}}{E_s}\right)^{\Delta\gamma}\right]\left[1 + \left(\frac{E}{E_s}\right)^{\Delta\gamma}\right]^{-1} & E > E_{\text{ankle}}. \end{cases}$$

The combined spectrum, the fit, and the parameters are given in fig. 2b. The combined spectrum shows a flattening above the ankle, which is found at $E_{\text{ankle}} = (5.08 \pm 0.06(\text{stat.}) \pm 0.8(\text{syst.})) \times 10^{18}$ eV, while the suppression is obtained at $E_s = (3.9 \pm 0.2(\text{stat.}) \pm 0.8(\text{syst.})) \times 10^{19}$ eV. The energy $E_{1/2}$ at which the integral spectrum drops by a factor of two below what would be the expected with no steepening is $E_{1/2} = (2.26 \pm 0.08(\text{stat.}) \pm 0.4(\text{syst.})) \times 10^{19}$ eV. The spectral indexes are: $\gamma_1 = 3.293 \pm 0.002(\text{stat.}) \pm 0.05(\text{syst.})$, $\gamma_2 = 2.53 \pm 0.02(\text{stat.}) \pm 0.1(\text{syst.})$ while $\Delta\gamma = 2.5 \pm 0.1(\text{stat.}) \pm 0.4(\text{syst.})$ [13].

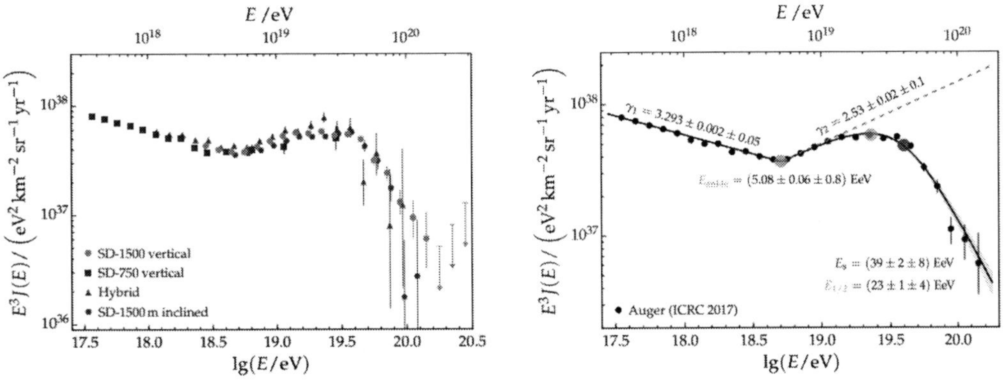

(a) (b)

Figure 2: (a) Energy spectra obtained from the three SD data sets and the hybrid data. The systematic uncertainty on the energy scale, which is common to all of them, is 14%. (b) Combined spectrum and fitted function. The optimal parameters are given in addition.

References

[1] A. Aab et al., Nuclear Instruments and Methods in Physics Research Section A: Accelerators, Spectrometers, Detectors and Associated Equipment **A798**, 172 (2015)

[2] J. Abraham *et al.* (Pierre Auger Collab.), Nucl. Instrum. Meth. **A613**, 29 (2010)

[3] D. Ravignani *et al.* (Pierre Auger Collab.), *Measurement of the energy spectrum of cosmic rays above* 3×10^{17} *eV using the AMIGA 750 m surface detector array of the Pierre Auger Observatory*, in *Proc. 33rd Int. Cosmic Ray Conf.* (Rio de Janeiro, Brazil, 2013)

[4] J. Abraham *et al.* (Pierre Auger Collab.), Nucl. Instrum. Meth. **A523**, 50 (2004)

[5] I. Allekotte et al. (Pierre Auger Collaboration), Nucl.Instrum.Meth. **A586**, 409 (2008)

[6] J. Abraham *et al.* (Pierre Auger Collab.), Nucl. Instrum. Meth. **A620**, 227 (2010)

[7] P. Abreu et al. (Pierre Auger Collaboration), Astropart. Phys. **34**, 368 (2011)

[8] X. Bertou et al. (Pierre Auger Collaboration), Nucl. Instrum. Meth. **A568**, 839 (2006)

[9] J. Hersil, I. Escobar, D. Scott, G. Clark, S. Olbert, Phys. Rev. Lett. **6**, 22 (1961)

[10] I. Valiño *et al.* (Pierre Auger Collab.), *The electromagnetic component of inclined air showers at the Pierre Auger Observatory*, in *Proc. 31st Int. Cosmic Ray Conf.* (Lódź, Poland, 2009)

[11] Rodriguez, G., *A measurement of the muon number in showers using inclined events detected at the Pierre Auger Observatory*, in *European Physical Journal Web of Conferences* (2013), Vol. 53 of *European Physical Journal Web of Conferences*, p. 07003

[12] R. Pesce *et al.* (Pierre Auger Collab.), *Energy calibration of data recorded with the surface detectors of the Pierre Auger Observatory: an update*, in *Proc. 32nd Int. Cosmic Ray Conf.* (Beijing, China, 2011)

[13] F. Fenu *et al.* (Pierre Auger Collab.), *The cosmic ray energy spectrum measured using the Pierre Auger Observatory*, in *Proc. 35th Int. Cosmic Ray Conf.* (Busan, South Korea, 2015)

Proton energy spectrum with the DAMPE experiment

Antonio De Benedittis[1,2,a] (on behalf of the DAMPE Collaboration)

[1] *Dip.to di Matematica e Fisica "E. De Giorgi", Università del Salento, Via per Arnesano, 73100, Lecce, Italy*
[2] *Istituto Nazionale di Fisica Nucleare (INFN), Sezione di Lecce, Via per Arnesano, 73100, Lecce, Italy*

Abstract. The DAMPE (DArk Matter Particle Explorer) experiment, in orbit since December 17th 2015, is a space mission whose main purpose is the detection of cosmic electrons and photons up to energies of 10 TeV, in order to identify possible evidence of Dark Matter in their spectra. Furthermore it aims to measure the spectra and the elemental composition of the galactic cosmic rays nuclei up to the energy of hundreds of TeV. The proton analysis and the flux with kinetic energy ranging from 50 GeV up to 100 TeV, at the end of two years of data taking, will be presented and discussed.

1 Introduction

Cosmic Rays (CRs) with energies up to ~ 4 PeV (the so called *knee* of the CRs spectrum [1]) are widely to be believed originated by the strong shock of Supernovae explosions in the Milky Way [2]. CRs are mainly composed of protons, therefore a very precise measurement of the proton flux up to energies of the order of PeV is crucial to understand the mechanism at the origin of their acceleration and the physical basis of the *knee*.

Starting from energies higher than 30 GeV the energy spectrum of CRs is expected to follow a featureless power-law according to the conventional acceleration models [3]. Recent measurements of proton flux carried out by baloon- and space-borne experiments (ATIC-2 [4], CREAM I-III [5], AMS-02 [6], PAMELA [7]) observed a spectral *hardening*, that is a deviation from the single power-law distribution. Moreover a spectral *softening* was observed at energies higher than 50 TeV (CREAM I-III [5], NUCLEON [9], ARGO-YBJ [8]).

One of the goals of DAMPE (DArk Matter Particle Explorer) is to investigate with high precision the *hardening* and the *softening* fitting a bridge between direct and indirect measurements.

2 DAMPE detector

The DAMPE detector [10] is composed by four sub-detectors from the top to the bottom: a Plastic Scintillator strip Detector (PSD), a Silicon-Tungsten tracKer-converter (STK), a BGO calorimeter and a NeUtron Detector (NUD).

The PSD measures the absolute value of the charge of incident particles (up to $Z \leq 28$), it provides also background rejection for gamma-ray detection. The STK reconstructs the trajectory and gives an independent charge measurement, it also acts as a gamma-ray converter using three tungsten foils

[a] e-mail: antonio.debenedittis@le.infn.it

interleaved in the first layers of silicon strips. The BGO calorimeter measures the energy of the particles and provides an electron/hadron discrimination, his depth (about 32 X_0) ensures that almost 100% of the energy of electrons and γ-rays is deposited in the calorimeter, and about 40% regarding nuclei [11]. The NUD gives a redundant electron/hadron discrimination using the neutrons produced in BGO by the hadronic showers.

3 Data analysis

Two years of data taking since the 1st January, 2016 to the 31st December, 2017 are taken in account in this analysis. The operational live time is about 75.5%, the lost time consists of three part: the instrumental dead time, the crossing time of the South Atlantic Anomaly (SAA) region, the time for the on-orbit calibration procedures.

3.1 Event selection

In this analysis some pre-selection cuts are applied to satisfy some basilar requests: to ensure that all sub-detectors work in good conditions, the SAA events are excluded; to avoid the effect due to the geomagnetic cut-off, only particles that release $E \geq 20$ GeV in the calorimeter are considered; the track of the particle has to be geometrically contained in the detector in order to select particles that are crossing all the sub-detectors.

After the pre-selection cuts, only the events passing the High Energy Trigger (HET), that foresees a deposited energy in the first four layers of BGO higher than 230 MeV (~ 10 MIPs). The HET efficiency (Fig. 1 *left side*) is estimated using the Unbias Trigger (which requires that the energy deposited in each of the first two BGO layers has to be ≥ 0.5 MIP):

$$\varepsilon_{trig} = \frac{N_{he|unb}}{N_{unb}} \tag{1}$$

where N_{unb} are the events passing the unbias trigger and $N_{he|unb}$ are those that satisfy also the HET.

Moreover only the events with at least a reconstructed track in the STK are selected, and the selected track has to cross all the PSD layers and has to satisfy the STK-BGO match. The track reconstruction efficiency (Fig. 1 *right side*) is estimated using a proton sample selected by an independent charge selection based on the reconstructed shower axis in the calorimeter, and it is defined as:

$$\varepsilon_{track} = \frac{N_{track|BGOtrack}}{N_{BGOtrack}} \tag{2}$$

where $N_{BGOtrack}$ are the events with a selected track in the calorimeter and $N_{track|BGOtrack}$ are those passing the STK track selection cuts.

Finally the proton candidates are selected by the information coming from the PSD charge distribution for different ranges of BGO energy. The charge recostruction efficiency (Fig. 2) is estimated independently for each PSD layer using the information coming from the first STK layer and from the other PSD layer.

3.2 Background

The background for protons is due mainly to mis-identified helium nuclei and electron events. The helium contamination is evaluated applying the template-fit method to the charge distributions obtained from PSD signals and using the MC samples of proton and helium nuclei (Fig. 3 *left side*). The

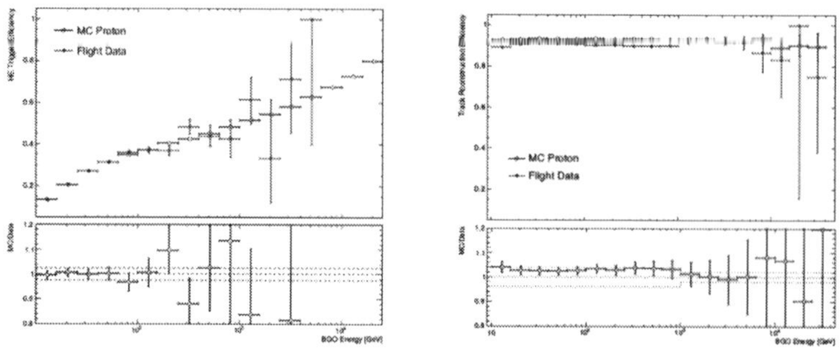

Figure 1. *Left* - The HET efficiency as a function of BGO energy. *Right* - The Track reconstruction efficiency as a function of BGO energy. In both plots the red points are for flight data and the blue points are for MC data.

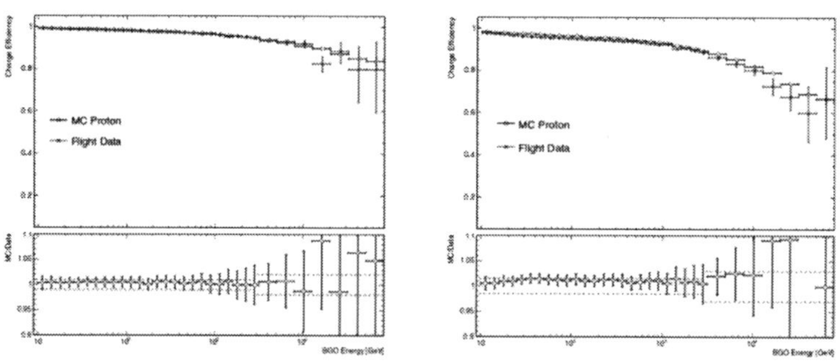

Figure 2. The charge reconstruction efficiency as a function of BGO energy, for PSD layer 1 (*Left*), for PSD layer 2 (*Right*). In both plots the red points are for flight data and the blue points are for MC data.

helium background is less than 2% up to energies of about 10 TeV; while the electron background, thanks to the high e/p discrimination power, is less than 0.1% (Fig. 3 *right side*). The helium and electron contamination is taken into account in the systematic uncertainties.

4 Results and conclusions

The effective acceptance (Fig. 4 *left side*) is evaluated using a MC sample of pure proton events. It is defined (for the *i*-th primary energy bin) as:

$$A_{eff,i} = A_{gen} \times \frac{N_{pass,i}}{N_{gen,i}} \quad (3)$$

where A_{gen} is the geometrical factor of the MC generation half-sphere, $N_{pass,i}$ and $N_{gen,i}$ are the number of generated events and surviving events respectively.

The *right side* of Fig. 4 shows the DAMPE preliminary proton flux as a function of kinetic energy, compared with the previous measurements of AMS-02, PAMELA, ATIC-2, CREAM-III. The spectral hardening at $E \simeq 200$ GeV is in agreement with the measurements of AMS-02 and PAMELA. Moreover a softening at energies higher than 10^4 GeV, according to ATIC and CREAM (within the systematic uncertainties) results, has been observed with unprecedent resolution.

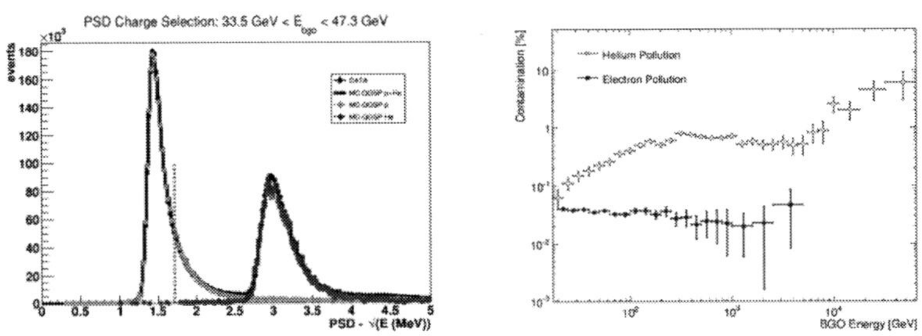

Figure 3. *Left* - The combined charge spectrum of PSD for protons and helium nuclei, for BGO energies between 33.5 GeV and 47.3 GeV. In this plot there are the on-orbit data (black), with the best fit templates of MC protons (green), helium nuclei (red) and their sum (blue). The vertical line shows the cut applied to select proton canditates. *Right* - The fraction of helium background (red) and electron background (blue) in proton candidate events as a function of BGO energy.

The data analysis and the evaluation of systematic uncertainties are still on-going. In the future, with the increase in data sample it is expected to obtain a proton flux up to energies higher than 100 TeV with high precision.

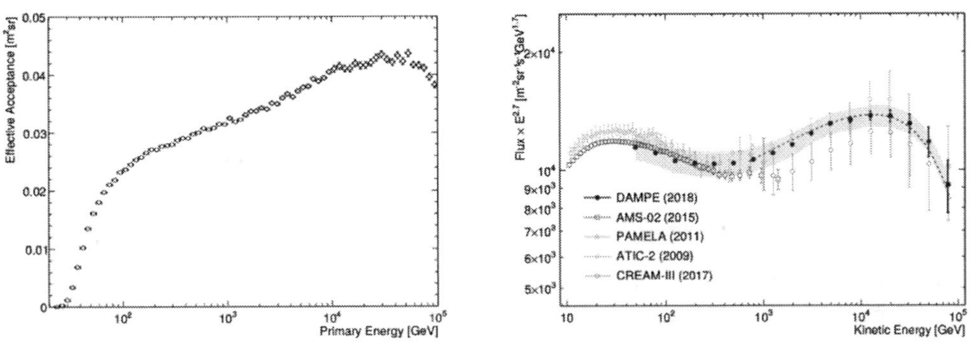

Figure 4. *Left* - The effective acceptance of protons as a function of primary energy. *Right* - The preliminary proton flux times $E^{2.7}$ as a function of primary energy (the shaded area shows the systematic errors).

References

[1] Hoerandel J., Astropart. Phys. **19**, 193–220 (2003).
[2] Baade W., Zwicky F., Phys. Rev., **46**, 76 (1934).
[3] Fermi E., Phys. Rev., **75**, 1169 (1949).
[4] Panov A. D. *et al.* (The ATIC Collaboration), Bull. Rus. Acad. Sci. Phys., **73**, 564 (2009).
[5] Ahn H. S. *et al.* (The CREAM Collaboration), Astrophys. J., **714**, L89 (2010).
[6] Aguilar M. *et al.* (The AMS Collaboration), Phys. Rev. Lett., **114**, 171103 (2015).
[7] Adriani O. *et al.* (The PAMELA Collaboration), Science, **332**, 69 (2011).
[8] Bartoli B. *et al.* (The ARGO-YBJ Collaboration), Phys. Rev. D **92**, 092005 (2015).
[9] Atkin E. *et al.* (The NUCLEON Collaboration), Astropart. Phys. **90**, 69-74 (2017).
[10] Chang J. *et al.* (The DAMPE Collaboration), Astropart. Phys. **95**, 6–24 (2017).
[11] De Mitri I., EPJ Web of Conferences, **136** , 02010 (2017).

EPJ Web of Conferences **209**, 01031 (2019) https://doi.org/10.1051/epjconf/201920901031
RICAP-18

Recoil Directionality Experiment

S. Sanfilippo[1,2,a], P. Agnes[3], M. Arba[4], M. Ave[5], E. Baracchini[6], A. Boiano[7], W. M. Bonivento[4], B. Bottino[8,9], M. Cadeddu[4], A. Caminata[9], M. Caravati[4,10], M. Cariello[9], M. Carpinelli[11,12], S. Catalanotti[7,13], V. Cataudella[7,13], R. Cereseto[9], C. Cicalò[4], G. Covone[7,13], A. de Candia[7,13], G. De Filippis[7,13], G. De Rosa[7,13], S. Davini[9], A. Devoto[4,10], C. Dionisi[14,15], D. D'Urso[11,12], G. Fiorillo[7,13], D. Franco[16], G. K. Giovannetti[17], C. Giganti[18], C. Galbiati[17,5], M. Gulino[19,12], G. Korga[3,20], M. Kuss[21], M. La Commara[7,13], L. La Delfa[4], M. Lissia[4], A. Mariani[6], S. M. Mari[1,2], C. J. Martoff[22], V. Masone[7], V. Oleynikov[23,7], M. Pallavicini[8,9], L. Pandola[12], A. Razeto[20], M. Rescigno[15], N. Rossi[15], D. Sablone[20,17], E. Scapparone[24], Y. Suvorov[7,13], G. Testera[9], M. Tuveri[4], H. Wang[25], and Y. Wang[25]

(The ReD Working Group)

[1] *INFN Roma Tre, Roma 00146, Italy*
[2] *Mathematics and Physics Department, Università degli Studi Roma Tre, Roma 00146, Italy*
[3] *Department of Physics, University of Houston, Houston, TX 7704, USA*
[4] *Istituto Nazionale di Fisica Nucleare, Sezione di Cagliari, Cagliari 09042, Italy*
[5] *Instituto de Física, Universidade de São Paulo, São Paulo 05508-090, Brazil*
[6] *Gran Sasso Science Institute, L'Aquila AQ 67100, Italy*
[7] *Istituto Nazionale di Fisica Nucleare, Sezione di Napoli, Napoli 80126, Italy*
[8] *Physics Department, Università degli Studi di Genova, Genova 16146, Italy*
[9] *Istituto Nazionale di Fisica Nucleare, Sezione di Genova, Genova 16146, Italy*
[10] *Physics Department, Università degli Studi, Cagliari 09042, Italy*
[11] *Chemistry and Pharmacy Department, Università degli Studi di Sassari, Sassari 07100, Italy*
[12] *Istituto Nazionale Fisica Nucleare, Laboratori Nazionali del Sud, 95123 Catania, Italy*
[13] *Physics Department, Università degli Studi Federico II, Napoli 80126, Italy*
[14] *Physics Department, Sapienza Università di Roma, Roma 00185, Italy*
[15] *Istituto Nazionale di Fisica Nucleare, Sezione di Roma, Roma 00185, Italy*
[16] *APC, Université Paris Diderot, CNRS/IN2P3, CEA/Irfu, Obs. de Paris, Sorbonne Paris Cité, Paris 75205, France*
[17] *Physics Department, Princeton University, Princeton, NJ 08544, USA*
[18] *LPNHE Paris, Université Pierre et Marie Curie, Université Paris Diderot, CNRS/IN2P3, Paris 75252, France*
[19] *Università di Enna KORE, Enna 94100, Italy*
[20] *INFN Laboratori Nazionali del Gran Sasso, Assergi (AQ) 67010, Italy*
[21] *Istituto Nazionale Fisica Nucleare, Sezione di Pisa, Pisa 56127, Italy*
[22] *Physics Department, Temple University, Philadelphia, PA 19122, USA*
[23] *Budker Institute of Nuclear Physics of Siberian Branch Russian Academy of Sciences (BINP SB RAS), Novosibirsk, 630090 Russian Federation*
[24] *Istituto Nazionale di Fisica Nucleare, Sezione di Bologna, Bologna 40126, Italy*
[25] *Physics and Astronomy Department, University of California, Los Angeles, CA 90095, USA*

Abstract. Directional sensitivity to nuclear recoils could provide a smoking gun for a possible discovery of dark matter in the form of WIMPs. A hint of directional dependence of the response of a dual-phase liquid argon Time Projection Chamber was found in the SCENE experiment. Given the potential importance of such a capability in the frame-

[a] e-mail: simone.sanfilippo@roma3.infn.it

work of dark matter searches, a new dedicated experiment, ReD (Recoil Directionality), was designed in the framework of the DarkSide Collaboration, in order to scrutinize this hint. This contribution will describe the performance of the detectors achieved during the first test-beam, the current status of ReD and the perspectives for physics measurements during the forthcoming beam-time.

1 Introduction

The hypothesis of dark matter was introduced more than 80 years ago to explain anomalous motions of galaxies gravitationally bound in clusters. Observational evidences had continued to accumulate since then, including rotation curves of galaxies and their clusters and discrepancies in the distributions of galaxy cluster mass estimated from luminosity vs. gravitational lensing. One of the most favored dark matter candidate is the Weakly Interacting Massive Particle (WIMP) [1, 2], which explains the current abundance of dark matter as a thermal relic of the big bang. Most models predict dark matter WIMP masses near the electroweak scale of 100's of GeV/c^2. However, dark matter particle masses ≤ 10 GeV/c^2 can also be compatible with experimental constraints if a significant asymmetry between dark matter and their anti-particles existed in the early universe.

2 The DarkSide program

DarkSide-50 (DS-50) is an experiment for the direct detection of WIMPs currently running at INFN Laboratori Nazionali del Gran Sasso (LNGS, Italy), at a depth of 3800 m.w.e. [3, 4]. The DarkSide Collaboration has recently published WIMP search results from about 500 days of data taking of DS-50. The results lead to the best WIMP-Argon cross section exclusion limits at present day, in a wide range of WIMP masses [5–7]. In order to further increase the sensitivity on WIMP-nucleus cross section with an argon based experiment, an enlarged DarkSide Collaboration (Global Argon Dark Matter Collaboration - GADMC) proposed a long term program for dark matter searches. The DarkSide-20k experiment (about 40 tonnes of active mass), in particular, aims at a significant improvement in the sensitivity for the direct detection of WIMPs in the mass range above a few hundred GeV/c^2, forecasting $\sigma = 1.2 \times 10^{-47}$ cm^2 for WIMPs of 1 TeV/c^2 mass [8]. Beyond this limit, however, demonstrating that a putative signal is due to WIMPs, is a major challenge. Several studies [9] show that the identification of recoil events due to particles coming from the galactic halo would give, for any WIMP model, a powerful tool to discriminate the signal from an isotropic, or fixed-location in the laboratory, background source. Direction-sensitivity is therefore a highly desirable characteristic for a direct detection experiment, which will become of paramount interest in case evidence is found for nuclear recoils that cannot be explained by any known backgrounds.

3 Recoil Directionality Experiment (ReD)

WIMP directional information is potentially available in a dual-phase Liquid Argon Time Projection Chamber (LAr TPC) by exploiting the recombination effect. Columnar recombination models [10] suggest that the magnitude of the recombination effect should vary with the angle between the field and the track direction. A difference in the electron-ion recombination effect is expected when the ionizing track is either parallel or perpendicular to the electric field. Such an effect were explored by the SCENE experiment [11], where a monoenergetic neutron beam were used to irradiate a small dual-phase LAr TPC with and without the application of an electric field. The collaboration achieved,

for the first time, the comparison of the light and charge yield of recoils parallel and perpendicular with respect to the electric field (directional sensitivity), in case of a S1 signal due to scintillation (S1, [11]). Unfortunately the same was not found for ionization signals (S2, [11]). The ReD experiment was proposed with the aim to improve the SCENE measurements as part of the DarkSide program, so it may provide a fully scalable technology to bring the future dark matter experiments based on LAr to the multi-ton scale, i.e. DarkSide-20k.
The main goal of the ReD experiment is to demonstrate the WIMP directional sensitivity by using a small scale, dual-phase LAr TPC explicitly designed for this purpose at University of California, Los Angeles (UCLA), in which the information is collected in form of optical signals by means of newly developed silicon photomultipliers (SiPMs), working at cryogenic temperature [12, 13]. The TPC

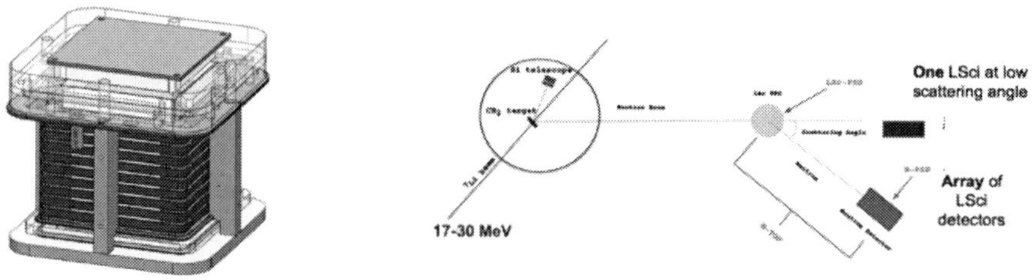

(a) (b)

Figure 1. (a) Schematic drawing of the ReD dual-phase LAr TPC. (b) The ReD experimental setup (drawing) at INFN Laboratori Nazionali del Sud (LNS).

has the dimensions of $5\times5\times6$ cm^3 in the inner part, and is enclosed by vertical acrylic-ESR sandwich reflection panels, and by two acrylic windows ITO and TPB coated, on the top (anode) and bottom (cathode) (fig. 1(a)). It is equipped with an array of 24x1 cm^2 FBK (Fondazione Bruno Kessler) SiPM with a 24 channels front-end board readout (custom-made for ReD by INFN Naples, in collaboration with INFN Bologna and INFN LNGS) on the top, and a 24x1 cm^2 FBK tile with a 4 channels front-end board readout on the bottom. Neutrons are produced by p(^{7}Li,n)^{7}Be reaction, available thanks to the 15 MV TANDEM accelerator of the INFN Laboratori Nazionali del Sud (LNS) in Catania, Italy, with energy of about 7 MeV and at a specific angle with respect to the beam line, in the direction of the LAr TPC. Neutron interactions producing 20-100 keV nuclear recoils in argon, were then recorded in the TPC, while the scattered neutrons were intercepted by a set of nine 3-inches neutron detectors (LSci) placed in coincidence with the TPC, allowing to close the kinematics and estimate the argon recoil direction (fig. 1(b)). All LSci are placed such to tag recoils having the same energy, i.e the same scattering angle with respect to the incident neutron, but different angle with respect to the drift field of the LAr TPC. One LSci detector is at small scattering angle and is devoted to the study of low-energy recoils. If a directional effect on recombination is present in LAr, it is expected to measure different scintillation and ionization responses for nuclear recoils of the same energy, but with tagged initial momentum in the parallel and perpendicular directions to the TPC electric drift field. The directional effect should also be dependent on the electric field, whose variation will provide a further signature to model the directionality [10]. The scatter plot of the amplitudes of a ΔE - E telescope, able to identify the charged particles (^{7}Be) which accompany the neutrons emitted towards the TPC, is displayed in fig. 2(a): it shows the ability of the detector to discriminate the charged products of the beam-target reactions, i.e. the main ^{7}Li band due to elastic scattering and the two ^{7}Be *loci* corresponding to the two solutions allowed by kinematics. As expected, coincident events between Si and TPC are mostly associated to one ^{7}Be *locus*. Prior to the beam start, the detectors were commissioned and tested

 https://doi.org/10.1051/epjconf/201920901031

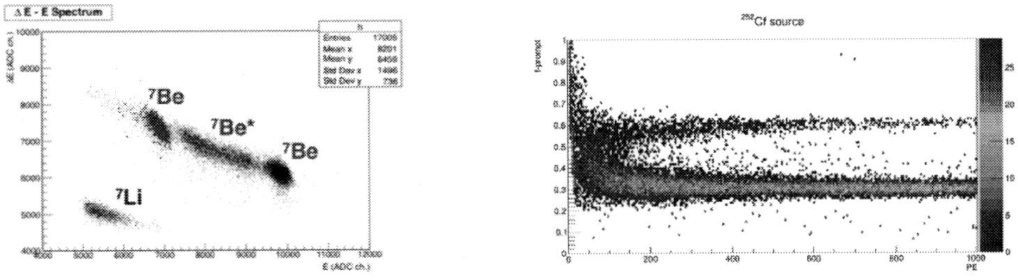

(a) (b)

Figure 2. (a) Scatter plot of the amplitudes of the ΔE and E Si detectors, placed at 5 deg with respect to the beam axis. The red markers are events having a coincident signal in the TPC. (b) The fraction of prompt light in the scintillation pulse (f-prompt) versus the total charge collected by the two SiPM arrays, expressed in units of photoelectrons (PE): are clearly visible the two different bands of neutrons (upper band) and γ-rays (lower band) that leave a signal inside the TPC.

individually, by using the laser and radioactive sources. Light yield and timing were checked by using ^{241}Am and ^{22}Na, respectively; a ^{252}Cf (fig. 2(b)) fission neutron source was used to characterize the pulse shape discrimination performance of the LAr TPC and the LSci.

4 Conclusions

Directional sensitivity to nuclear recoils would provide a smoking gun for a possible discovery of dark matter in the form of WIMPs. A hint of directional dependence of the response of a dual-phase LAr TPC was found in the SCENE experiment. Given the potential importance of such a capability in the framework of dark matter searches, a new dedicated experiment, ReD (Recoil Directionality), was designed in the framework of the DarkSide Collaboration, in order to scrutinize this hint. Here we gave a preliminary look on the potentiality of such a technology by the successful installation and running of the ReD Experiment at INFN Laboratori Nazionali del Sud in Catania (Italy).

References

[1] G. Steigman G. and M.S. Turner, Nucl. Phys. B **253**, (375) 1985
[2] G. Bertone, D. Hooper and J. Silk, Phys. Rep. **405**, (279) 2005
[3] P. Agnes et al. (The DarkSide Collaboration), Phys. Lett. B **743**, 456-466 (2015)
[4] P. Agnes et al. (The DarkSide Collaboration), Phys. Rev. D **93**, 081101(R) (2016)
[5] P. Agnes et al. (The DarkSide Collaboration), Phys. Rev. Lett. **121**, 111303 (2018)
[6] P. Agnes et al. (The DarkSide Collaboration), arXiv:1802.07198 (accepted to Phys. Rev. D) (2018)
[7] P. Agnes et al. (The DarkSide Collaboration), Phys. Rev. Lett. **121**, 081307 (2018)
[8] C. E. Aalseth et al. (The DarkSide Collaboration), Eur. Phys. J. Plus **133**, 131 (2018)
[9] R. Morgan, A. Green, N. J. C. Spooner, Phys. Rev. D **71**, 103507 (2005)
[10] G. Jaffè, Ann. Phys. **42**, 303 (1913) ; G. Jaffè, Phys. Rev. **58**, 968 (1940)
[11] Cao H. et al. (The SCENE Collaboration), Phys. Rev. D **91**, 092007 (2015)
[12] C. E. Aalseth et al., (The DarkSide Collaboration), J. Instr. **12**, P09030 (2017)
[13] F. Acerbi et al., IEEE Trans. Electr. Dev. **64**, 521-526 (2017)

New solar modulation modeling of the galactic proton flux measured by the AMS02 and PAMELA experiments

E. Fiandrini[1,a], B. Bertucci[1], N. Tomassetti[1], and B. Khiali[2]

[1] *Dipartimento di Fisica e Geologia, University of Perugia and INFN Perugia, Italy*
[2] *INFN Perugia, Italy and ASI Space Science Data Center (SSDC), Roma, Italy*

Abstract. A thorough understanding of solar effects on the galactic cosmic rays is relevant both to infer the local interstellar spectrum characteristics and to investigate the dynamics of charged particles in the heliosphere. We present a newly developed numerical modulation model to study the transport of galactic protons in the heliosphere. The model was applied to the 27-day averaged galactic proton flux recently released by the PAMELA and AMS02 experiments, covering an extended time period from mid-2006 to mid-2017.

1 Introduction

Nearly all direct observations of Cosmic Rays (CRs) are made inside the heliosphere. As a result, the CR spectrum observed at Earth can differ significantly from the Local Interstellar Spectrum (LIS) due to many physical processes that modify the spectrum. This is particularly true for CRs with kinetic energies below ~20 GeV, which can be efficiently diffused, curved and de-accelerated adiabatically during their propagation from heliosphere to the Earth. The changes in the Galactic CR (GCR) intensities and energy spectra are related to the solar activity, and are referred to as CR solar modulation. A thorough understanding of solar effects on the GCR propagation is therefore relevant both to infer the LIS and to investigate the dynamics of charged particles in the heliosphere. It is also important for astronauts and the electronic components radiation hazard during long-duration missions.

The new precise data from AMS02 [1] and PAMELA [2] experiments offer a unique possibility to study the solar modulation over a long period of time.

2 The proton model

Propagation in the heliosphere is described by the Parker's equation[3];

$$\frac{\partial f}{\partial t} = \nabla \cdot [\mathbf{K} \cdot \nabla f] - \vec{V} \cdot \nabla f - < \vec{V_D} > \cdot \nabla f + \frac{1}{3}(\nabla \cdot \vec{V}) \frac{\partial f}{\partial (lnp)} \quad (1)$$

Here t is the time, p is the CR particle momentum, $\vec{V}$ is the solar wind speed, $\vec{V}_D$ is the averaged guiding center speed for a pitch angle-averaged distribution function f and $\mathbf{K}$ is the diffusion tensor. In

[a] e-mail: emanuele.fiandrini@pg.infn.it

this work, a 2D numerical integration with $\partial/\partial t = 0$ is made using a Stochastic Differential Equation approach with a backward time integration, based on [4].

The heliosphere is a dynamic void in the local interstellar medium where Sun activity dominates. In this work the heliosphere has a spherical symmetry with the boundary located at r = 122 AU, where local interstellar spectrum is given as boundary condition, with the termination shock at r = 85 AU and the Earth position at r = 1 AU in the equatorial plane. The Heliospheric Magnetic Field (HMF) is wounded up in a modified Parker's spiral [5, 6], as

$$B = B_0(\frac{1AU}{r})^2\sqrt{1 + tan^2(\psi) + (\frac{r\delta(\theta)}{r_\odot})^2} \tag{2}$$

where B_0 is the HMF value at Earth position, $r_\odot$ is the Sun radius and $\delta(\theta) = 8.7 \times 10^{-5}/\sin(\theta)$ if $\theta < 1.7^o$ or $\theta > 178^o$ and = 0 otherwise. A major corotating structure of importance to CR modulation is the wavy Heliospheric Current Sheet (HCS), which divides the solar magnetic field into hemispheres of opposite polarity. The HCS is parametrized by using its tilt angle α and is well correlated to solar activity. The waviness of the HCS is one of the best proxy for solar activity relevant to CR modulation and is widely used in numerical approaches. The tilt angle has been observed at the Wilcox solar observatory since 1976.

Solar wind velocity is radially directed from close to the Sun, up to the inner heliosheath. During periods of minimum solar activity the flow is latitude dependent, changing from an average of 450 kms^{-1} in the equatorial plane to 800 kms^{-1} in the polar regions. This significant effect disappears with increasing solar activity, when the average speed is ~400 kms^{-1} at all the latitudes. The latitudinal dependence of the solar wind speed is parametrized as in [7] and was made dependent on the solar activity by increasing the polar angle at which the solar wind speed changes from slow to fast as a function of the tilt angle.

Diffusion is described by a diagonal diffusion tensor with three diffusion coefficients, namely parallel diffusion $K_\parallel$, the transverse radial, $K_{\perp r}$, and transverse polar diffusion coefficient, $K_{\perp\theta}$. The parallel diffusion can be computed using quasi linear theory [8]. The diffusion coefficient parallel to the background field, $K_\parallel$, was parametrized as in [10]

$$K_\parallel = K_0\beta\frac{B_0}{B}\left(\frac{R_0}{R}\right)^a\left[\frac{(R/R_0)^h + (R_k/R_0)^h}{1 + (R_k/R_0)^h}\right]^{(b-a)/h} \tag{3}$$

with β = v/c, the ratio of the particle's speed to the speed of light. K_0 is a constant in units of 10^{23} cm^2s^{-1}, with R_0 = 1 GV, B the HMF magnitude and B_0 the local field value, such that the units are in K_0. Here a and b are a power indexes that determine the slope of the rigidity dependence, respectively, below and above a rigidity with the value R_k , whereas $h = 3.0$ determines the smoothness of the transition. The value R_k determines the rigidity where the break in the power law occurs. The parameters K_0, a, b depend on the solar activity, therefore vary in time and are free parameters that can be used to fit measured CR flux at a given epoch. Perpendicular diffusion in the radial direction is parametrized as $K_{\perp r} = 0.02 \times K_\parallel$, which a widely used assumption [9], while the polar perpendicular diffusion was parametrized as $K_{\perp\theta} = 0.02 \times g(\theta) \times K_\parallel$, where $g(\theta)$ is a function that enhances the diffusion by a factor d = 3 near the poles, as defined by Eqn. (8) in [10].

The motion in the large-scale average HMF is given by pitch-angle averaged velocity V_D. In this 2D modeling, the HCS is described as a region with an extension that depends on the rigidity of CR particles and on the value of the tilt angle. In this approach, the drift velocity is given by

$$< V_D >= f(\theta)\nabla\times\left(K_A\frac{\vec{B}}{B}\right) + \left(\frac{\partial f(\theta)}{\partial\theta}\right)\frac{K_A}{r}\vec{e}_\theta\times\frac{\vec{B}}{B} \tag{4}$$

The first term describes the gradient and curvature drifts, while the second one describes the motion in the region affected by the HCS. $f(\theta)$ is a function that describes the transition between the regions inside and outside the HCS, given in [6], $\vec{e}_\theta$ is the unit vector along the polar direction and K_A is the anti-symmetric part of the diffusion tensor [11], given by

$$K_A = K_{A0}\frac{Q}{|Q|}\frac{\beta R}{B}\left[\frac{(R/R_{Ak})^2}{1+(R/R_{Ak})^2}\right] \tag{5}$$

Here Q is the CR charge, R_{Ak} is the break in K_A and K_{A0}=1. The local interstellar spectrum (LIS) of galactic CR is an input for any modulation models. Fluxes are assumed isotropically distributed at heliosphere boundary, in a steady-state configuration. In the current model the proton LIS flux that was used is taken from [13].

2.1 Application to AMS02 and PAMELA data

The numerical model was applied to the Bartel-rotation averaged AMS02 data from May 2011 to May 2017 [1] and to the Bartel-rotation averaged PAMELA data from June 2006 to January 2014 [2]. The model has six free parameters that can be varied to fit the measured data sets. They are: tilt angle $\alpha(t)$, HMF at Earth $B_0(t)$, the HMF polarity $A(t)$, the parallel diffusion normalization coefficient $K_0(t)$, the two spectral indexes of the diffusion coefficient, $a(t)$ and $b(t)$. The first three define the average status of heliosphere at a given epoch, the last three fit the data sets and define the modulation of CR at the same epoch. The break R_k is fixed at R_k = 3 GV and R_{Ak}=0.3 GV, since it was seen that can be treated as constant in time. As a first approximation, to take into account the time-varying status of the heliosphere, we take the averaged tilt angle and HMF over a time period of 1 year preceding the selected BR, using a Backward Moving Average (BMA). To fit the measured data sets of CR of AMS02 and PAMELA, a six dimensional discrete grid of the model parameters vector $\vec{q} = (\alpha, B_0, A, K_0, a, b)$ was built with the parameters values shown in the Table 1.

Table 1. Model parameters ranges and step used to build the proton model flux grid.

	Min	Max	Step
α	5	75	5
B_0/5 nT	0.7	1.5	0.1
A	-1	1	1
K_0	0.1	0.9	0.1
a	0.45	1.4	0.05
b	0.45	1.65	0.05

For each of the 1.215×10^6 points in the grid, a proton flux $J_m(E, \vec{q})$ was simulated. Then, for each data set $J_d(E, t)$, a χ^2 was defined as

$$\chi^2(\vec{q}) = \sum_i \frac{[J_d(E_i, t) - J_m(E_i, \vec{q})]^2}{\sigma^2(E_i, t)} \tag{6}$$

The errors are given by $\sigma^2(E_i, t) = \sigma_d^2(E_i, t) + \sigma_{SDE}^2(E_i, t)$. Here $\sigma_d^2(E_i, t)$ are the experimental errors for i-th energy bin and $\sigma_{SDE}^2(E_i, t)$ are the errors due to the limited statistics of the Stochastic Differential Equation (SDE) MonteCarlo approach and the uncertainties in the LIS flux. These errors are ~10% at 20 MeV and become constant at ~2% above 1 GeV. For each BR-averaged data set

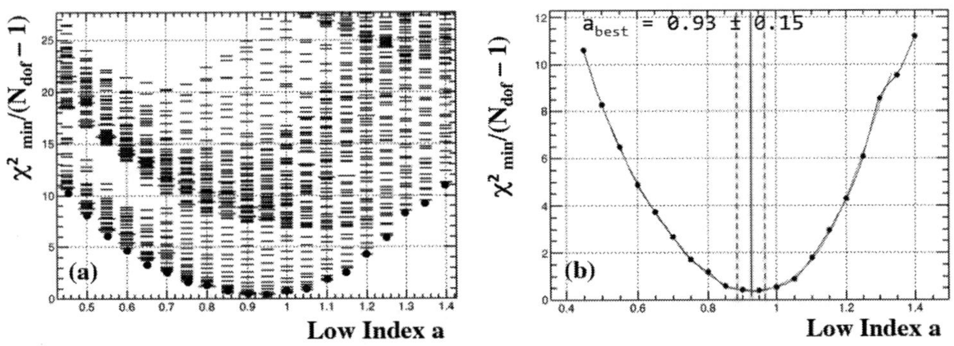

Figure 1. (a) Distribution of the χ^2 of the parameter a. The black dots indicate the χ^2_{min} curve. (b) χ^2_{min} curve for the same data set as in (a). x_{best} and errors are shown as solid and dashed lines, respectively.

at a given epoch, defined as the time t of the beginning of the corresponding BR, the first three parameters, $\alpha(t)$, $B_0(t)$ and A(t) fix the average status of the heliosphere. For $x = K_0(t)$, $a(t)$ and $b(t)$), the minimum chi-squared curve, $\chi^2_{min}(x)$, is found, regardless of the values of all the other parameters, as shown in Figure 1a for the spectral index below the break a for one data set, where the $\chi^2_{min}(x)$ is represented by the black dots. Then, an interpolation is done on the $\chi^2_{min}(x)$ to find its miminum and the value of the parameter x at minimum, x_{best}. The errors on the parameters are estimated as $\sigma_x = max(|x_- - x_{best}|, |x_+ - x_{best}|)$, where $x_\pm$ is the parameter value such that $\chi^2_{min}(x_\pm) = \chi^2_{min}(x_{best}) + 1$ above and below x_{best}, as shown in 1b, where the solid line indicates x_{best} and the dashed lines $x_\pm$. The model was applied to the Bartel-Rotation averaged proton fluxes measured by the PAMELA and AMS02 experiments [1, 2]. The data time-series cover the solar minimum from 2006 to 2009, the ascending phase to solar maximum, occurred on ~ February 2013, when the Sun polarity A reversed, and the following descending phase until May 2017. The comparison to the data, $\sum \frac{|J_d - J_{best}|}{\sigma}$, shows that the difference between any data set and the corresponding best flux model are all well below 1. In Figure 2 are shown the model parameters K_0, a and b as a function of time for the data sets of AMS02 and PAMELA. It is clearly visible a correlation to the solar activity measured by the averaged Sun Spot Number (SSN) with a time lag between the SSN and the model parameters. To find the time lag between the model parameters and the SSN we compare $K_0(t)$ with SSN(t - Δt) varying Δt and find the Δt that maximizes the correlation. It turns out to be $\Delta t = 1.04 \pm 0.1$ years. K_0 is anti-correlated with SSN when time lag is taken into account, the slopes a and b are less clearly correlated with SSN but a variation can be seen between solar max and solar min.

3 Conclusions

A new modeling of the proton flux has been implemented. It successfully describes AMS02 and PAMELA data covering 11 years of CR measurements. A clear correlation between solar activity and CR modulation is found. The model array can be used to find the modulation parameters at any epoch.

References

[1] M. Aguilar et al., Phys. Rev. Lett. 121, 051101, 2018
[2] Adriani et al., The Astroph. J., 765 (2013) 91

https://doi.org/10.1051/epjconf/201920901032

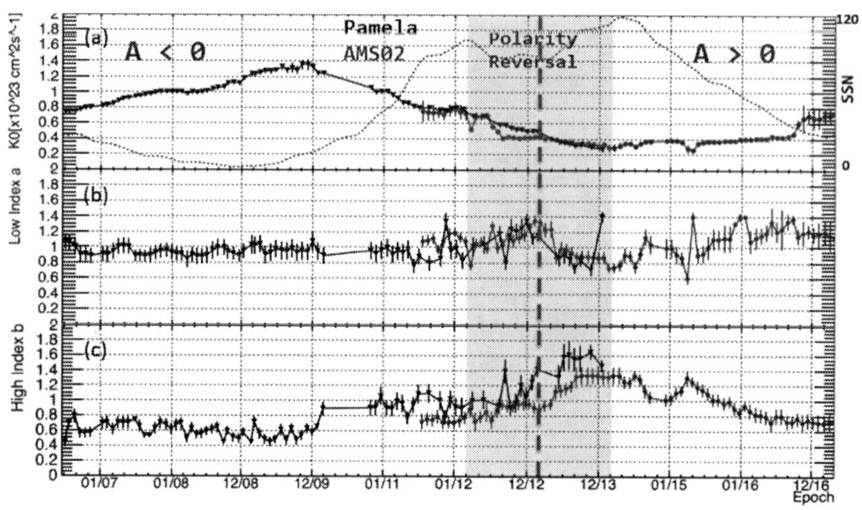

Figure 2. Results for (a) K_0, (b) spectral index a and (c) spectral index b for all the data sets as a function of time. Dotted line in (a) displays the averaged SSN.

[3] Parker, E. N. 1965, Planet. Space Sci., 13, 9
[4] Kappl, Comp. Phys. Comm. 207 (2016) 386?399
[5] Parker, E. N. 1958, ApJ, 128, 664
[6] Bobik *et al.* The Astroph. J., 745:132 (21pp), 2012
[7] Potgieter, M. S., Vos, E. E., Boezio, M., et al. 2014, SoPh, 289, 391
[8] Jokipii, J. R. 1966, ApJ, 146, 480
[9] le Roux, J. A. , Fichtner, H. 1999, J. Geophys. Res., 104, 4709
[10] M.S. Potgieter et a., Solar Phys (2014) 289:391?406 DOI 10.1007/s11207-013-0324-6
[11] Burger, R. A., Hattingh, M. 1995, Ap SS, 230, 375
[12] Heber, B., Potgieter, M.S.: 2006, Space Sci. Rev. 127, 117
[13] Boschin et al., Astrophysical Journal, 840:115 (2017)

EPJ Web of Conferences **209**, 01033 (2019) https://doi.org/10.1051/epjconf/201920901033
RICAP-18

Particle propagation in clumpy supernova remnants

Silvia Celli[1,a]

[1]*Gran Sasso Science Institute, Viale F. Crispi 7, 67100, L'Aquila, Italy*

Abstract. The presence of dense clumps in the environment where a supernova remnant expands might have a strong impact in shaping the observed hadronic gamma-ray spectrum. A detailed numerical study about the penetration of relativistic protons into clumps which are engulfed by a supernova remnant shock is here presented, taking into account the magneto-hydrodynamical properties of the background plasma. This has strong implications for the formation of the spectrum of hadronic gamma rays, which does not reflect anymore the acceleration spectrum of protons, resulting substantially modified by propagation effects. A hadronic scenario including dense clumps inside the remnant shell is shown to adequately reproduce the broadband gamma-ray spectrum of the Galactic supernova remnant RX J1713-3946.7 from GeV to TeV energies.

1 Introduction

During the latest years, numerous detections of supernova remnants (SNRs) in TeV gamma rays have confirmed the theoretical predictions that SNRs can operate as powerful CR accelerators. However, the details of how particles propagate within the SNR, escape the shock and are released into the interstellar medium (ISM) are poorly understood, though these processes strongly influence the very-high-energy emissions observed. In the following, the effects of the presence of dense non homogeneities in the circumstellar medium (CSM), where the shock expands, are investigated. In particular, dense molecular clouds constitute potential sources of both gamma rays and neutrinos, in that the protons accelerated at the SNR shock might interact with the dense gas contained in clumps (in the so-called pp interaction) and produce both neutral pions and charged pions. The methods developed for the computation of the spectrum of protons penetrated inside the clump, from both the point of view of the background plasma and of the accelerated particles, are described in Sec. 2. This spectrum results much harder than that in the diffuse inter-clump medium: as a consequence, the gamma-ray spectrum resulting from pp interactions will reflect this hardening. The cumulative contribution of a uniform clump distribution within the remnant shell is able to reproduce the hard GeV spectrum observed in the SNR RX J1713-3946.7, as explained in Sec. 3. Finally, conclusions are derived in Sec. 4.

2 Shocks expanding in non-homogeneous media

Observations of the ISM have revealed a strong non homogeneity, particularly inside the Galactic Plane. For instance, the environment where type II SNe explode is likely populated by molecular clumps: in fact, given their fast evolution, the explosion happens in an environment rich of molecular

[a]e-mail: silvia.celli@gssi.infn.it

clouds, the same that generated the star. Moreover, given their massive progenitor, strong winds in the giant phase of the star evolution accelerate the fragmentation of clouds into clumps, while creating a large cavity of hot and rarified gas around them. If the mass of the cool clouds does not exceed the Jeans mass, clouds can remain in pressure equilibrium with the surrounding hot phase. The physical size of these ISM inhomogeneities results from MHD simulations which include interstellar cooling, heating and thermal conduction. These have been conducted by [1], who found that the characteristic length scale of clumps amounts to $R_c \simeq 0.1$ pc, corresponding to the smaller scale where thermal instability is effective. Such a scale will be fixed in the following. This scenario could be similar to the one in which the remnant of RX J1713.7-3946 is evolving. In the following, a simplified scenario is realized, where a single clump much denser than the circumstellar plasma is engulfed by a shock. In such a configuration, two different kinds of processes amplify the magnetic field. Around the clump, the regular field results to be compressed reaching a value up to $\sim$ 10 times the original value, in a layer that is found to be about half of the clump size, i.e. 0.05 pc, similar to the results obtained by [1]. This region will be referred to as the *clump magnetic skin*. In addition, in the region behind the clump, turbulence develops and the associate vorticity further amplifies the magnetic field. It is also shown that if the density contrast between the clump and the surrounding medium is very large, the clump can survive for long time before evaporating, even longer than the SNR age.

2.1 MHD simulations with PLUTO

The description of the thermal properties of a classical fluid follows from the solution of the Navier-Stokes equations, coupled to the induction equation for the time evolution of the background magnetic field $\mathbf{B_0}$. In order to simulate a shock discontinuity expanding in the presence of a clump, a numerical approach has been adopted, through a three dimensional simulation in cartesian coordinates within the PLUTO code [2]. A magnetic field of intensity $B_0^{\rm up} = 5\mu$G is set in the region upstream of the shock, inclined by 45° with respect to the shock normal, with $\mathbf{B_0} = (B_{0x}, 0, B_{0z})$ and $B_{0x} = B_{0z}$. In order to investigate a situation as much similar as that of high-mass star SN explosion, like RX J1713.7-3946, the upstream region is simulated as a low density medium with $\rm n_{up} = 10^{-2}$ cm^{-3}, which gets compressed by the shock in the downstream region up to $\rm n_{down} = 4 \times 10^{-2}$ cm^{-3}. A strong shock is moving in the direction of the z-axis, with a sonic Mach number $M = v_s/c_s \simeq 37$, as expected for this remnant, given its measured shock speed of $v_s = 4.4 \times 10^8$ cm s^{-1} and an upstream temperature of $T = 10^6$ K, which is typical for bubbles inflated by stellar winds. As initial condition, the clump is set in the upstream with a density as high as $\rm n_c = 10^3$ cm^{-3}. Therefore, a density contrast $\chi = n_c/n_{\rm up} = 10^5$ is assumed: if the shock speed is $\mathbf{v} = v_s\widehat{z}$, it propagates inside the clump with a velocity $\mathbf{v_{s,c}} = v_{s,c}\widehat{z}$ equal to $v_{s,c} = v_s/\sqrt{\chi} = 1.4 \times 10^6$ cm s^{-1} [1]. The interesting time interval for the evolution of the background plasma is about 300 years from the first shock-clump interaction, as for larger times clumps will have been crossed by the contact discontinuity and thus they will soon be emptied of CRs. Within this time interval, results from MHD simulations can be summarized as follows: i) the shock slows down inside the clump; ii) the clump maintains its density contrast, although the density distribution tends to smoothens; iii) the regular magnetic field is wrapped around the clump surface in a region with a typical size of $R_c/2$: it results compressed in the clump skin, such that the magnetic energy density is $\sim$ 100 times larger than outside; iv) in the region immediately behind the clump, a long tail develops where the plasma becomes turbulent: as a consequence, the vorticity amplifies the magnetic field in a turbulent dynamo-like process, by a factor up to 10. Note that the fact that the magnetic field around the clump is mainly directed along the tangential direction implies that it is difficult for accelerated particles to diffuse orthogonally to the clump surface.

2.2 Particle transport

The propagation of accelerated particles into a magnetized ISM is described by the transport equation, which regulates the temporal and spatial evolution of the CR density function in the phase space $f(\mathbf{x}, \mathbf{p}, t)$. In in the shock reference frame, this equation reads as

$$\frac{\partial f}{\partial t} + \mathbf{v} \cdot \nabla f = \nabla \cdot [D \nabla f] + \frac{1}{3} p \frac{\partial f}{\partial p} \nabla \cdot \mathbf{v} + Q_{\rm CR} \quad (1)$$

Here $D(\mathbf{x})$ is the particle diffusion coefficient, $\mathbf{v}$ is the plasma velocity field, p is the particle momentum and $Q_{\rm CR} = f_{\rm inj} v_s \delta(z - z_s)$ is the injection flux acting at the shock position $z = z_s$. Given the system symmetry, Eq. (1) will be solved in cylindrical coordinates, through a finite difference method (see [3] for further details). The initial condition of the simulation includes the presence of a shock precursor in the upstream region. A stationary isotropic Bohm diffusion is assumed through all the space: its space dependency is dictated by a magnetic field $B_0(\mathbf{x})$, which defined four regions in the space: i) the unshocked CSM, in the far upstream, ii) the shocked CSM, where $B_0 \equiv B_{\rm CSM} = 10\,\mu$G, iii) the clump interior, with a size of $R_{\rm c} = 0.1$ pc, where diffusion is expected to be not efficient because of ion-neutral damping, such that $B_0 \equiv B_c = 1\,\mu$G, and iv) the clump skin, with a size of $R_{\rm s} = 0.5 R_{\rm c}$ around the clump itself, where the amplification of the magnetic field is realized such that $B_0 \equiv B_s = 100\,\mu$G. Following the test-particle approach of DSA, a p^{-4} power-law spectrum is set at the shock position: an exponential cut-off in momentum $p_{\rm cut}$ is introduced in order to take into account the maximum attainable energy. Results are shown in Fig. 1(a), where $p_{\rm cut} = 70$ TeV/c was set in order to reproduce the very-high-energy gamma-ray data of RX J1713.7-3946. The spectrum of particles from younger clumps appears much harder than the one accelerated at the shock: in fact, low-energy CRs are prevented from penetrating into the clump, due to both the amplified magnetic field at the skin and the linear dependency of the diffusion coefficient with the particle momentum. Thus, the entrance of low-energy CRs into the clump is delayed, while high-energy particles can freely propagate in it.

3 The case of RX J1713.7-3946

The case of RX J1713.7-3946 is of special interest to this respect. This remnant has been considered for long time the best candidate for an efficient acceleration scenario, mainly due to its high gamma-ray flux. The detection of gamma-ray emission in the energy range [1 – 300] GeV by the Fermi-LAT satellite [4] has shown an unusually hard spectrum which, at a first glance, seems to be in a better agreement with a leptonic scenario. Nevertheless, a deeper analysis shows that neither the hadronic nor the leptonic scenarios, taken in their simplest form, can unequivocally explain the observations [5–7]. To address this issue, the presence of dense inhomogeneities can be invoked: in fact, the magnetic field can result amplified all around the clump both because of field compression and because of hydrodynamical instabilities developed in the shock-clump interaction. Consequently, the gamma-ray spectrum would be harder than the parent proton spectrum accelerated at the forward shock. Results are shown in Fig. 1(b), where a uniform distributions of clumps inside the remnant shell was assumed (for a total mass in clumps equal to 45 $M_\odot$). An efficiency of the pressure conversion mechanism from bulk motion to accelerated particles equals to $\eta \simeq 2\%$ was also set. Two different models are presented, either with $B_c = 1\,\mu$G or with $B_c = B_{\rm CSM} = 10\,\mu$G: both are shown to adequately describe the hardening observed in multi-GeV gamma rays.

4 Conclusions

The presence of dense molecular clumps where a shock propagates strongly affects the plasma properties, in such a way that the large scale magnetic field around the clumps result amplified. As a

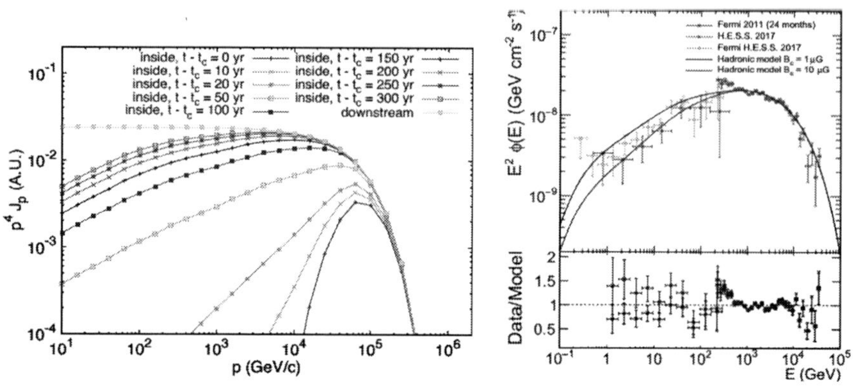

(a) (b)

Figure 1. (a) Proton spectrum inside clumps of different ages. The downstream particle energy density, which is constant in time, is also reported. (b) Gamma-ray flux from SNR RX J1713.7-3946. The points are Fermi-LAT data (pink), H.E.S.S. data (violet) and H.E.S.S. analysis of Fermi-LAT data (light blue). The hadronic models (solid lines) refer to the configuration with a magnetic field inside the clump equal to $B_c = 1\,\mu$G (black) and to $B_c = 10\,\mu$G (blue). The field in the clump skin is fixed to 100μG in both models, to mimic the amplification.

consequence, low-energy accelerated particles needs more time to penetrate the clump compared to high-energy ones, resulting into a significant hardening of the particle energy spectrum inside clumps with respect to the acceleration spectrum. The cumulative contribution of clumps embedded between the contact discontinuity and the current shock position is able to reproduce the observed GeV hardening observed in RX J1713.7-3946. Remarkably, for the gas density inside the clump assumed here ($n_c = 10^3$ cm^{-3}) the evaporation time is much longer than the SNR age: hence, the clumps crossed by the forward shock do not produce significant thermal X-ray emission, in agreement with observations. This scenario would also naturally account for the fast variability in non-thermal X-rays reported in some hot-spots inside RX J1713.7-3946: the electrons entering the regions of amplified magnetic field would rapidly lose their energy because of synchrotron emission. An independent signature on the clump origin of the gamma-ray emission could be revealed by means of morphological studies by CTA, as a large spatial fluctuation of the gamma-ray flux is expected, unlike a scenario where the SNR is expanding into a uniform medium. Finally, the detection of neutrinos would constitute an unambiguous probe of the hadronic origin of the radiation observed: for a 100% hadronic origin, a significant detection appears realistic within one decade of operation of KM3NeT.

References

[1] T. Inoue, R. Yamazaki, S.i. Inutsuka, Y. Fukui, ApJ **744**, 71 (2012), 1106.3080
[2] A. Mignone et al., ApJSS **170**, 228 (2007), astro-ph/0701854
[3] S. Celli, G. Morlino, S. Gabici, F. Aharonian, ArXiv e-prints (2018), 1804.10579
[4] A.A. Abdo et al., ApJ **734**, 28 (2011), 1103.5727
[5] G. Morlino, P. Blasi, E. Amato, APh **31**, 376 (2009), 0903.4565
[6] V.N. Zirakashvili, F.A. Aharonian, ApJ **708**, 965 (2010), 0909.2285
[7] S. Gabici, F.A. Aharonian, EPJWC **121**, 04001 (2016), 1502.00644

The Limadou-HEPD Segmented Calorimeter

Vincenzo Vitale[1,a] for the CSES-Limadou Collaboration

[1] *INFN sez. Roma Tor Vergata, Via della Ricerca Scientifica 1, 00133 Roma, Italy*

Abstract. The core of the High-Energy Particle Detector (HEPD) on board of the China Seismo-Electromagnetic Satellite (CSES) is a segmented calorimeter, which is composed with an upper tower of plastic scintillator counters and a bottom array of LYSO large crystals. Electrons with energy below 100 MeV, protons and light nuclei, below few hundreds of MeV/nucleon are fully contained within this calorimeter. Mainly the LYSO array (density 7.3 g/cm^3, thickness around 29.2 g/cm^2) extends the HEPD energy range, allowing those measurements (solar energetic particles, low-energy cosmic rays) which are more related to astroparticle physics topics. Two identical copies of HEPD, and then of its calorimeter, exist: the Flight (FM) and the Qualification (QM) models. While the FM has achieved the orbit on board of the CSES satellite in February 2018, the Qualification Model, is used, at ground, for tests and calibrations. A report on the characterization of this compact particle space detector and on preliminary studies and results, will be given.

1 HEPD

The High-Energy Particle Detector (HEPD, [1]) is a space detector for high-energy particles (electrons from 3 to 100 MeV, protons from 30 to 300 MeV, light nuclei up to hundreds of MeV/n), developed and built by a consortium of Italian institutions. Figure 1 it is shown a model of the entire High-Energy Particle Detector with its sub-detectors, obtained with a GEANT4 based MC code, and a picture of the apparatus. HEPD is one of the instruments on board of the Chinese Seismo-Electromagnetic Satellite (CSES, [2]). This is a CNSA (Chinese National Space Administration) and ASI (Italian Space Agency) mission, focused to study the near-Earth electromagnetic, plasma and particle environment, to search for electromagnetic seismo associated disturbances. Two HEPD copies were built: the Qualification Model (QM) and the Flight model (FM). These had intense test and qualification campaigns with proton and electron beam tests (INFN-LNF-BTF and ProtonTerapy Center in Trento), space qualification tests (thermo-vacuum, pyro-shock, thermal cycles, SERMS lab. in Terni). CSES was successfully launched February the 2nd 2018. HEPD FM completed its commissioning phase in June 2018. HEPD QM is used, at ground, for further characterization of the detector, such as a beam test with light nuclei (H, He, C, O, at 62MeV/n) performed during July 2018 at INFN-LNS. Apart from the searches for Geo-space anomalies related to powerful seismic events, with HEPD several astroparticle topics can be addressed. The main are: - the study of the solar energetic particles [3]; - the measurement of the low-end of the Galactic Cosmic Rays spectrum [4]; - the characterization of the orbit radiation environment [5].

[a] e-mail: vincenzo.vitale@roma2.infn.it

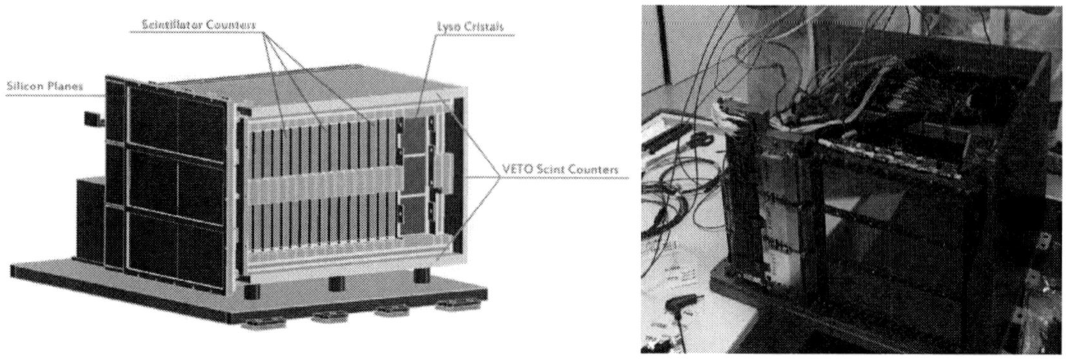

Figure 1. On the left panel is shown the High-Energy Particle Detector: - in blue, the silicon tracker (two silicon planes of 213×213 mm^2), with 192 micron read-out (p and n) and resolution of 40-50 micron); - below the tracker there is the trigger plane (one thin plane of 6 counters with dimen-sion 200×30×5 mm^3 of plastic scintillator EJ-200 and read out by PMTs Hamamatsu R9880U.); - in light blue 16 plastic counters of the upper calorimeter and on the right are visible three of the LYSO crystals of the lower calorimeter (see text for details); - around the calorimeter three counters of the VETO can be seen (the VETO has four lateral and one bottom counters similar to the CALO ones; - in green the power supply and electronics. On the right panel a picture of the HEPD FM during the detector integration.

2 The Detector

The core of HEPD consists of a segmented calorimeter, which is divided in two sections. Its upper part is a tower made of 16 plastic scintillator EJ-200 counters. Each counter measures 150×150×10 mm^3 , and is read out by two PMTs at two opposite corners.The lower calorimeter consists of an array of 3×3 crystals of heavy inorganic scintillator (LYSO, density 7.3 g/cm^3). Each crystal has dimensions 48×48×40 mm^3 and is read out by one PMT. This array is located at the bottom of the tower. In Figure 2 pictures of the tower and LYSO array are shown. These two devices constitute two different radiator homogeneus layers. The main departures from homogeneity are given by the mechanical structures, which had to be included between each pair of tower counters, to the gaps (few mm) between the LYSO crystals, as also to the mechanical structure between tower and array. A fraction of the incident particles will be fully contained within the calorimeter. For them it is possible to measure the total energy (to have which also the energy release within the trigger should be included). The particle identification can be obtained with the dE/dX vs E_{Tot} method, using the energy release within the trigger (or the tracker) as dE/dX. For contained events, the particle range can also be obtained. Another fraction of particles pass through the calorimeter and exit from the bottom veto. For these passing through particles it is possible to measure the differential energy loss within each calorimeter element, and then a study of the event is possible with a comparison with an energy deposition model (passing through analysis).

3 Energy Calibration

A preliminary energy calibration of the plastic tower was performed by means of the electron (30 MeV) and proton (51, 70, 100 and 125 MeV) irradiations. For each of the PMT channels, the pedestal and the associated rms were determined with special runs, during which a high-frequency artificial

Figure 2. On the right panel: the HEPD-QM calorimeter. The tower of 16 plastic scintillator counters is visible. On the bottom is located the 9 LYSO crystals array. On the left panel: the HEPD-QM LYSO array. In both pictures also a part of the mechanical structures are visible)

trigger was sent to the detector. The PMTs light responses to a MIP-like particle were set to a conventional value, by means of acquisitions of atmospheric muons. Beam tests data were required: a) to give signal only in one trigger bar (trigger multiplicity = 1); b) to be contained within the upper calorimeter (no hit in the veto counters and no hit in the LYSO crystals); c) to have more than a minimum number of hit planes, to reject events with anomalous low energy loss. Then we calculated the sum of the signal of trigger and tower as $S_{Tot} = S_{Tower} + \frac{1}{2} S_{Trigger}$, with S_{Tower} the sum of the signals of all the tower PMTs, $S_{Trigger}$ the sum of the signal of the trigger PMTs over threshold. The trigger bars have half of the thickness of a tower counter. An energy response function was built, by associating the S_{Tot} peak-position to the related particle energy. The energy losses of the particles before arriving to the trigger (air, tracker, etc) were taken into account, by means of MC simulations. The LYSO array calibration was performed with a different method, because all the beam particles which reach the LYSO array have already interacted with the HEPD elements above it (tracker, trigger, tower, structures). To build the energy response function S_{Tot} peak-position was associated to the expected energy deposition within the LYSO crystals. This deposition was estimated with Monte Carlo simulations. The LYSO array was calibrated with 60 and 90 MeV electrons, 174, 202 and 228 MeV protons. The obtained energy response functions for the tower and the LYSO array are shown in Figure 3, together with an example of energy reconstruction, e.g. BTF 60 MeV electrons, which loose energy both in tower and LYSO array.

4 Outlook

A preliminary energy calibration of the HEPD calorimeter has been performed with the use of the electron and proton beams data. Studies of secondary effects such as larger charges, mechanical structures effects, etc, are ongoing. After the calibration, further data analysis steps are possible, as

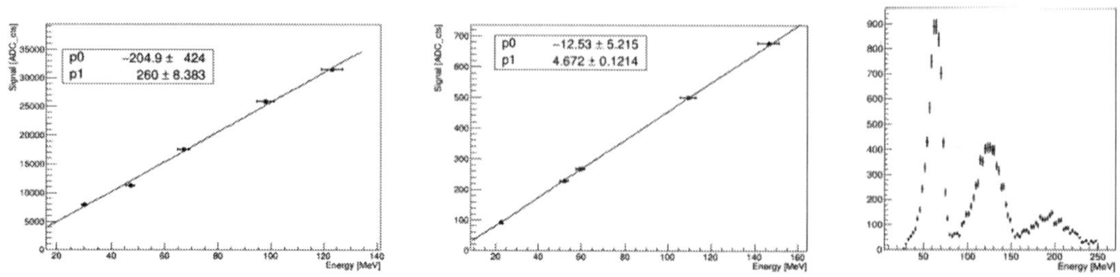

Figure 3. On the left: tower Energy Repsonse Function. The peak position of the signal distribution is associated with the beam energy minus the losses in the tracker and air. On the centre : LYSO array Energy Response Function.The peak position of the signal distribution is associated to the expected energy deposition within the LYSO array, obtained with MC simulations. On the right: Energy Reconstruction for the BTF 60 MeV electrons. Events were required to have more than 15 active tower counters. These particles reach both tower and array, The second and third peaks are produced by the pile-up of 2 or 3 simultaneous electrons.

the implementation of the particle identification methods (dE/dX vs E_{Tot}, passing through particles analysis).

References

[1] Ambrosi, G. et al. "The HEPD particle detector of the CSES satellite mission for investigating seismo-associated perturbations of the Van Allen belts" Sci. China Technol. Sci.(2018) 61: 64 DOI:10.1007/s11431-018-9234-9

[2] Shen, X., et al. "The state-of-the-art of the China Seismo-Electromagnetic Satellite mission." (2018) Sci. China Technol. Sci. (2018) 61: 634 DOI:10.1007/s11431-018-9242-0

[3] "The High-Energy Particle Detector (HEPD) on board of the CSES mission", Proceedings of the International Astronomical Union, IAU Symposium (2018) IAUS 335, 365V, DOI:10.1017/S1743921317008328

[4] Vitale V., et al., for the CSES/HEPD Collab. "The High-Energy Particle Detector on board of the CSES mission.", (2018). EPJ Web of Conferences 170, 01007 doi:10.1051/epjconf/201713601007

[5] Vitale V., et al., for the CSES/HEPD Collab. "Measurement of the Low-Energy Charged Particle Background with the space detector HEPD", (2018). Proceedings of the Symposium "A Decade of AGILE", LYNC-D-18-00180, submitted

The Science Case for a Southern Wide Field of View Detector

Di Sciascio Giuseppe[1,a]

[1] *INFN - Roma Tor Vergata*

Abstract. EAS arrays are survey instruments able to monitor continuously all the overhead sky. Their sensitivity in the sub-TeV/TeV energy domain cannot compete with that of Cherenkov telescopes, but the wide field of view (about 2 sr) is ideal to complement directional detectors by performing unbiased sky surveys, by monitoring variable or flaring sources such as Active Galactic Nuclei (AGN) and to discover transients or explosive events (GRBs). Arrays are well suited to study extended sources, such as the Galactic diffuse emission, and to measure the spectra of Galactic sources at the highest energies (near or beyond 100 TeV). An EAS array is able to detect at the same time events induced by photons and charged cosmic rays, thus studying the connection between these two messengers of the non-thermal Universe. Therefore, these detectors are, by definition, multi-messenger instruments. All EAS arrays presently in operation or under installation are located in the Northern hemisphere. The scientific potential of a next-generation survey instrument in the Southern Hemisphere will be presented and briefly discussed.

1 Introduction

The riddle of the origin of Cosmic Rays (CR) is unsolved since more than one century. Charged cosmic rays, gammas and neutrinos are strongly correlated in CR sources where hadronic accelerators are at work. Their integrated study is one of the most important and exciting fields in the *'multi-messenger astronomy'*, the exploration of the Universe through combining information from different cosmic messengers: electromagnetic radiation, gravitational waves, neutrinos and cosmic rays.

The study of CRs is based on two complementary approaches:

(1) Measurement of energy spectrum, elemental composition and anisotropy in the CR arrival direction distribution, the three basic observables crucial for understanding origin, acceleration and propagation of the radiation.

(2) Search of their sources through the observation of neutral radiation (photons and neutrinos), which points back to the emitting sources not being affected by the magnetic fields.

The identification of the galactic sources able to accelerate particles beyond PeV (=10^{15} eV) energies, the so-called *'PeVatrons'*, is certainly one of the main open problems of high energy astrophysics. In fact, even there is no doubt that galactic CR are accelerated in SuperNova Remnants (SNRs), the capability of SNRs to accelerate CRs up to the *'knee'* of the all-particle energy spectrum ($\sim 3\times10^{15}$

[a]e-mail: giuseppe.disciascio@roma2.infn.it

eV) and above is still under debate. The determination of the maximum energy at which protons are accelerated inside their sources (the *'proton knee'*), as well as the measurement of the evolution of the heavier component across the knee, are the key for understanding acceleration mechanisms and the propagation processes in the Galaxy, and to investigate the transition from Galactic to extra-galactic CRs.

In a hadronic interaction the secondary photons have an energy a factor of 10 lower than the primary proton. Therefore, the quest for CR sources able to accelerate particles up to the knee energy region and above requires to survey the γ-ray sky above 100 TeV. In addition, the Inverse Compton scattering at these energies is strongly suppressed by the Klein-Nishina effect. Therefore, the observation of a γ-ray power law spectrum extending up to the 100 TeV range would be a strong indication of the hadronic nature of the emission.

So far no photons above 100 TeV have been observed from any source, and only a few sources have data above 30 TeV. Their spectra is, however, only known with large uncertainties, being the sensitivity of the current instruments at the highest energies not enough to determine clearly the spectral shape.

To open the 100 TeV range to observations a detector with a very large effective area, operating with high duty-cycle, is required. The most sensitive experimental technique for the observation of γ-rays at these energies is the detection of EAS via large ground-based arrays.

EAS arrays are also extremely valid instruments in the sub-TeV/TeV energy domains. Their sensitivity in this energy range cannot compete with that of the future generation Cherenkov telescopes, as CTA (that also have a better energy and angular resolution) [1], but the wide FoV offers the opportunity to continuously monitor a large fraction of the sky, looking for unpredictable transient phenomena. In fact, most current Cherenkov telescopes can work only during clear moonless nights, with a total observation time of about 1000–1500 hours per year (depending on the location), with a typical duty cycle of $\sim$10–15%, and a FoV of a few degrees radius. This implies that they can observe only one source at the time, and only in the season of the year when the source culminates during night time.

By contrast, an EAS detector every day can observe a large fraction of the celestial sphere (spanning 360° in right ascension and about 90° in declination, in a declination interval depending on the geographic location). Sources located in this portion of the sky are in the FoV of the detector, either always, or for several hours per day, depending on their celestial declination. This situation is ideal to perform sky surveys, discover transients or explosive events, such as GRBs, and monitor variable or flaring sources such as AGN.

Another important topic that can be successfully addressed by a wide FoV detector is the extended γ-ray emission and in particular the study of the galactic diffuse gamma emission. This radiation could trace the location of the CR sources and the distribution of interstellar gas. In addition, the observation of a knee in the energy spectrum of the diffuse emission, at an energy of a few hundred TeV, corresponding to the knee of the CR all-particle spectrum, would provide a complementary way to investigate the origin of the knee [2]. A new wide FoV experiment should also be able to map the diffuse emission in different regions of the Galactic Plane to observe a location dependence of the γ-ray spectral index [3] and/or of the knee energy.

Wide FoV telescopes are also important for a multi-messenger study of the Gravitational Wave events due to their capability to survey simultaneously all the large sky regions identified by LIGO and VIRGO, looking for a possible correlated γ-ray emission. As an example, the event GW170817 was localized within a 3 by 10 degree region, well beyond the typical FoV of a Cherenkov telescope and required multiple pointings to H.E.S.S. to cover the area [4].

The only wide FoV detector currently in data taking to study gamma-ray astronomy in the sub-TeV/multi-TeV energy domain is HAWC, located in central Mexico at an elevation of 4100 m a.s.l.

Two new wide FoV detectors, LHAASO and TAIGA-HiSCORE, are under installation. All these experiments are located in the Northern hemisphere. Differential sensitivity (multiplied by E^2) to a Crab-like point gamma ray source of different experiments and projects is shown in Fig. 1.

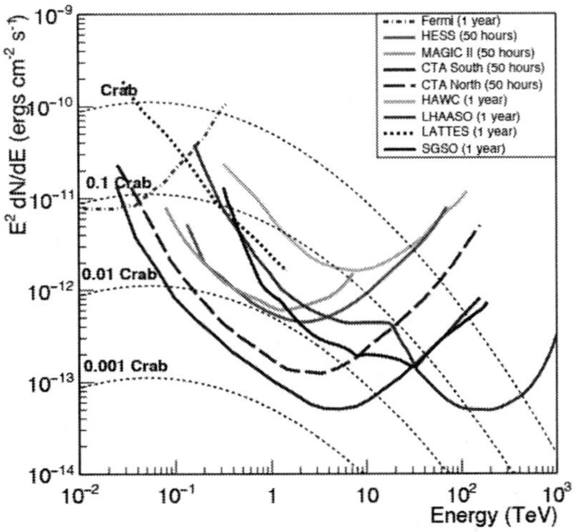

Figure 1. Differential sensitivity (multiplied by E^2) to a Crab-like point gamma ray source of different experiments and projects. The best fit of the Crab Nebula data obtained by different detectors [17] is reported as a reference, extrapolated to 1 PeV

The construction of a new survey instrument at sufficiently Southern latitude to continuously monitor the Galactic Center and the Inner Galaxy should be a high priority. Such an array could be also a 'finder' telescope for the future CTA-South experiment.

1.1 The strong case for a wide FoV telescope in the Southern hemisphere

The observation of the diffuse emission by AGILE [5] and Fermi [6] shows that the distribution of CRs above the GeV energy range is smoothly distributed throughout the disc of the Galaxy but the intensity is higher in the Inner Galaxy and falls off towards the outer disc. Accordingly, the majority of γ-ray sources discovered in the last years above 100 GeV have been detected by the Cherenkov telescopes of the H.E.S.S. experiment located in Namibia. A continuous monitoring of the Inner Galaxy is expected to allow the detection of a large number of new sources.

In addition to the reasons mentioned in the Introduction, the opportunity to observe the Inner Galaxy offers a lot of additional motivations to push the construction of a wide FoV in the Southern hemisphere.

As mentioned, H.E.S.S. recently claimed the possible detection of a PeVatrons in the Galactic Center, most likely related to a supermassive black hole [7], thus opening new perspectives for the observations of CR sources other than SNRs. In fact, the diffuse emission around J1745-290 (positionally compatible with SgrA*) extends up to ~50 TeV, thus suggesting acceleration of CR protons up to the PeV energy range. Combined with the Fermi-LAT observations of a harder spectrum in the Galactic center region [8] and the discovery of the so-called Fermi bubbles, we expect that a lot of high-energy particle accelerators are at work in this region that can be studied by a future wide FoV telescope.

Some authors found that the bulk of the Galactic Ridge emission can be naturally explained by the interaction of the diffuse steady-state Galactic CR sea with the gas present in the central molecular zone. Although they confirm the presence of a residual radial-dependent emission associated with a central source, the relevance of the large-scale diffuse component prevents claim solid evidence of a PeVatrons in the Galactic Center to explain the H.E.S.S. data [8]. A wide FoV detector may confirm this scenario observing the emission from larger region centered on the Galactic Center.

Another interesting opportunity for a wide FoV telescope located in the Southern hemisphere is to investigate the so-called *'IceCube spectral anomaly'*. Recently it was suggested that different IceCube datasets are not consistent with the same power law spectrum of cosmic neutrinos, thus suggesting that they are observing a multicomponent spectrum. The passing-muon data (from the Northern sky) agree with isotropy and E^{-2} distribution up to very high energies. The HESE events (mostly from the Southern sky and shower-like) suggest a softer distribution at low energies. This North-South asymmetry is the IceCube spectral anomaly. It strengthens the likelihood that a Galactic neutrino component exists, mainly observable from the Southern hemisphere. This Galactic component of neutrinos implies the existence of a Galactic component of very high energy gamma rays [9]. A measurement of the γ-ray flux from the Southern sky in the same energy domain of neutrinos (above 30 TeV) could clarify the origin of the emission.

Recently, some authors emphasized the existence of an extended 'hot' region of the Southern gamma sky where the cumulative sources contribution dominates over the diffuse component [10]. This region is located in the Inner Galactic Plane and could be also an important source of HE neutrinos. Interestingly, this region approximately coincides with the portion of the galactic plane from which a $\sim 2\sigma$ excess of showers is observed in the HESE IceCube data sample. If photons are not absorbed, we expect that the high energy neutrino sky is strongly correlated with the high energy gamma sky. A wide FoV detector will allow to study the gamma/neutrino correlation and will provide us a handle to perform a detailed multi-messenger study of the Galactic Plane.

With EAS array some aspects of fundamental physics can be also studied, in particular the existence of multi-TeV Dark Matter particles. The assumption of a single astrophysical power-law flux to explain the IceCube 6-year HESE events yields an anomalous large spectral index (γ^{6yr}=2.92). Adopting a spectral index in the range [2.0, 2.2], which is compatible with the one deduced by the analysis performed on the 6-year up-going muon neutrinos data, the latest IceCube data show an up to 2.6σ excess in the number of events in the energy range 40–200 TeV. Such an excess can be interpreted as a decaying Dark Matter signal [11]. DM particles with masses in the 100 TeV energy range can be investigated only by large wide FoV detectors [12].

Finally, a wide FoV detector is an ideal instrument to deepen the observations of the CR anisotropy carried out in the Southern hemisphere by IceCube/IceTop experiments [13] as a function of the particle rigidity.

2 Design consideration for a Southern Hemisphere Wide FoV detector

A new Southern Hemisphere wide FoV detector, to be fully complementary to CTA-South and to carry out an integrated study of gamma and nuclei induced showers, needs

1. an energy threshold of $\sim$100 GeV, to be a transient factory;
2. a sensitivity at few percent Crab flux level below the TeV, to have high exposure for flaring activity;
3. an angular resolution of $\sim 1°$ at the threshold, to reduce source confusion in the Inner Galaxy;

4. to survey the gamma sky at 100 TeV with a capability to discriminate the background at a level of 10^{-5} to observe the knee in the energy spectrum of the gamma diffuse emission in different regions of the Galactic Plane;

5. to be able to discriminate different primary masses in the knee energy range to measure the proton knee and investigate the maximum energy of accelerated particles in CR sources and to observe the CR anisotropy as a function of the particle rigidity.

Is this possible ?
The main parameters to push down the sensitivity to gamma-ray sources are [14]: (1) the energy threshold; (2) the angular resolution; (3) the gamma/hadron relative trigger efficiency; (4) the effective area for photon detection; (5) the background rejection capability.

Two different experimental techniques have been applied in the last two decades in ground-based survey instruments: (1) water Cherenkov (Milagro); (2) Resistive Plate Chambers (ARGO-YBJ).

The benefits in the use of RPCs in ARGO-YBJ were [50]: (1) high efficiency detection of low energy showers (energy threshold ~300 GeV at 4300 m asl) by means of the high density sampling of the central carpet (~92% coverage); (2) unprecedented wide energy range investigated by means of the digital/charge read-outs (~300 GeV $\rightarrow$ 10 PeV); (3) good angular resolution ($\sigma_\theta \approx 1.66°$ at the threshold, without any lead layer on top of the RPCs) and unprecedented details in the core region by means of the high granularity of the different read-outs.

RPCs allowed to study also charged CR physics (energy spectrum, elemental composition and anisotropy) up to about 10 PeV. By contrast, the capability of water Cherenkov facilities in extending the energy range to PeV and in selecting primary masses must be investigated.

In both experiments (Milagro and ARGO-YBJ) the limited capability to discriminate the background was mainly due to the small dimensions of the detectors. In fact, in the new experiments, like HAWC and LHAASO, the discrimination of the CR background is made studying shower characteristics far from the shower core (at distances R> 40 m from the core, the dimension of the Milagro and ARGO-YBJ detectors).

The key to lower the energy threshold is to locate the detector at extreme altitude (about 5000 m asl for a threshold in the 100 GeV range). But the energy threshold, as well as the angular resolution, depends also on the coverage (the ratio between the detection area and the instrumented one), on the granularity of the read-out, on the particular type of detector and on the trigger logic.

The ARGO-YBJ experiment, combining the full coverage approach at high altitude with a high granularity of the read-out (about 15,000 strips 7×62 cm^2 wide), sampled 100 GeV γ-induced showers with an efficiency of about 70% at 4300 m a.s.l. . The median energy of the first multiplicity bin (20-40 fired pads) for photons with a Crab-like energy spectrum was 340 GeV [16]. The granularity of the read-out at cm level allowed to sample events with only 20 fired pads, out of 15,000, with a background-free topological-based trigger logic. A water Cherenkov based facility will hardly be able to lower the energy threshold below ~500 GeV.

Therefore, a full coverage approach based on the RPC technology is one of the most interesting solution for a new survey instrument in the South. A hybrid detector with RPCs coupled to a HAWC/LHAASO - like water Cherenkov detector, will allow to lower the energy threshold at 100 GeV level, to measure the arrival direction with two independent techniques, to exploit the HAWC/LHAASO approach to reject the background, and to study also CR physics with nuclei induced events up to 10 PeV.

3 Conclusions

Open problems in cosmic ray physics push the construction of new generation EAS arrays to study, in the $10^{11} - 10^{18}$ eV energy range, at the same time photon- and charged-induced events. In the next decade CTA-North and LHAASO [51] are expected to be the most sensitive detectors to study gamma-ray astronomy in the Northern hemisphere from about 20 GeV up to PeV.

A survey instrument to monitor the Inner Galaxy and the Galactic Center with high sensitivity should be a high priority. Extreme altitude (~5000 m asl), high coverage coupled to a high granularity of the read-out are the key to improve the angular resolution and the sensitivity to gamma-ray sources. A detector able to sample shower up to particle densities of $\approx 10^4$ - 10^5 p/m^2 will allow to study primary particles up to ~ 10 PeV.

The ARGO-YBJ Collaboration demonstrated that bakelite RPCs can be safely operated at extreme altitudes for many years. The benefits in the use of RPCs in ARGO-YBJ were: (1) high efficiency detection of low energy showers (energy threshold ~300 GeV) by means of the dense sampling of the central carpet; (2) unprecedented wide energy range investigated by means of the digital/charge read-outs (~300 GeV $\rightarrow$ 10 PeV); (3) good angular resolution and unprecedented details in the core region by means of the high granularity of the read-outs.

The ARGO-like RPCs should be an important element of a future experiment in the South, possibly coupled to HAWC/LHAASO - like water Cherenkov detectors to exploit the added values of two experimental approaches.

Science case and experimental solutions for a survey instrument in the South are under discussion in the framework of the Southern Gamma-ray Survey Observatory (SGSO) alliance [19]. Preliminary calculations of the sensitivity of a water Cherenkov based array is shown in Fig. 1.

References

[1] B.S. Acharya et al., preprint: arXiv:1709.07997.
[2] S. Vernetto and P. Lipari, Pos(ICRC2017) 718.
[3] B. Bartoli et al., ApJ **806** (2015) 20.
[4] H. Abdalla et al., preprint:arXiv:1710.05862.
[5] A. Giuliani et al., Mem SAIt Suppl. **5** (2004) 135.
[6] A.A. Abdo et al., Phys. Rev. Lett. **103** (2009) 251101.
[7] A. Abramovski et al., Nature **531** (2016) 476.
[8] D. Gaggero et al., Phys. Rev. Lett. **119** (2017) 031101.
[9] A. Palladino et al., JCAP **12** (2016) 45.
[10] G. Pagliaroli and F.L. Villante, preprint arXiv:1710.01040.
[11] M. Chianese et al., preprint arXiv:1707.05241
[12] G. Di Sciascio and A. Morselli, PoS(ICRC2015) 296.
[13] M.G. AArtsen et al., ApJ **826** (2016) 220.
[14] G. Di Sciascio et al., PoS(ICRC2017) 781.
[15] G. Di Sciascio, Int. J. Mod. Phys. D **23** (2014) 1430019.
[16] B. Bartoli et al. ApJ **798** (2015) 119.
[17] H.H.He et al., 34th ICRC Proc., 2015
[18] G. Di Sciascio, in *CRIS 2015 proceedings*, Preprint arXiv:1602.07600.
[19] www.sgso-alliance.org

Multimessenger Probes of High-energy Sources

Dafne Guetta[1,a]

[1] *ORT Braude College, Karmiel, Israel*

Abstract. Multimessenger observations may hold the key to learn about the most energetic sources in the universe. The recent construction of large scale observatories opened new possibilities in testing non thermal cosmic processes with alternative probes, such as high energy neutrinos and gravitational waves. We propose to combine information from gravitational wave detections, neutrino observations and electromagnetic signals to obtain a comprehensive picture of some of the most extreme cosmic processes. Gravitational waves are indicative of source dynamics, such as the formation, evolution and interaction of compact objects. These compact objects can play an important role in astrophysical particle acceleration, and are interesting candidates for neutrino and in general high-energy astroparticle studies. In particular we will concentrate on the most promising gravitational wave emitter sources: compact stellar remnants. The merger of binary black holes, binary neutron stars or black hole-neutron star binaries are abundant gravitational wave sources and will likely make up the majority of detections. However, stellar core collapse with rapidly rotating core may also be significant gravitational wave emitter, while slower rotating cores may be detectable only at closer distances. The joint detection of gravitational waves and neutrinos from these sources will probe the physics of the sources and will be a smoking gun of the presence of hadrons in these objects which is still an open question. Conversely, the non-detection of neutrinos or gravitational waves from these sources will be fundamental to constrain the hadronic content.

1 Introduction

With the contemporary operation of the IceCube and Antares, Advanced LIGO and Advanced Virgo facilities, we are now able to observe the universe using two new, distinct astrophysical messengers. In addition to photons from radio waves to gamma-rays, we can now, for the first time, simultaneously observe the sky with neutrinos and gravitational waves (GWs).

This new era of multi-messenger astrophysics offers a unique view of the universe and provide powerful insights into the workings of some of the most energetic and enigmatic objects in the cosmos.

High-energy cosmic neutrinos observed by the IceCube experiment reveal a deep view of the universe at energies where the sky is opaque to photons. With more than one thousand times the energy of the most energetic neutrinos produced with earthbound accelerators, cosmic neutrinos collected by IceCube also also exceed by a factor of one billion the energy of the neutrinos detected from the supernova explosion in the Large Magellanic Cloud detected in February 1987, the only astrophysical neutrinos observed from outside the solar system prior to IceCube's breakthrough. An immediate

[a] e-mail: dafneguetta@braude.ac.il

inference made about the large neutrino flux observed by IceCube, which is predominantly of extragalactic origin, is that the total energy density of high-energy neutrinos in the universe is similar to that of gamma-rays. This is worthy of a closer look. Astrophysical neutrinos are mostly the decay products of pions. Protons accelerated in regions of high magnetic fields near neutron stars or supermassimve black holes (like in jet og Active Galactic Nuclei) may interact with the radiation or dust surrounding them to produce these pions and kaons that decay into neutrinos.

Together with charged pions generating neutrinos, neutral pions are produce, which promptly decay into two gamma-rays. These gamma-rays in the matching 10 to 10,000 TeV energy range of IceCube neutrinos can not reach Earth, as they interact with extragalactic (mostly microwave) photons. This interaction triggers a cascade process that leads to photons in the GeV-TeV energy range, which are detected.

The matching energy densities of the extragalactic gamma-ray flux detected by Fermi and the high-energy neutrino flux measured by IceCube would be a natural consequence of having been originated from common sources. However, the majority of the extragalactic, unresolved gamma-ray flux is thought to originate from blazars, which alone cannot explain the observed cosmic neutrinos[1]. A possible explanation that does not violate the gamma-ray flux limit nor the source emission constraints is that neutrino sources or their environment absorb gamma-rays; i.e. they are electromagnetically hidden [2].

Neutrino astronomy represents a unique tool within multi-messenger astrophysics to probe the most extreme cosmic processes. IceCube plans to extend our reach for neutrino sources by instrumenting 10 km^3 of glacial ice at the South Pole, improving IceCube's sensitive volume by an order of magnitude [3]. The new facility will increase the event rates for the highest energy neutrinos from thens to hundreds per year over several years.

On the other side of the world from IceCube is the ANTARES telescope, in the Mediterranean sea. ANTARES is currently the only deep sea high energy neutrino telescope that is operating in the Northern hemisphere. The telescope covers an area of about 0.1 km^2 on the sea bed, at a depth of 2475 m, 40 km off the coast of Toulon, France. ANTARES is planned to be followed by a multi-kilometer detector in the Mediterranean sea called KM3NeT in the next few years. KM3NeT is the future generation of under water neutrino telescopes. The infrastructure will consist of three so-called building blocks, each made of 115 strings of 18 optical modules, that have 31 photo-multiplier tubes each. KM3Net is made of KM3NeT/ARCA (Toulon, France) and KM3NeT/ORCA (Capo Passero, Sicily)[4]. The realization of next generation high energy detectors like CTA for TeV photons, KM3Net and IceCube-Gen2 for higher energy neutrinos and the improving sensitivity of GW detectors will open a new era in multi-messenger astrophysics that we propose to exploit.

Beyond neutrinos, another major pillar of multi-messenger probes are GWs. Advanced LIGO [5] recently made the first direct detection of GWs [6]. On September 14th, 2015, the LIGO detectors recorded a signal that was soon reconstructed to have come from the merger of two black holes, each with $\sim 30\,M_{\odot}$, about 1.3 billion light years away. This accomplishment was not only the beginning of GW observations that can probe binary black holes system and their gravitational interaction in a unprecendented ways; it also started a new chapter for multi-messenger astronomy.

Combining information from GW detections with electromagnetic and neutrino observations allows us to gain a better understanding of some of the most extreme cosmic processes [7]. GWs are indicative of source dynamics, such as the formation, evolution and interaction of compact objects. These compact objects are anticipated to play an important role in astrophysical particle acceleration and high-energy emission, making a direct link between the GW observation and high-energy neutrino emissionand in general high-energy astroparticle studies.

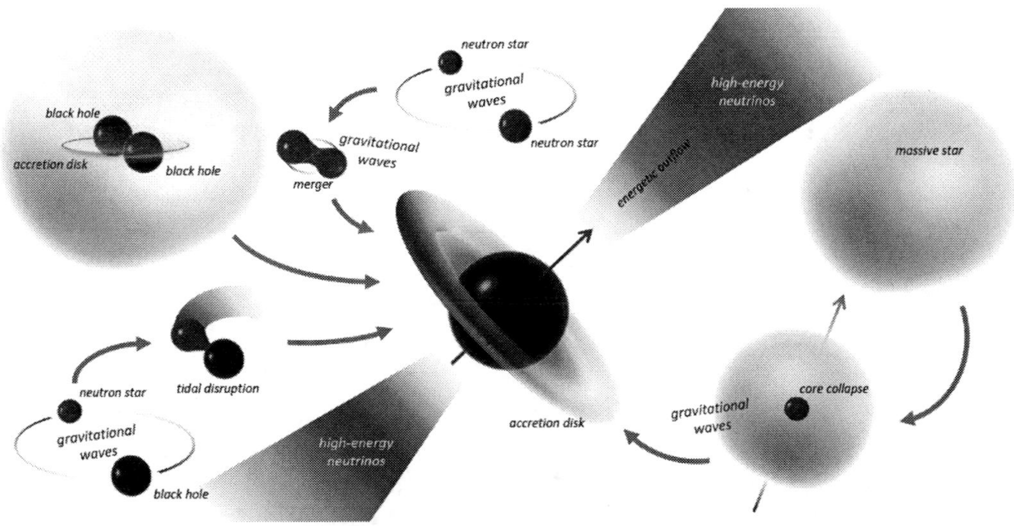

Figure 1. Illustration of multi-messenger sources of GWs and high-energy neutrinos.

The most promising GW emission is related to compact stellar remnants. The merger of binaries of black holes and neutron stars are abundant GW sources and will likely make up the majority of detections [8]. Stellar core collapse with a rapidly rotating core may also be a significant GW emitter [9–11], while slower rotating cores may be detectable at closer distances [12]. If the core collapse results in a protoneutron star, fallback accretion can significantly increase the angular momentum thus resulting in increased GW production (along with electromagnetic emission), improving detection prospects [13]. Other GW sources include rotating neutron stars [14], and plausibly magnetar flares [15–17]. See Fig. 1 for an illustration of some of the main GW sources that may also produce high-energy neutrino emission.

With the onset of GW observations, there has been a significant effort to search for electromagnetic and neutrino emission from GW sources [18]. Started during the operation of initial LIGO and Virgo [19, 20], electromagnetic follow-up efforts now include a large number of partner observatories from radio to gamma-rays [21]. A significant number of these telescopes searched for counterparts of the first observed GW, GW150914, but no obvious one was found [21].

Joint GW+neutrino searches started with initial GW detector operations as well. These searches adopted a ± 500 s time window around the gravitational wave event [22], and most of them were based on the same baseline GW+neutrino search technique [23]. The first observational constraints for joint sources were derived using initial LIGO-Virgo and the IceCube detector [24], which was soon followed by searches with ANTARES [25] and IceCube [26]. Most recently, high-energy neutrino searches were carried out for the first GW detection, GW150914, with the ANTARES and IceCube detectors [27] and the Pierre Auger Observatory [28], while the KamLAND detector [29] and IceCube [27] searched for MeV neutrino counterpart. No significant temporally and directionally coincident neutrinos were found by these searches.

2 Multi-messenger Detection of IceCube Alert IceCube-170922A

IceCube detects muon neutrinos above a threshold of approximately 10 GeV, resulting in one well-reconstructed upgoing muon track event every few minutes. These events are dominated by low-energy atmospheric neutrinos. In 2016, IceCube installed an online filter that selects from this sample, in real time, very high energy events that are potentially cosmic in origin, typically having a probability greater than 50% of coming from outside the earth's atmosphere. These are reconstructed within less than one minute, and the energy and arrival direction of the neutrino are communicated to the Gamma-ray Coordinate Network (GCN) for follow-up by astronomical telescopes. The tenth such alert [30], IceCube-170922A, on September 22, 2017, reported a well-reconstructed muon neutrino with an energy of 290 TeV and, therefore, with a high probability of originating in an astronomical source.

What makes this alert special is that, for the first time, telescopes detected enhanced gamma-ray activity from a flaring blazar aligned with the cosmic neutrino to within less than 0.06 degree. The source is a known blazar, TXS0506+056, and its redshift has been subsequently measured to be $z \sim 0.34$ [31]. Originally detected by NASA's Fermi [32] and Swift [33] satellite telescopes, the alert was followed up by the MAGIC air Cherenkov telescope [34]. MAGIC detected the emission of gamma rays with energies exceeding 400 GeV. With a redshift of 0.34, we can conclude that the source is a TeV blazar. Several other telescopes subsequently observed the flaring blazar.

It is important to realize that nearby blazars like the Markarian sources are at a redshift that is ten times smaller, and therefore TXS0506+056, with a similar flux despite the greater distance,is between the 3that multiple attempts have not found a correlation between the arrival directions of cosmic neutrinos previously observed by IceCube and the various Fermi blazar catalogues that are dominated by "vanilla" nearby sources; we can infer to this flare peculiar luminosy and duration as well as a particular hadron activity.

Given where to look, IceCube searched its archival neutrino data up to and including October 2017, for evidence of neutrino emission at the location of TXS0506+056. When searching the sky for point sources of neutrinos, two analyses have been routinely performed: one looking for a steady emission and one that searches for flares over a variety of timescales. Evidence was found for a spectacular burst of 14 high-energy neutrinos in 110 days. It dominates the neutrino flux in coincidence with the source position over the last 9.5 years for which we have data. It is interesting to note that a subset of blazars, around 1% – 10% of all blazars, bursting once in 10 years at the levels of TXS, can accommodate the diffuse cosmic neutrino flux observed by IceCube. The energy of the neutrino flux potentially generated by the flaring blazars is at the same level as the flux in extragalactic cosmic rays, the Waxman-Bahcall bound.

The coincident observation of Fermi and IceCube, the significance of the 2014 neutrino flare, and the detection of the TeV emission by MAGIC, puts the discovery of the first comic ray accelerator beyond question.

Studying the relation between the diffuse neutrino flux and point source flux of TXS, we derive that the efficiency of the proton beam for producing the pions that are the parents of the observed neutrinos must be close to unity. We will investigate the above facts with detailed modeling of this blazar jet. Early results based on the interaction of the accelerated protons with the dominant 10 eV blue photons in the galaxy, known as "the blue bump," are particularly promising.

In summary, we intend to model the rich multiwavelength data provided by this event. The challenge is that the blazar jet must have a sufficiently dense photon target to produce the neutrinos seen by IceCube and, at the same time, be transparent to the TeV photons implied by the MAGIC observation and the large redshift of the source.

3 Astrophysical modeling to optimize multi-messenger searches

How effective multi-messenger searches will be in yielding astrophysical insights strongly depends on how well we understand the sources we are searching for, and how much information on these sources we incorporate in searches. Current multi-messenger searches [21, 27] typically build on minimal source assumptions, primarily operating with a coincidence time window. We will examine the constraints and uncertainties of the emission properties for some of the primary multi-messenger sources [7], and design a strategy that optimizes search sensitivity based on our understanding of the emission. In particular, we will study compact binary mergers and core collapse supernovae with rapidly rotating cores. We will focus on (i) constraints on the progenitor from the GW signal, (ii) relative time of emission, and (iii) relative flux of GWs, neutrinos and electromagnetic counterparts.

An important goal is to understand the role of a delay between the merger and the jet launching events. The longer the delay between the merger and the jet launching, the larger and more massive is the polluted region [35, 36]. Understanding the consequence of the jet-ejecta interaction on the radiation properties can potentially set constraints on the length of the delay. This can give valuable constraints to GW searches as well as improving our understanding of the launching of relativistic jets from compact binary merger remnants, i.e. newly formed black holes or long-lived supramassive neutron stars.

We expect that the sources associated with GW detections may produce neutrinos in the IceCube energy range. Moreover these sources are expected to be at are low to intermediate redshift, it is not unreasonable to surmise that meaningful neutrino detections or upper limits will be provided by the IceCube experiment [37]. We will estimate the expected relative luminosities of GWs, neutrinos and electromagnetic counterparts for different source models. This will allow multi-messenger searches to use the observed flux in one messenger to calculate a flux prior in another messenger.

4 Hidden cosmic accelerators: gravitational wave and neutrino emission scenarios

We will revise the expected high-energy neutrino flux, along with GW emission models, from hidden sources that are opaque in gamma-rays and therefore are consistent with the IceCube-Fermi constraints. In particular we they are exwill consider transient neutrino sources, choked GRBs [38] and low luminosity GRBs [39].

We will also consider emission in jets driven by compact binary mergers that interact with sub-relativistic dynamical/wind ejecta from the merger, which can result in attenuated gamma emission prior to the jet burrowing through the slower ejecta, while trans-ejecta neutrinos can escape [40].

Merging binary systems containing two compact objects, i.e. a double neutron star, or a stellar-mass black hole and neutron star are hypothesized to be the progenitors of short GRBs. These mergers are also powerful GW sources within the LIGO sensitive band.

The ratio of short GRBs to GW events depends on the fraction of mergers that produce gamma-rays, but also on the beaming factor of gamma-ray emission with respect to the more isotropic GW emission. The observational connection between observed GW events and short GRBs therefore can provide constraints on the physical mechanism and nature of the GRB jetted emission.

5 Conclusions and plan for the future

The study of common sources of GWs, neutrinos and gamma-rays requires a broad understanding of the emission processes and detection technique. It is fundamental that multimessenger observations are performed following a more clear theoretical framework, instead of combined from different studies that have a diverse set of assumptions. How effective multimessenger searches will be in yielding astrophysical insights strongly depends on how well we understand the sources we are searching for, and how much information on these sources we incorporate in searches.

The studies described in this paper will build on observations by the Advanced LIGO [5], Advanced Virgo [41], KAGRA [42], IceCube [37], ANTARES [43], Fermi [44, 45] and Swift [46] detectors, and potentially others, as well as information from future detectors, namely KM3NeT [4], ISS-TAO [47], and ULTRASAT [48].

We plan to derive quantitative models for some of the most promising multi-messenger transient sources, namely binary neutron star and neutron star–black hole mergers (presumably short GRBs) and core-collapse supernovae with rapidly rotating cores (long GRBs), delayed neutrino emission from core-collapse supernovae.

We plan to renew the model of hidden or faint gamma-ray sources, in particular choked/low-luminosity/trans-ejecta GRBs. We will consider the possibility of choked short GRBs by examining the prospects of these sources for joint GW+neutrino observations.

We will also expand on the recent models of neutrino production in jets of binary neutron star mergers [40, 49–51], in particular by studying the effect of the sub-relativistic ejecta and viewing angle on detectability. This scenario became particularly interesting since the multi-messenger discovery of binary neutron star merger GW170817 [49, 52, 53]. Future observations of similarly nearby mergers whose jets contain a significant fraction of hadrons and that are beamed towards the Earth are expected to be detectable by IceCube for source directions on the northern hemisphere [40].

We will revise neutrino emission models from these sources and derive their detectability with km-scale neutrino detectors. We will combine this information with the detectability of GWs from the same sources to quantitatively estimate the prospects of multi-messenger observations.

References

[1] K. Bechtol, M. Ahlers, M. Di Mauro, M. Ajello, J. Vandenbroucke, ArXiv e-prints (2015), 1511.00688
[2] K. Murase, D. Guetta, M. Ahlers, Physical Review Letters **116**, 071101 (2016), 1509.00805
[3] IceCube-Gen2 Collaboration, :, M.G. Aartsen, M. Ackermann, J. Adams, J.A. Aguilar, M. Ahlers, M. Ahrens, D. Altmann, T. Anderson et al., ArXiv e-prints (2014), 1412.5106
[4] U.F. Katz, Nuclear Instruments and Methods in Physics Research A **567**, 457 (2006), astro-ph/0606068
[5] J. Aasi, B.P. Abbott, R. Abbott, T. Abbott, M.R. Abernathy, K. Ackley, C. Adams, T. Adams, P. Addesso, et al., CQG **32**, 074001 (2015), 1411.4547
[6] B.P. Abbott, R. Abbott, T. Abbott, M.R. Abernathy, F. Acernese, K. Ackley, C. Adams, T. Adams, P. Addesso, et al., Phys. Rev. Lett. **116**, 061102 (2016), 1602.03837
[7] S. Ando, B. Baret, I. Bartos, B. Bouhou, E. Chassande-Mottin, A. Corsi, I. Di Palma, A. Dietz, C. Donzaud, D. Eichler et al., Reviews of Modern Physics **85**, 1401 (2013), 1203.5192
[8] B.P. Abbott, R. Abbott, T.D. Abbott, M.R. Abernathy, F. Acernese, K. Ackley, C. Adams, T. Adams, P. Addesso, R.X. Adhikari et al., Living Reviews in Relativity **19** (2016), 1304.0670
[9] I. Bartos, P. Brady, S. Márka, Classical and Quantum Gravity **30**, 123001 (2013), 1212.2289
[10] K. Kashiyama, K. Murase, I. Bartos, K. Kiuchi, R. Margutti, Astrophys. J. **818**, 94 (2016), 1508.04393
[11] K. Murase, K. Kashiyama, K. Kiuchi, I. Bartos, Astrophys. J. **805**, 82 (2015), 1411.0619
[12] C.D. Ott, Classical and Quantum Gravity **26**, 063001 (2009), 0809.0695
[13] A.L. Piro, E. Thrane, Astrophys. J. **761**, 63 (2012), 1207.3805
[14] L. Bildsten, Astrophys. J. Lett. **501**, L89 (1998), astro-ph/9804325
[15] B. Abbott, R. Abbott, R. Adhikari, P. Ajith, B. Allen, G. Allen, R. Amin, S.B. Anderson, W.G. Anderson, M.A. Arain et al., Physical Review Letters **101**, 211102 (2008), 0808.2050
[16] A. Corsi, B.J. Owen, Phys. Rev. D. **83**, 104014 (2011), 1102.3421
[17] D. Murphy, M. Tse, P. Raffai, I. Bartos, R. Khan, Z. Márka, L. Matone, K. Redwine, S. Márka, Phys. Rev. D. **87**, 103008 (2013), 1302.3915
[18] M.W.E. Smith, D.B. Fox, D.F. Cowen, P. Mészáros, G. Tešić, J. Fixelle, I. Bartos, P. Sommers, A. Ashtekar, G. Jogesh Babu et al., Astroparticle Physics **45**, 56 (2013), 1211.5602
[19] J. Piscionere, S. Marka, P.S. Shawhan, J. Kanner, T.L. Huard, D.C. Murphy, *LOOC UP: Locating and Observing Optical Counterparts to Unmodeled Pulses in Gravitational Waves*, in *American Astronomical Society Meeting Abstracts* (2007), Vol. 39 of *Bulletin of the American Astronomical Society*, p. 910
[20] J. Kanner, T.L. Huard, S. Márka, D.C. Murphy, J. Piscionere, M. Reed, P. Shawhan, Classical and Quantum Gravity **25**, 184034 (2008), 0803.0312
[21] B.P. Abbott, R. Abbott, T.D. Abbott, M.R. Abernathy, F. Acernese, K. Ackley, C. Adams, T. Adams, P. Addesso, R.X. Adhikari et al., Astrophys. J. Lett. **826**, L13 (2016), 1602.08492
[22] B. Baret, I. Bartos, B. Bouhou, A. Corsi, I.D. Palma, C. Donzaud, V.V. Elewyck, C. Finley, G. Jones, A. Kouchner et al., Astroparticle Physics **35**, 1 (2011), 1101.4669
[23] B. Baret, I. Bartos, B. Bouhou, E. Chassande-Mottin, A. Corsi, I. Di Palma, C. Donzaud, M. Drago, C. Finley, G. Jones et al., Phys. Rev. D. **85**, 103004 (2012), 1112.1140
[24] I. Bartos, C. Finley, A. Corsi, S. Márka, Physical Review Letters **107**, 251101 (2011), 1108.3001

[25] S. Adrián-Martínez, I.A. Samarai, A. Albert, M. André, M. Anghinolfi, G. Anton, S. Anvar, M. Ardid, T. Astraatmadja, J.J. Aubert et al., JCAP **6**, 008 (2013), 1205.3018

[26] M.G. Aartsen, M. Ackermann, J. Adams, J.A. Aguilar, M. Ahlers, M. Ahrens, D. Altmann, T. Anderson, C. Arguelles, T.C. Arlen et al., Phys. Rev. D. **90**, 102002 (2014), 1407.1042

[27] S. Adrián-Martínez, A. Albert, M. André, M. Anghinolfi, G. Anton, M. Ardid, J.J. Aubert, T. Avgitas, B. Baret, J. Barrios-Martí et al., Phys. Rev. D. **93**, 122010 (2016), 1602.05411

[28] The Pierre Auger Collaboration, A. Aab, P. Abreu, M. Aglietta, I.A. Samarai, I.F.M. Albuquerque, I. Allekotte, A. Almela, J. Alvarez Castillo, J. Alvarez-Muñiz et al., ArXiv e-prints (2016), 1608.07378

[29] KamLAND Collaboration, A. Gando, Y. Gando, T. Hachiya, A. Hayashi, S. Hayashida, H. Ikeda, K. Inoue, K. Ishidoshiro, Y. Karino et al., ArXiv e-prints (2016), 1606.07155

[30] C. Kopper and E. Blaufuss on behalf of the IceCube Collaboration, GCN 21916 (2017)

[31] S. Paiano, R. Falomo, A. Treves, R. Scarpa, Astrophys. J. Lett. **854**, L32 (2018), 1802.01939

[32] Y.T. Tanaka, S. Buson, D. Kocevski on behalf of the Fermi-LAT Collaboration, ATel #10791 (2017)

[33] P.A. Evans, A. Keivani, J.A. Kennea, D.B. Fox, D.F. Cowen, J.P. Osborne, and F.E. Marshall on behalf of the Swift-IceCube collaboration, ATel #10792 (2017)

[34] R. Mirzoyan for the MAGIC Collaboration, ATel #10817 (2017)

[35] A. Murguia-Berthier, G. Montes, E. Ramirez-Ruiz, F. De Colle, W.H. Lee, Astrophys. J. Lett. **788**, L8 (2014), 1404.0383

[36] J. Granot, D. Guetta, R. Gill, Astrophys. J. Lett. **850**, L24 (2017), 1710.06407

[37] A. Karle, J. Ahrens, J.N. Bahcall, X. Bai, T. Becka, K.H. Becker, D.Z. Besson, D. Berley, E. Bernardini, D. Bertrand et al., Nuclear Physics B Proceedings Supplements **118**, 388 (2003), astro-ph/0209556

[38] P. Mészáros, E. Waxman, Physical Review Letters **87**, 171102 (2001), astro-ph/0103275

[39] K. Murase, K. Ioka, S. Nagataki, T. Nakamura, Astrophys. J. Lett. **651**, L5 (2006), astro-ph/0607104

[40] S.S. Kimura, K. Murase, I. Bartos, K. Ioka, I.S. Heng, P. Mészáros, ArXiv e-prints (2018), 1805.11613

[41] F. Acernese, M. Agathos, K. Agatsuma, D. Aisa, N. Allemandou, A. Allocca, J. Amarni, P. Astone, G. Balestri, G. Ballardin et al., Classical and Quantum Gravity **32**, 024001 (2015), 1408.3978

[42] Y. Aso, Y. Michimura, K. Somiya, M. Ando, O. Miyakawa, T. Sekiguchi, D. Tatsumi, H. Yamamoto (The KAGRA Collaboration), Phys. Rev. D **88**, 043007 (2013)

[43] M. Ageron, J.A. Aguilar, I. Al Samarai, A. Albert, F. Ameli, M. André, M. Anghinolfi, G. Anton, S. Anvar, M. Ardid et al., Nuclear Instruments and Methods in Physics Research A **656**, 11 (2011), 1104.1607

[44] C. Meegan, G. Lichti, P.N. Bhat, E. Bissaldi, M.S. Briggs, V. Connaughton, R. Diehl, G. Fishman, J. Greiner, A.S. Hoover et al., Astrophys. J. **702**, 791-804 (2009), 0908.0450

[45] M. Ackermann, M. Ajello, A. Albert, A. Allafort, W.B. Atwood, M. Axelsson, L. Baldini, J. Ballet, G. Barbiellini, D. Bastieri et al., Astrophys. J. Supp. Ser. **203**, 4 (2012), 1206.1896

[46] N. Gehrels, *The Swift Gamma-Ray Burst Mission*, in *5th INTEGRAL Workshop on the INTEGRAL Universe*, edited by V. Schoenfelder, G. Lichti, C. Winkler (2004), Vol. 552 of *ESA Special Publication*, p. 777

[47] J. Camp, S. Barthelmy, L. Blackburn, K.G. Carpenter, N. Gehrels, J. Kanner, F.E. Marshall, J.L. Racusin, T. Sakamoto, Experimental Astronomy **36**, 505 (2013), 1304.3705

[48] I. Sagiv, A. Gal-Yam, E.O. Ofek, E. Waxman, O. Aharonson, S.R. Kulkarni, E. Nakar, D. Maoz, B. Trakhtenbrot, E.S. Phinney et al., Astron. J. **147**, 79 (2014), 1303.6194

[49] A. Albert, M. André, M. Anghinolfi, M. Ardid, J.J. Aubert, J. Aublin, T. Avgitas, B. Baret, J. Barrios-Martí, S. Basa et al., Astrophys. J. Lett. **850**, L35 (2017), 1710.05839

[50] D. Biehl, J. Heinze, W. Winter, Mon. Not. Roy. Astron. Soc. **476**, 1191 (2018), 1712.00449

[51] S.S. Kimura, K. Murase, P. Mészáros, K. Kiuchi, Astrophys. J. Lett. **848**, L4 (2017), 1708.07075

[52] LIGO and Virgo Collaborations, Phys. Rev. Lett. **119**, 161101 (2017)

[53] B.P. Abbott et al., Astrophys. J. Lett. **848**, L12 (2017)

Measurement of the cosmic ray Moon shadow with the ANTARES detector.

Matteo Sanguineti[1,2,a] and Tommaso Chiarusi[3,b] on behalf of ANTARES collboration

[1] *INFN - Sezione di Genova, Via Dodecaneso 33, 16146 Genova, Italy*
[2] *Dipartimento di Fisica dell'Universita di Genova, Via Dodecaneso 33, 16146 Genova, Italy*
[3] *INFN - Sezione di Bologna, Viale Berti-Pichat 6/2, 40127 Bologna, Italy*

Abstract. The ANTARES detector is the largest neutrino telescope currently in operation in Mediterranean sea. One of the main goals of the ANTARES telescope is the search for point-like neutrino sources, so both the pointing accuracy and the angular resolution of the detector need a proper direct estimation. One possibility to evaluate the pointing performance of the detector is to analyse the shadow of the Moon, i.e. the deficit in the atmospheric muon flux in the direction of the Moon induced by absorption of cosmic rays. The ANTARES data taken between 2007 and 2016 shows a Moon shadow evidence of 3.5σ significance. This is the first measurement of the ANTARES angular resolution and absolute pointing for atmospheric muons using a celestial calibration source. The presented results confirm the good pointing performance of the detector as well as the predicted angular resolution.

1 Introduction

The study of the Universe using cosmic neutrino as messanger opened a new era of astrophysics. Neutrinos are chargeless, weakly interacting particles; they can cross radiation fields or dense matter without being absorbed for cosmological distance. Neutrino are a complementary probe with respect to other messengers such as multi-wavelength electromagnetic radiation, charged cosmic rays and gravitational waves allowing thus the observation of the interior of the astrophysical sources and the far Universe. The ANTARES detector (Astronomy with a Neutrino Telescope and Abyss environmental RESearch) [1] is the largest neutrino telescope currently operating in sea water. The pointing accuracy and the angular resolution of the neutrino telescopes are critically important for the detection of point-like sources and a proper way to evaluate these performances is needed. Figures of merit depend on the relative delays among the optical sensors, but also on the instantaneous shape and absolute orientation of the detector lines.

In this measurement the Moon absorption of the cosmic rays is exploited to evidence a deficit in the event density of the atmospheric muon flux corresponding to the direction of the Moon disk in the ANTARES detector. The detection of this deficit - the so-called Moon shadow - and its position in the sky provides the measurements of the detector angular resolution for atmospheric down-going

[a] e-mail: matteo.sanguineti@ge.infn.it
[b] e-mail: tommaso.chiarusi@bo.infn.it

muons and the detector absolute pointing capability. The energy threshold for muons detectable at the depth of the ANTARES telescope is about 500 GeV when they are at the sea surface level, most of them with energy above 1 TeV. Primaries which are progenitors of such highly energetic muons are practically not affected by the Earth geomagnetic field. This assumption of large rigidity holds also for the secondary muons detected by the ANTARES detector, thus they can be exploited in the study of the Moon shadow without introducing any bias.

2 Moon shadow analysis optimization

The optimization of the selection criteria applied to the reconstructed muon tracks has been performed using a dedicated Monte Carlo (MC) simulation with the MUPAGE code [2], it includes both single and bundle muon events. The Monte Carlo comprises the generation of Cherenkov light emitted by the muon tracks, its propagation up to the detector, the optical background caused by bioluminescence and the radioactive isotopes and the detector response. Finally the hits detected by the telescope are used to reconstruct the atmospheric muon tracks with a robust track fitting procedure based on a maximisation likelihood method [3].

The Moon shadow effect is simulated rejecting the muons generated within the Moon disk, having an angular radius $R_{Moon} = 0.26°$. In this analysis two different Monte Carlo simulation sets were performed: one considering the shadowing effect of the Moon (H_1 hypothesis) and the other without this effect (H_0 hypothesis). In order to measure the shadowing effect, the region around the Moon centre is divided in concentric rings with increasing radius. The event density of each ring is computed as the number of events detected in that sector over the surface of the ring. Event tracks reconstructed as down-going and detected when the Moon is above the Horizon are selected. A test statistic function λ is defined to evaluate the statistical significance of the Moon shadow effect:

$$\lambda = \chi^2_{H_1} - \chi^2_{H_0} \quad with \quad \chi^2_H = 2\sum_{i=1}^{N_{bin}} \left[N_{i,H} - n_i + n_i \ln \frac{n_i}{N_{i,H}} \right], \tag{1}$$

where n_i stands for the measurement in the i-th bin to be compared with the expectations $N_{i,H}$ under the H_0 and H_1 hypotheses.

Using the two Monte Carlo sets mentioned above, 10^6 pseudo-experiments were generated in order to obtain the distribution of the variable λ in the case of Moon shadow effect or vice versa (Figure 1).

The dashed area below the dotted curve represents the fraction of pseudo-experiments where the Moon shadow hypothesis is correctly identified; the filled-coloured area below the smooth curve corresponds to a p-value equal to 3.6×10^{-4}, or equivalently $3.4\,\sigma$. This is the expected median significance of the Moon shadow effect with the optimized selection criteria.

3 Results

The measured muon event density as a function of the angular distance δ from the Moon is reported below for the data sample of the years 2007-2016, selecting $9.6 \cdot 10^5$ events out of the pre-cuts collection about three times larger. (Figure 2). Assuming a Gaussian shape for the detector point spread function, it is possible to evaluate the detector angular resolution fitting the event density with the formula:

$$\frac{dn}{d\delta^2} = k(1 - \frac{R^2_{Moon}}{2\sigma^2_{res}} e^{-\frac{\delta^2}{2\sigma^2_{res}}}). \tag{2}$$

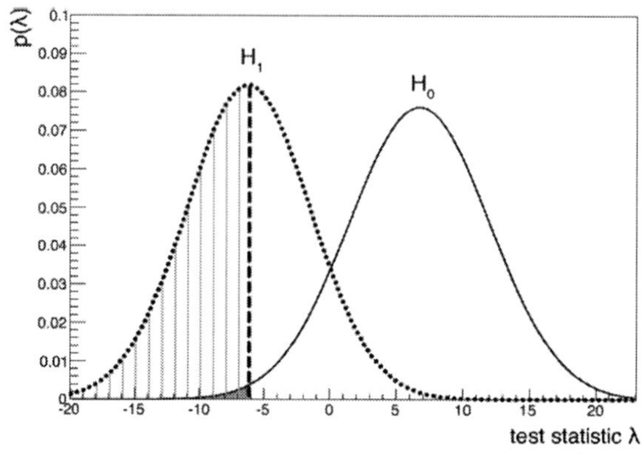

Figure 1. The test statistics λ distribution for the "Moon shadow" hypothesis H_1 (dotted curve) and the "no Moon shadow" hypothesis H_0 (smooth curve). The dashed area corresponds to the 50% of the pseudo-experiments where the Moon shadow hypothesis is correctly identified. The shaded area quantifies the expected median significance (here 3.4 σ) to observe the Moon shadow.

where δ is the angular distance from the Moon centre. The two free parameters are k, the average muon event density in the off-source region, and σ_{res}, the angular resolution. The Moon radius R_{Moon} is fixed to 0.26°.

The angular resolution for downward-going atmospheric muons resulting from the fit is σ_{res} = 0.73°±0.14°. The significance of the shadowing is evaluated using a χ^2 test comparing the measured event density with the flat distribution $\frac{dn}{d\delta^2} = k$. Such χ^2 test leads to a significance of the Moon shadow effect of 3.3 σ.

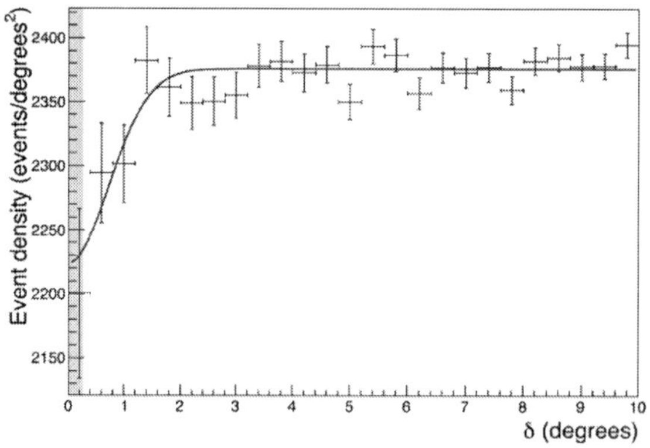

Figure 2. Measured muon event density as a function of the angular distance δ from the Moon. The smooth line is the best fit according to equation (2); the shaded area corresponds to the apparent radius of the Moon (0.26°).

The procedure for evaluating the pointing accuracy of the Moon shadow is based on determining the statistical significance of the selected data set under the assumption of the Moon in a given direction. In this study the field of view is limited in both x and y within the range $[-10°, 10°]$, and it is subdivided in a grid of $0.2° \times 0.2°$ squared bins. The used test statistic is again λ as reported in equation (1), but now the sum is evaluated on all 100×100 square bins. For each bin in the field of view, the value of the test statistic λ is minimised finding the best fit of a bi-dimensional Gaussian point spread function (Figure 3-left) [4]. The smallest value of λ is found equal to -17.05 in the bin of coordinates $(0.5°, 0.1°)$, this value corresponds to a significance of 3.5 σ of Moon shadow effect. The interval corresponding to a desired confidence level (CL) (Figure 3-right) can be directly obtained from Figure 3-left accoding to [5].

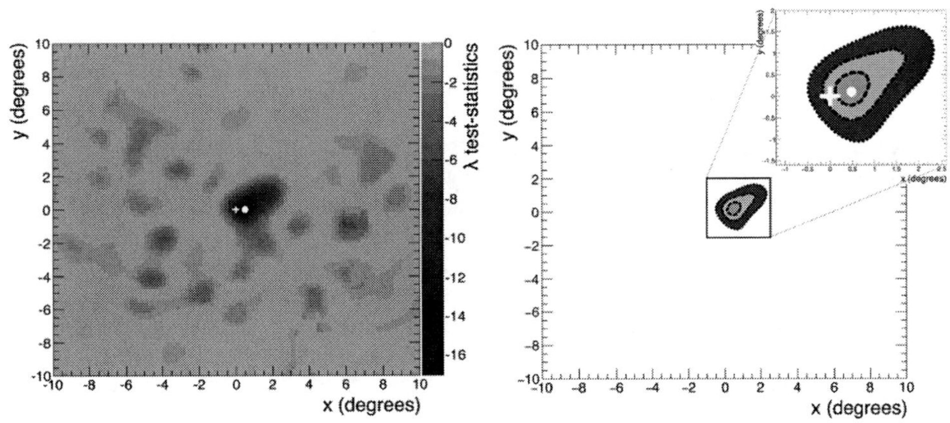

Figure 3. Left: measured distribution of the test statistic λ from equation (1) in the field of view around the Moon nominal position $O \equiv (0°, 0°)$, indicated by a white cross; the white dot refers to the coordinates $(0.5°, 0.1°)$ where the test statistics reaches the minimum. Right: contour plots corresponding to different confidence levels (cyan/dashed: 68.27%; green/dot-dashed: 95.45%; red/dotted: 99.73%); in the zoom the dot represents the position in the FoV where λ is minimum, the cross indicates the nominal position of the Moon.

4 Conclusions

The pointing performance of the ANTARES telescope has been estimated using the Moon shadow effect. The analysis of the ANTARES data sample of the years 2007-2016 allowed the identification of the Moon shadow effect with 3.5σ significance and the detector angular resolution for downward-going atmospheric muons of $0.73° \pm 0.14°$. The pointing accuracy of the detector is consistent with the expectations.

References

[1] M. Ageron et al. (ANTARES), Nucl. Instrum. Meth. **A656**, 11 (2011), 1104.1607
[2] G. Carminati, A. Margiotta, M. Spurio, Comput. Phys. Commun. **179**, 915 (2008), 0802.0562
[3] S. Adrian-Martinez et al. (ANTARES), JCAP **1303**, 006 (2013), 1302.6750
[4] A. Albert et al., The European Physical Journal C **78**, 1006 (2018)
[5] G. Cowan, *Statistical Data Analysis* (Clarendon (Oxford), Oxford, 1998 (ISBN: 0-19-850156-0))

The Cherenkov Telescope Array

Science Goals and Current Status

The CTA Consortium[1,a], represented by Rene A. Ong[2,b]

[1] *See https://www.cta-observatory.org/consortium_authors/authors_2018_09.html*
[2] *Department of Physics and Astronomy, University of California, Los Angeles CA, 90095, USA*

Abstract. The Cherenkov Telescope Array (CTA) is the major ground-based gamma-ray observatory planned for the next decade and beyond. Consisting of two large atmospheric Cherenkov telescope arrays (one in the southern hemisphere and one in the northern hemisphere), CTA will have superior angular resolution, a much wider energy range, and approximately an order of magnitude improvement in sensitivity, as compared to existing instruments. The CTA science programme will be rich and diverse, covering cosmic particle acceleration, the astrophysics of extreme environments, and physics frontiers beyond the Standard Model. This paper outlines the science goals for CTA and covers the current status of the project.

1 Introduction

The field of very high-energy gamma-ray (VHE, $E > 20\,\mathrm{GeV}$) astronomy has come of age in recent years, due to a wealth of information from both ground-based and space-borne telescopes. Prior to 2000, the general expectation for the field was to detect the accelerators of the high-energy cosmic rays, most likely supernova remnants (SNRs). However, the reality over the last two decades has been the discovery of an astonishing variety of VHE gamma-ray sources, both within the Milky Way, e.g. SNRs, pulsars, pulsar wind nebulae (PWNe), cosmic ray bombarded molecular clouds, stellar binaries, massive stellar clusters, and the supermassive black hole Sgr A*, and outside, e.g. BL Lac objects, flat-spectrum radio quasars, radio galaxies, starburst galaxies, Milky Way satellites, and gamma-ray bursts (GRBs), see [1]. Most of these discoveries have been made by the major ground-based imaging atmospheric Cherenkov telescopes (IACTs), H.E.S.S., MAGIC, and VERITAS, that have been joined in recent years by the air shower array HAWC. Important contributions have also come from the space instruments Fermi-LAT and AGILE.

The recent results have been exciting and have established the field, but there are indications that they represent the tip of the iceberg and that more source classes and many more sources remain to be discovered. In addition, we know that the potential of the imaging atmopsheric Cherenkov technique has not yet been fully exploited and that significant improvements in sensitivity and resolution over existing instruments can be achieved with an array of Cherenkov telescopes that is both larger (to

[a]e-mail: info@cta-observatory.org
[b]e-mail: rene@astro.ucla.edu

provide more views of the shower) and more capable (e.g. with wider field-of-view and higher resolution cameras). Thus, there is both strong scientific and technical motivation for a next-generation observatory that further develops the imaging atmospheric Cherenkov technique.

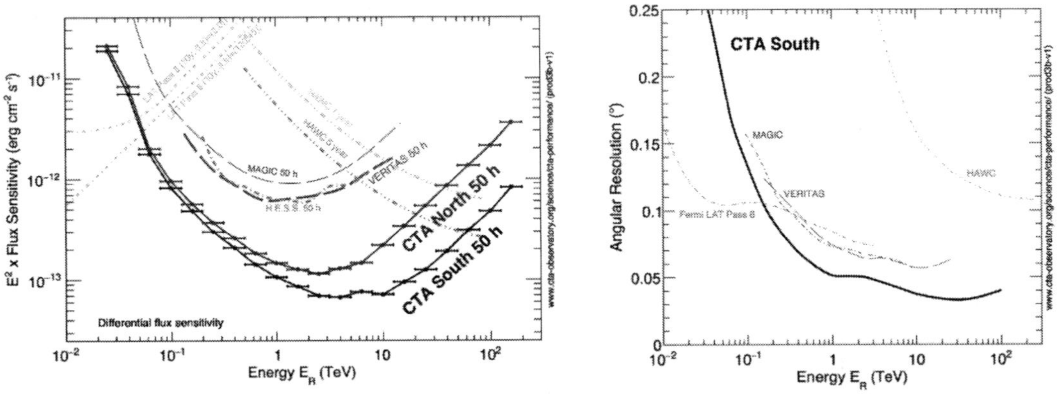

Figure 1. Left: Differential gamma-ray flux sensitivity (multiplied by E^2) for CTA-South and CTA-North as a function of energy. Also shown are curves for the existing IACTs (H.E.S.S., MAGIC and VERITAS), the air-shower experiment HAWC, and the Fermi-LAT space telescope. **Right:** Gamma-ray angular resolution for CTA-South as a function of energy, compared to existing instruments. For details on the sensitivity and angular resolution calculations and comparisons, see [2].

The Cherenkov Telescope Array (CTA) will transform our understanding of the VHE universe and will address important questions in fundamental physics, such as the nature of dark matter. It will also be an important member of the suite of experiments and observatories participating in the expanding areas of multi-wavelength and multi-messenger astronomy. Compared to the existing IACTs, CTA will: 1) cover a wider energy range, 2) have a significantly larger field-of-view, and 3) achieve up to an order of magnitude improvement in sensitivity (see Figure 1). Angular resolution and energy resolution will also be improved and full-sky coverage will be ensured by arrays in both the southern and northern hemispheres. CTA will also be the first open observatory to operate in the VHE waveband, with approximately 50% of the observing time set aside for guest observer proposals and all high-level data available to the public after a proprietary period (typically one year).

CTA has been in development for more than a decade. The concept was originated by the CTA Consortium, currently consisting of around 1,400 scientists from 31 countries. The Consortium has also developed the core science programme and has led the prototyping of telescope hardware. The CTA Observatory (CTAO) has also been established; it is the legal entity that will oversee the construction and operation of the observatory. Given a project the size and scope of CTA, international partnerships and good cooperation between the different stakeholders involved in the project (e.g. funding agencies, national laboratories, and scientific communities) are required.

2 CTA Key Science

The core programme of science for CTA consists of major legacy studies, called Key Science Projects (KSPs), to be carried out by the Consortium using guaranteed time (comprising approximately 40% of the observing time during the first ten-year period). The KSPs have been developed and refined by Consortium members, with extensive input from scientists from outside of CTA. The programme will be refined further as the observatory moves closer to its operational phase. The currently proposed KSPs are described in an extensive document that has been published as a book [3].

The core programme comprises three main scientific themes: ***1) Cosmic Particle Acceleration:*** How and where are particles accelerated, how do they propagate and what is their impact on the environment? ***2) Probing Extreme Environments:*** Understanding the physical processes close to neutron stars and black holes, characterizing relativistic jets, winds and explosions, and exploring radiation fields and magnetic fields in cosmic voids. ***3) Physics Frontiers:*** What is the nature of dark matter and how is it distributed? Do axion-like particles exist? Is the speed of light a constant for high-energy photons? These themes are developed fully within nine KSPs and the Dark Matter Programme. Given space limitations, only highlights of a few KSPs are discussed here; see [3] for more complete information.

2.1 Dark Matter Programme

The existence of dark matter is well established from a variety of astrophysical and cosmological measurements. CTA will search for dark matter via the indirect detection technique, looking for gamma-ray signatures from the annihilation or decay of weakly interacting massive particles (WIMPs). The astrophysical targets for CTA are the Galactic centre halo, dwarf spheroidal galaxies, the Large Magellanic Cloud and galaxy clusters. The programme strategy focuses on maximizing the possibility of a detection. Accordingly, the primary target will be the Galactic centre halo that houses the largest concentration of dark matter in our Galaxy, with secondary data taken on the other targets. With a deep 500 h observation of the Galactic centre halo, CTA will achieve a sensitivity level below the benchmark relic cross-section over a wide WIMP mass range that extends into the multi-TeV region (see Figure 2). Thus, CTA will be an important player in the world-wide search for dark matter, providing unique and complementary information to direct-detection experiments and the Large Hadron Collider.

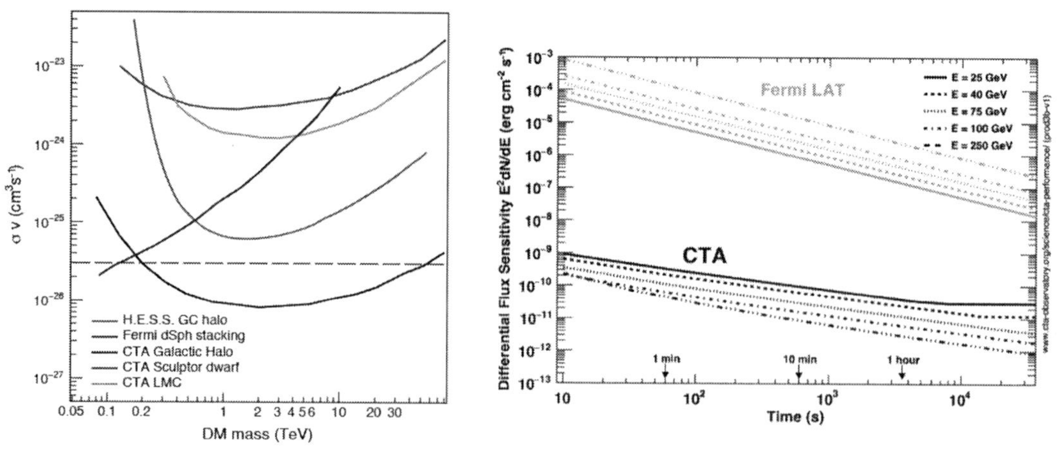

Figure 2. Left: Predicted CTA sensitivity on the velocity-weighted cross section ($\langle \sigma v \rangle$) for dark matter annihilation for various targets, as indicated in the legend. Results from H.E.S.S. on the Galactic centre halo [4] and from Fermi-LAT on dwarf spheroidal galaxies [5] are also shown; see [3] for details. **Right:** Flux sensitivities for CTA and Fermi-LAT at selected energies as a function of observing time; see [2] for details.

2.2 Transients KSP

CTA is a powerful instrument for exploring the transient universe. With its very large collection area and high signal to noise, CTA has outstanding instantaneous sensitivity over a broad energy range, giving it a distinct advantage over Fermi-LAT at lower energies and HAWC at higher energies. For transient time scales of a day or less, CTA is at least several orders of magnitude more sensitive than Fermi-LAT (see Figure 2). CTA's disadvantage is that it has a limited field-of-view compared to Fermi-LAT or HAWC; hence, reaction time for CTA is critical and its transients programme is focused on follow-up observations. The Transients KSP is an integrated programme that encompasses a variety of multi-wavelength and multi-messenger alerts, including explosive transients such as GRBs and gravitational wave events, active galactic nuclei (AGN) flares, Galactic transients, and high-energy neutrino alerts. Rapid dissemination to the wider community of the VHE properties of transients observed by CTA is an important component of the KSP.

2.3 Galactic Plane Survey KSP and Cosmic Ray PeVatrons KSP

Surveys of the Galactic plane are key legacy programmes for all major observatories. Previous surveys in the VHE gamma-ray band have been carried out by both ground-based and space-borne instruments. The CTA Galactic Plane Survey (GPS) will be the first high sensitivity, high angular resolution survey carried out at TeV energies. Using 1600 h of observing time, the GPS KSP will cover the entire Galactic plane with a graded-exposure strategy, emphasizing key regions of the Galaxy in both the southern and northern hemispheres. From simulations and our (incomplete) knowledge of the VHE source population, we expect the GPS to discover many hundreds of new sources, especially PWNe, SNRs, and binaries. The GPS will also have great potential for the discovery of new, unexpected phenomena and will provide a detailed view of diffuse VHE gamma-ray emission. Figure 3 shows a simulated image of the GPS results.

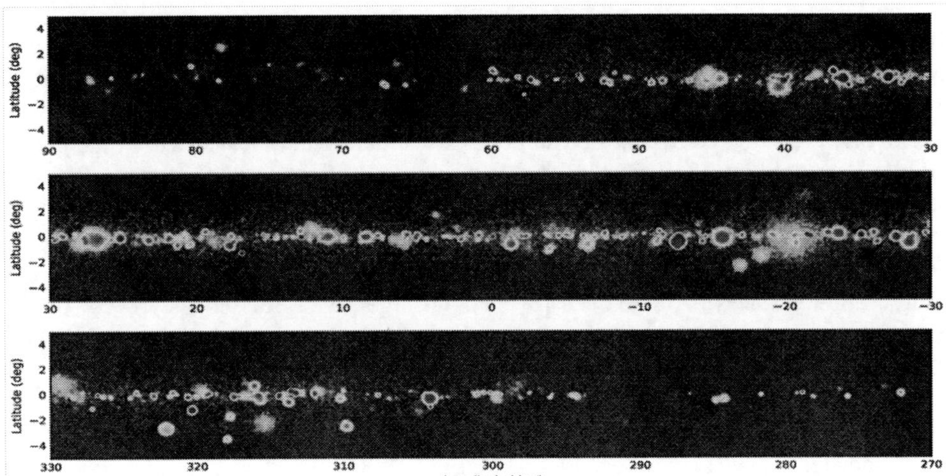

Figure 3. Simulated results from the CTA Galactic Plane Survey in very high-energy gamma rays for half of the plane. The simulation incorporates the proposed GPS observation strategy and a source model containing SNR and PWN populations, as well as diffuse emission.

A long-standing issue in high-energy astrophysics is to understand what sources accelerate cosmic ray nuclei (hadrons) to the knee in the energy spectrum. The standard paradigm involves diffusive shock acceleration in SNRs, but so far only a handful of SNRs provide good evidence for hadronic acceleration and their energy spectra reach only to ~ 10 TeV. The goal of the Cosmic Ray Pevatrons KSP is to determine the sources capable of accelerating hadrons to PeV energies. For this, the GPS will be used as a finder and the five brightest sources with spectral extension out to ~ 100 TeV will be targeted for deep follow-up observations. For any PeVatrons detected, correlation with data from other wavebands will be crucial for clear identification of the physical sources.

2.4 Extragalactic Survey KSP

The motivation for a survey of the extragalactic sky is similar to that for the GPS; we wish to observe a large portion of the sky in an uniform manner to get an understanding of the dominant source population (in this case AGN) and to have sensitivity to completely unexpected phenomena. In the Extragalactic Survey KSP, one-quarter of the sky will be surveyed to a limiting sensitivity of 0.6% of the (standard candle) Crab nebula flux. The survey will be carried out by both CTA-South and CTA-North and will connect to the GPS and cover the important regions of Coma, Virgo, Cen A, and the northern Fermi Bubble. The key science targeted by the survey includes the unbiased determination of the luminosity function for AGN, the discovery of VHE emission from as yet undetected source classes such as Seyfert galaxies and ultraluminous infrared galaxies, and the detection of GRBs in the prompt phase. Studies are ongoing to understand the potential of carrying out the survey using a divergent pointing strategy (in which various telescopes are pointed to sky positions slightly offset from the target position), which would allow for larger instantaneous field-of-view.

3 CTA Design and Current Status

3.1 CTA Design

CTA will consist of two large arrays of imaging atmospheric Cherenkov telescopes. The larger southern array, CTA-South, will be constructed in the Atacama Desert of Chile at an elevation of 2,100 m, close to the existing Paranal Observatory of the European Southern Observatory. CTA-North, will be built at an elevation of 2,200 m on the island of La Palma, Spain, located within the existing Observatorio del Roque de los Muchachos of the Instituto de Astrofisica de Canarias. Two important administrative/scientific centers will be located in Europe: the CTA Headquarters in Bologna, Italy and the CTA Science Data Management Center in Zeuthen, Germany.

In order to optimize the scientific capability within a constrained funding envelope, CTA will employ three different sizes of telescopes [6]. Four Large-Sized Telescopes (LSTs), with mirror diameters of 23 m, will provide the sensitivity in the lowest energies of the CTA waveband (down to 20 GeV), where the Cherenkov photon density in gamma-ray air showers is low (~ $1/\mathrm{m}^2$). Medium-Sized Telescopes (MSTs), with mirror diameters of 9-12 m, will be distributed around the LSTs to provide sensitivity in the core energy range of 100 GeV - 10 TeV. Small-Sized Telescopes (SSTs), with mirror diameters of ~ 4 m, will extend the size of CTA-South and provide sensitivity to energies of 300 TeV and above. SSTs are only envisioned for the southern array for which the entire inner regions of the Galactic plane will be visible. For both the MSTs and the SSTs, prototypes are being developed that use either the conventional single-mirror Davies-Cotton optical design or the newly developed dual-mirror Schwarzschild-Couder optical design. In the baseline CTA design, CTA-South will comprise 4 LSTs, 25 MSTs, and 70 SSTs, covering an area of ~ 5 km^2, and CTA-North will comprise 4 LSTs and 15 MSTs, covering an area of ~ 0.5 km^2. The proposed telescope array layouts, shown in Figure 4, were derived through a lengthy and comprehensive simulation process [7].

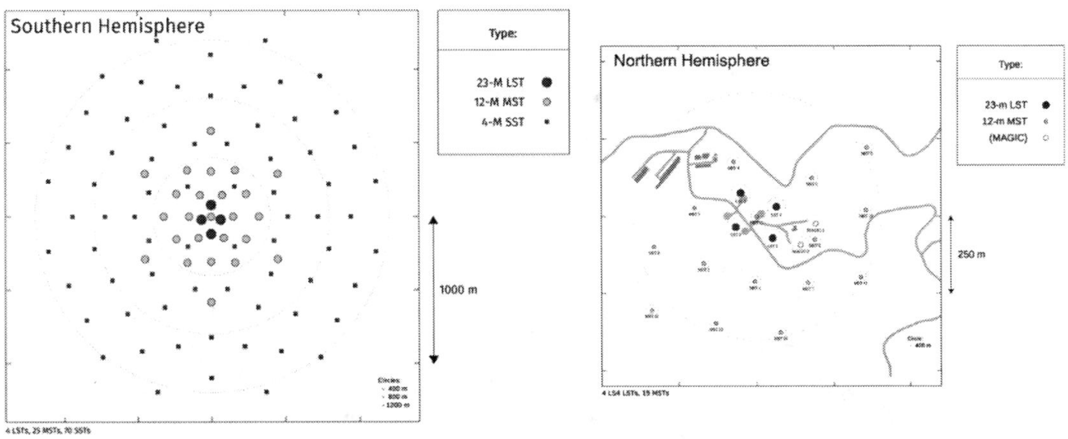

Figure 4. Left: Proposed layout for the three sizes of telescopes (LSTs, MSTs, and SSTs) in CTA-South (Paranal, Chile). **Right:** Proposed layout for the two sizes of telescopes (LSTs and MSTs) in CTA-North (La Palma, Spain). The roads are shown in grey and the buildings in purple. The current MAGIC telescopes are indicated by the open red circles.

3.2 CTA Status

Initial planning for CTA started as early as 2005, when the community came together at Palaiseau, France to discuss the future of ground-based gamma-ray astronomy [8]. Since then, an enormous amount of development work has been carried out and CTA has been heavily reviewed and endorsed by the major roadmap committees in particle physics, astroparticle physics, and astronomy (see, e.g., [9–11]). Most recently, CTA was promoted to Landmark status in the European Strategy Forum on Research Infrastructure (ESFRI) 2018 Roadmap [12]. The management of CTAO has been established. The CTA Council takes care of overall governance and the Project Office manages the construction project. An interim legal entity (CTAO gGmbH), composed of shareholders from eleven countries and associate members from two countries, has been established for the preparation of the implementation of CTA.

The development of telescope hardware is being carried out by the Consortium institutes, as in-kind contributions to the observatory. Substantial software development has also started, both at the Project Office and within Consortium institutes. As shown in Figure 5, six full prototype telescopes exist and are being actively tested. Work is now focused on design optimization, with the goal to reduce the number of telescope and camera options. Site hosting agreements for both CTA sites have been signed and site preparatory work is ongoing. In this context, CTA-North is more advanced since much of the necessary infrastructure already exists. The first telescope to be located at a CTA site is the prototype LST that was inaugurated at the CTA-North site in October 2018 and is now undergoing commissioning. Current plans are: start of construction in 2019, first pre-production telescopes on site in 2020 and completion of construction in 2025. Since the CTA design is quite modular, initial operations can start as early as 2023, using a portion of the array available at either site.

References

[1] TeVCAT: http://tevcat.uchicago.edu/

Figure 5. Concept drawings and photographs of CTA protoype telescopes, as of 2018. Clockwise (from lower left): SST-ASTRI at Mt. Etna, Italy, SST-1M at Krakow, Poland, SST-GCT at Meudon, France, SCT at Whipple Observatory, AZ, USA, MST at Adlershof, Germany and LST-1 at La Palma, Spain.

[2] https://www.cta-observatory.org/science/cta-performance/

[3] The CTA Consortium, *Science with the Cherenkov Telescope Array* (World Scientific Publishing, 2019), ISBN 978-981-3270-08-4, arXiv: 1709.07997, DOI: 10.1142/10986

[4] H. Abdallah, A. Abramowski, F. Aharonian, F. Ait Benkhali, A.G. Akhperjanian, E. Angüner, M. Arrieta, P. Aubert, M. Backes, A. Balzer et al., Physical Review Letters **117**, 111301 (2016), 1607.08142

[5] M. Ackermann, A. Albert, B. Anderson, W.B. Atwood, L. Baldini, G. Barbiellini, D. Bastieri, K. Bechtol, R. Bellazzini, E. Bissaldi et al., Physical Review Letters **115**, 231301 (2015), 1503.02641

[6] B.S. Acharya, M. Actis, T. Aghajani, G. Agnetta, J. Aguilar, F. Aharonian, M. Ajello, A. Akhperjanian, M. Alcubierre, J. Aleksić et al., Astroparticle Physics **43**, 3 (2013)

[7] The CTA Consortium, Astroparticle Physics (2019), submitted

[8] B. Degrange, G. Fontaine, eds., *Proceedings, 7th Workshop on Towards a Network of Atmospheric Cherenkov Detectors 2005*, Ecole Polytechnique (Ecole Polytechnique, Palaiseau, France, 2005)

[9] The AStroParticle ERAnet (ASPERA) roadmap: http://www.aspera-eu.org/

[10] The U.S. Astronomy and Astrophysics Decadal Survey (ASTRO2010): http://www.nationalacademies.org/astro2010

[11] The U.S. HEPAP P5 Panel: http://usaparticlephysics.org/p5/

[12] The ESFRI 2018 Roadmap: http://roadmap2018.esfri.eu/

CaloCube and "Tracker In Calorimeter" projects for the direct measurement of high energy charged astro-particles and gamma rays.

Lorenzo Pacini[1,2,*] and *Nicola* Mori[2,**] for the CaloCube and TIC collaborations

[1]Institute of Applied Physics (IFAC-CNR), Via Madonna del Piano, 10, 50019, Italy
[2]INFN Firenze, via B. Rossi 1, I-50019 Sesto Fiorentino (Firenze), Italy

Abstract. Measurements of high energy cosmic rays in the "knee" region (about 10^{15} eV) are currently available only with ground detectors: new observations of cosmic particles up to these energies with direct measurements are one of the main goals of the next generation space experiments. To achieve those aims, a large acceptance, good energy resolution and particle identification are needed. CaloCube is the design of a space borne calorimeter which is capable to accept particles coming from any direction, increasing the acceptance with respect to traditional telescopes. A good performance for both hadronic and electromagnetic showers is achieved with a 3-D sampling capability: the basic picture of CaloCube is a cubic homogeneous calorimeter which consists of cubic scintillating crystals. MC simulations, concerning different materials and geometrical configurations, and several beam tests with different versions of the CaloCube prototype have been employed to optimize both the detector design and the data analysis method. Taking advantage of the CaloCube project, the space experiment HERD ("High Energy Cosmic Radiation Detection") will include a large acceptance cubic calorimeter with cubic LYSO crystals. It will be installed on-board of the Chinese space station around 2025.
Beside the charged particle observations, high energy gamma-rays provide direct information about the galactic cosmic ray sources. A new project named "Tracker In Calorimeter" (TIC) was approved by INFN in 2017 with the main purpose of the optimization of the calorimeter design for the reconstruction of the gamma-ray direction, without the requirement of additional not homogeneous pre-shower detector. A TIC prototype was recently assembled and tested at the PS-CERN and SPS-CERN accelerators.

1 Introduction

Accurate measurements of high energy astro-particles are needed to provide new information about the origin of cosmic rays (CR), but these are also relevant for the search of dark matter signatures. For example, the presence of possible CR sources nearby the Earth can be investigated observing the CR electron spectrum up to tens of TeV, while the models related to the CR acceleration and propagation can be tuned with new observation of proton and nuclei spectra up to the "knee" region, i.e. about 10^3 TeV per nucleon. Actually CR

*e-mail: l.pacini@ifac.cnr.it
**e-mail: nicola.mori@fi.infn.it

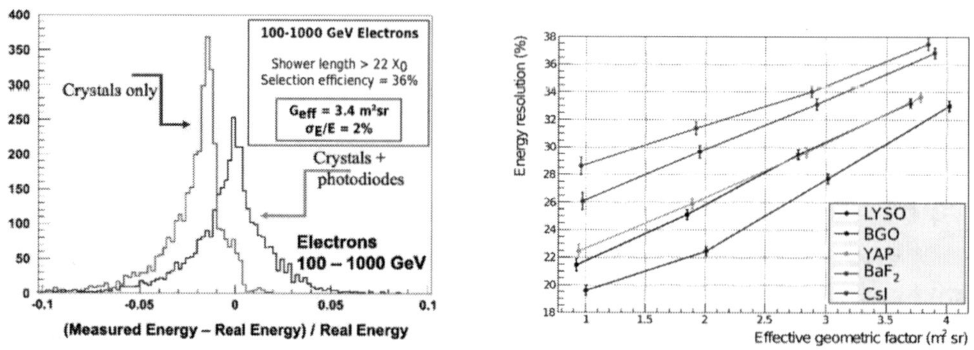

Figure 1. Left: fraction of energy deposited in the calorimeter by electrons from 100 GeV to 1 TeV, taking into account the energy deposited in crystals only (red histogram) or also the contribution of direct ionization on photodiodes (black histogram). Right: energy resolution for protons at 1 TeV by employing different materials and different event selections, corresponding to effective GFs.

observations in these energy ranges are available thanks to indirect measurements with ground detectors. These experiments exploit large geometrical factors (GF) but they are affected by large systematic errors due to the modeling of shower development in the atmosphere.
Space detectors are capable to provide precise direct measurements but they are limited by small acceptances with respect to ground experiments. A GF of about 3 m^2sr with an observation time of 5 years is required to measure protons up to the knee region. Furthermore an energy resolution for protons better than 40% and a good charge resolution, of about 0.3, for nuclei are needed. Finally an high dynamic range of about 10^7 is essential to measure the energy deposit from MIP particle to high energy shower.

2 The CaloCube project.

The goal of the CaloCube project[1] is the optimization of the design of a calorimeter which is capable to fulfill the requirements for next generation CR detectors. CaloCube is an homogeneous cubic calorimeter which detects of incoming particles from all the directions thanks to a 3-D segmentation: this design features a larger acceptance with respect to a typical CR telescope, which can detect particles hitting the detector on top surface only. The project base-line consists of 20x20x20 cubic CsI(Tl) scintillating crystals, with a total thickens of about 30 X_0. The crystal sides are equal to 1 Moliere radius, i.e. 3.6 cm, thus allowing for a good 3-D shower reconstruction. The scintillation light is read-out with a couple of photodiodes (PD) with different active area and connected to a double gain front-end electronics to extend the dynamic range of the detector.
By using the FLUKA package, accurate MC simulations[2] were developed to optimize the design. These take into account instrumental effects like electronic noise and direct ionization of the active area of the PDs. The left panel of figure 1 shows the energy deposited in the calorimeter by electrons simulated in the energy range between 100 GeV and 1 TeV. The red histogram takes into account the energy deposited in the crystals while the black one includes also the signals due to the direct ionization of the PDs. In both cases an energy resolution of about 2% is obtained. A simple event selection is employed in order to include only electrons

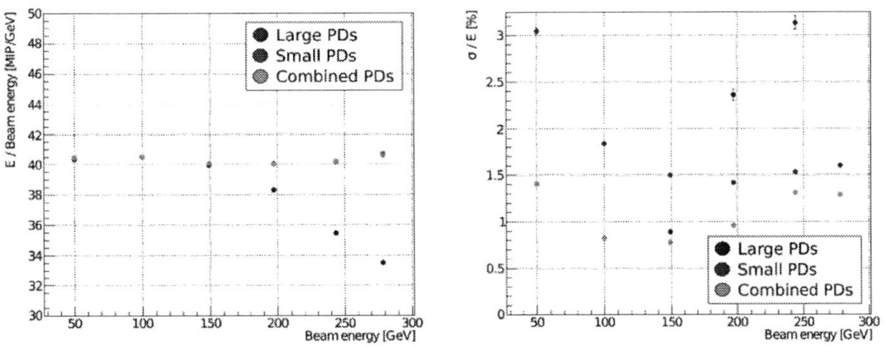

Figure 2. Fraction of the energy deposit in the calorimeter (left) and energy resolution (right) for electrons at different energies.

well contained in the calorimeter: the effective GF, which is the nominal GF multiplied by the selection efficiency, is 3.4 m^2sr for this analysis. The CaloCube performance for hadronic showers has been studied using different materials for the scintillating crystal (CsI(Tl), BaF_2, YAP(Yb), BGO, LYSO(Ce)). For each material, the number of crystals is adjusted in order to keep the total calorimeter weight equal to about 1.6 tons. The energy resolution for protons at 1 TeV obtained with different event selections, corresponding to different effective GFs, is shown in the right panel of figure 1. LYSO is the best candidate mainly thanks to its short hadronic interaction length.
To confirm the MC simulation results a prototype of CaloCube was developed and continuously upgraded[4]. The final version consists of 5x5x18 CsI(Tl) crystals, the scintillation light is read-out by two PDs (VTH2090 and VTP9421) with different active area. The front end-electronics consists of CASIS chips, which are low noise, low consumption, high dynamic range charge amplifiers developed by the INFN[3]. The performance of the prototype was studied with several tests at SPS-CERN by using electron and nuclei beams. Figure 2 shows the mean energy deposit in the calorimeter divided by the beam energy (left panel) and the energy resolution (right panel) obtained with electron beams from 50 GeV to 280 GeV using the large and the small PDs. The green points are obtained using the large PD where when its signal is not saturated and the small PD otherwise. The saturation of the large PDs starts at about 150 GeV but by using the combined information, a resolution better than 1.5% is achieved in the entire energy range, as expected from the FLUKA MC simulations of the prototype. Finally the energy resolution for protons obtained with beam test data is better than 40%.

3 The Tracker In Calorimeter project.

In parallel with observations of high energy charged particle, also gamma-rays detection is relevant: the direction of incoming photons points the CR sources, providing direct and unique information about these. The standard approach of a typical gamma-ray detectors is based on a named tracker-converter with silicon tracking layers interleaved by tungsten layers, e.g. the Fermi-LAT detector[5]. This design achieve a good angular resolution, since it exploits the photon conversion in tungsten, but it limits the acceptance of the detector with respect to the calorimeter GF, as shown in left panel of figure 3. Furthermore the tungsten layers are thick passive material, thus when using the same detector to detect charged CR,

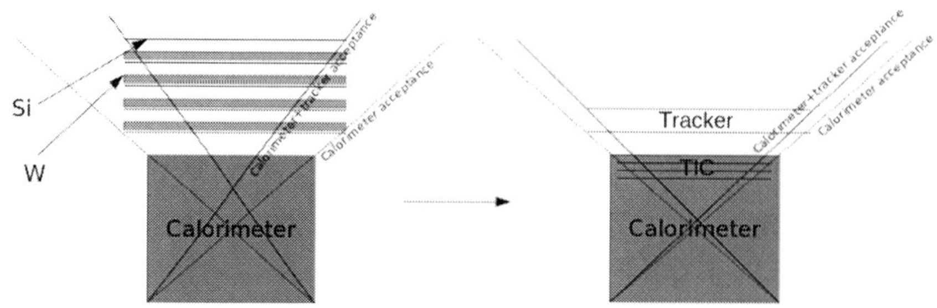

Figure 3. Schematic picture of the standard configuration of a tracker-converter (left) and the TIC basic idea (right). For simplicity only the top face of the instrument is considered in this illustration.

they degrade the energy resolution for low energy particles and the charge resolution for nuclei due to fragmentation in the passive material. Tracker In Calorimeter (TIC) is an *R&D* project approved by INFN in 2017; its main scientific goal is to adding to a detector for charged CRs the possibility to measure also gamma-rays, without compromising the performance for charged particles. The basic idea is to insert silicon micro-strip detectors inside CaloCube and make use of the scintillating crystals as an active converter. By combining the signals of the cubes and the silicon planes the early stage of the electromagnetic shower generated by gamma-rays can be precisely sampled, and the sampling can then be used to reconstruct the arrival direction of the primary photons giving origin to the shower. Some conventional tracking layers are needed outside of the calorimeter to track charged particles and measure the particle charge. This external tracking device can have a lower profile than that of a conventional tracker-converter; furthermore, tungsten is no longer needed. Preliminary MC simulations of several TIC configurations feature an angular resolution better than 0.5° for gamma-rays above 10 GeV. A TIC prototype have been realized integrating several silicon layers inside the CaloCube prototype: prototype has been tested in 2018 at PS-CERN and SPS-CERN accelerators with several electron beams from 1 GeV to 100 GeV. The data analysis is still ongoing.

4 Prospectives of the CaloCube and TIC projects

The idea of a 3-D deep isotropic calorimeter was investigated by the CaloCube collaboration and the good performance of this design is consistent with the one required for the next generation CR experiments. The future space experiment HERD ("High Energy Cosmic Radiation Detection") has chosen the 3-D large acceptance calorimeter design with cubic LYSO crystals. Furthermore if the ongoing data analysis of TIC beam test will confirm competitive performance with that of the tracker-converter design then it will be proposed for being used in HERD.

References

[1] Elena Vannuccini et al., NIM, Section A, 2017, **845** 421-424
[2] O. Adriani et al., Astroparticle Physics, 2017, **96** 11–17
[3] V. Bonvicini et al., IEEE Transactions on Nuclear Science, 2010, **57**, 2963–2970
[4] L. Pacini et al., Journal of Physics: Conference Series, 2017, **928** p 012013
[5] Atwood W et al., The Astrophysical Journal, 2009, **697** 1071

The future of the high energy cosmic ray detection: HERD

Chiara Perrina[1,a] on behalf of the HERD Collaboration

[1] *DPNC, Université de Genève, 24 quai Ernest-Ansermet, CH-1211 Genève 4, Switzerland*

Abstract. The High Energy cosmic-Radiation Detection (HERD) facility will be one of the space astronomy payloads on board the future Chinese space station. The ambitious aim of HERD is the direct detection of cosmic rays towards the "knee" region ($\sim$ 1 PeV), with a detector able to measure electrons, photons and nuclei with an excellent energy resolution (1% for electrons and photons at 200 GeV and 20% for nuclei at 100 GeV - PeV), an acceptance 10 times the one of present generation missions ($\sim$ 1 m^2 sr), and long lifetime ($>$ 10 years). The primary objectives of HERD are the indirect search for dark matter particles and the precise measurement of energy distribution and composition of cosmic rays from 30 GeV up to a few PeV, determining the origin of the "knee" structure of the spectrum. Furthermore, HERD will monitor the high energy gamma-ray sky from 500 MeV, observing gamma-ray bursts, active galactic nuclei, galactic microquasars, etc. HERD will be composed of a homogeneous calorimeter, surrounded by a particle tracker and a plastic scintillator detector. Two possible trackers are under study: a 5-side tracker made of silicon strip detectors and a 4-side scintillating fiber tracker with a silicon strip top tracker. The total volume of HERD will be $(2.3 \times 2.3 \times 2.6)$ m^3 with a weight of about 4 t. The HERD design, perspectives, expected performances in terms of energy sensitivity and acceptance will be presented in this contribution.

1 Introduction

From 2025, the expected launch year, and for at least 10 years, HERD will be observing the high energy sky from the Chinese space station (CSS). Thanks to its unprecedented geometric acceptance at PeV energies, HERD represents the future of the direct high energy cosmic ray detection. An international scientific collaboration is growing around the HERD project: at present, it involves researchers from Chinese institutions lead by CSU and IHEP, Italy, Switzerland and Spain, but more institutes have shown interest and will eventually join the effort.

2 The objectives of HERD

The main objectives of HERD are:

1. The search for signatures of annihilation and decay products of dark matter (DM) both in the energy spectrum and in anisotropy of high energy electrons from 10 GeV to 100 TeV and in the gamma-ray spectrum from 500 MeV to 100 TeV. HERD will have an unprecedented sensitivity

[a] e-mail: chiara.perrina@unige.ch

Table 1. Main requirements for the HERD payload

Energy range (e/γ)	10 GeV - 100 TeV
Low energy range for γ	500 MeV - 10 GeV
Energy range (nuclei)	30 GeV - 3 PeV
Angular resolution (e/γ)	0.1° at 10 GeV
Charge resolution (nuclei)	10% - 15% for $Z = 1 - 26$
Energy resolution (e/γ)	< 1% at 200 GeV
Energy resolution (p)	20% at 100 GeV - PeV
e/p separation power	> 10^5
Geometric factor (e)	> 3 m^2sr at 200 GeV
Geometric factor (p)	> 2 m^2sr at 100 GeV

in searching for DM signals thanks to its large geometric acceptance, excellent energy resolution and long duration. The current generation of missions such as AMS-02 [1] and DAMPE [2] have observed some unexpected features in the positron fraction and in the total electron spectrum that could be related to DM, but could also be produced by nearby pulsars. HERD will perform high statistics and precise measurements in the same energy range and beyond, to more than 10 TeV, of both spectral features and anisotropy, which will definitely resolve the question whether the features are due to nearby sources. Fig. 1 (left) illustrates the expected HERD capability to distinguish if the excess in the total electron flux ($e^- + e^+$) is due to an astrophysical source, like a pulsar wind nebula (PWN), or to DM. Another particularly interesting detection channel is provided by γ-rays. Two-body DM annihilation or DM decay could be observed as a sharp peak over the high energy diffuse γ-ray background. The γ-ray line feature is considered as the "smoking gun" signal of DM. HERD will cover the γ-ray line searches from 10 GeV to 10 TeV with the best sensitivity ever achieved.

2. The measurement of energy spectrum and composition of cosmic rays (protons and nuclei) from 30 GeV to a few PeV that carry the information on the production, acceleration and propagation of these particles, which is fundamental to understand the formation and the evolution of stars and galaxies, and the extreme energetic processes such as supernova explosions, pulsar jets, and neutron star and black hole merging. HERD will measure the spectra of different chemical elements in cosmic rays, from proton to iron, with an unprecedented precision, up to the "knee" region where ground-based experiments have observed a change of spectral index in the all-particle cosmic ray flux, which could indicate the confinement limit of high energy particles by the galactic magnetic field, and the existence of extra-galactic sources. HERD will perform these measurements directly in space, avoiding large systematic effects associated with ground-based experiment, thus helping to elucidate another great puzzle in cosmic ray physics. Fig. 1 (center) shows the simulation of the proton flux as measured by HERD after 5 years.

3. A wide field of view monitoring of γ-rays from 500 MeV to study gamma-ray bursts, active galactic nuclei and galactic microquasars. The monitoring of high energy γ-ray sources in space is very crucial in the era of the multi-messenger and multi-wavelength astronomy, as has been demonstrated by the recent detection by Fermi [3] of the counterpart of the gravitational wave from a neutron star merger detected by LIGO. HERD will play a dominant role in the GeV-TeV γ-ray astronomy in 2025-2035. Without HERD there will be a large gap in the coverage of the electromagnetic spectrum. The sky map expected from HERD after 1 year of activity is illustrated in Fig. 1 (right).

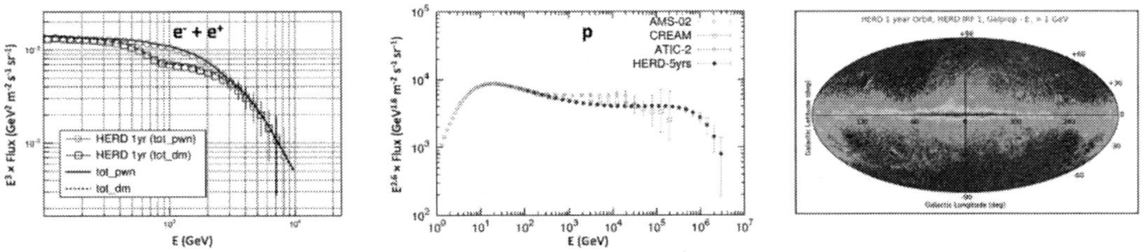

Figure 1. Expected e^- + e^+ flux as measured by HERD with one year data accumulation. HERD will clearly distinguish between a smooth spectrum, due to a PWN, or a spectrum with a "kink", due to DM (left). Expected proton flux as measured by HERD with five years data accumulation. HERD can reach the PeV region with unprecedented statistics (center). Expected HERD's one-year map of the gamma-ray sky (right).

In Table 1 the requirements for HERD in terms of energy range, energy, angular and charge resolution, electron/proton separation and geometric factors are listed. Moreover it is important to note that the geometric factors are $\sim$ 10 times the geometric factors of present detectors, with a field of view of $\pm 70°$.

3 The HERD detector

Fig. 2 shows the exploded view of the HERD detector in which it is possible to see the different sub-detectors. From inside to outside, the HERD detector consists of a calorimeter (CALO) for the measurement of the energy of traversing particles and for the electrons/protons discrimination, a 5-side particle tracker which will serve to reconstruct the particle trajectories and to identify the nuclei charge, a plastic scintillator detector (PSD) for the measurement of low energy γs and for a redundant measurement of the nuclei charge and a transition radiation detector (TRD) to perform the energy calibration. The option with a 4-side scintillating fiber tracker (FIT) and a top silicon strip tracker (STK) is depicted.

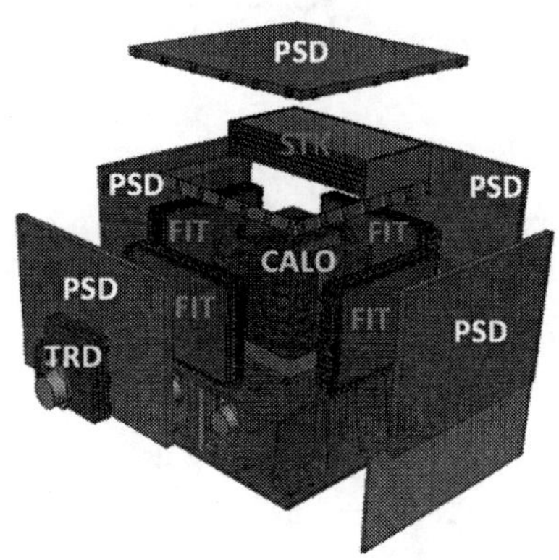

Figure 2. Exploded view of the HERD detector. The whole apparatus will have a weight < 4 t and a volume of $(2.3 \times 2.3 \times 2.6)$ m^3.

3.1 The calorimeter (CALO)

The calorimeter comprises three sub-systems: the crystal array, the intensified scientific CMOS (IsC-MOS) camera and the trigger sub-system. The crystal array (Fig. 3 left) is made of about 7500 cubic crystals with 3 cm side, corresponding to ~ 55 radiation lengths and ~ 3 nuclear interaction lengths in all directions. Cosmic ray measurements up to PeV energy is then made feasible with 20% energy resolution and an e/p separation power of at least 10^5. A novel method of reading the LYSO signals by wavelength shifting fibers (WLSFs) and IsCMOS is adopted for its compactness, simple electronics and external power consumption. For the realization of larger dynamic range, each crystal is coupled with three WLSFs, reshaped into spirals (Fig. 3 center) to get the largest contact area with the crystal. Two of them are for high and low range IsCMOS cameras and the other one for the trigger sub-system. The crystal and fibers are coated with reflectors to increase light output (Fig. 3 right). The complete MIP peak shall be visible in the spectrum: the starting point of the dynamic range of the camera is 1/3 MIP (1 MIP $\simeq$ 30 MeV in a LYSO cube), while the maximum energy deposition in one crystal, calculated on simulation, is about 20 TeV which is created by a PeV hadronic shower. Therefore the dynamic range of the readout system should be about 10^7, thus two different image intensifiers with two different gain settings are needed. All the 7500 trigger fibers are routed to the PMTs of the trigger sub-system, dedicated to providing common trigger signals to all the other instruments after veto and coincidence logic. A beam test of a HERD calorimeter prototype consisting of 250 LYSO crystals was conducted at the CERN SPS H2 and H4 beam lines in 2017. Preliminary results show the good energy linearity of the calorimeter. The energy resolution is better than 1.3% for 200 GeV/c electrons.

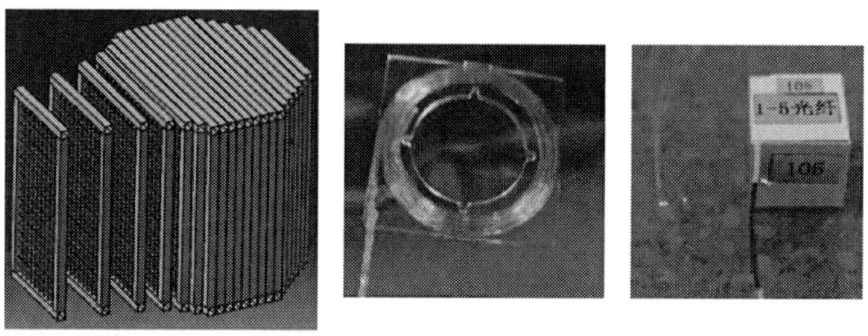

Figure 3. Sketch of the crystal array of the HERD calorimeter. It will consist of 21 layers of cubic LYSO crystals with 3 cm sides (left). Three WLSFs reshaped into spirals are coupled to one side of the crystal (center). The crystal sides are covered with reflectors to increase the light output (right).

3.2 The tracking detector

Two possible tracker options are under study: a 5-side tracker made of silicon strip detectors (SSDs) and a 4-side scintillating fiber tracker with a silicon strip tracker on the top.

3.2.1 STK

The Silicon Tracker (STK) is designed to perform particle track reconstruction and particle charge measurement up to Z = 26. The top STK (Fig. 4 panel 2) is composed of six double planes of silicon detectors, each plane has an active area of 133 cm × 133 cm and is equipped with 28 Type A ladders made of 7 single-sided SSDs each. Ladders on alternate planes are placed orthogonally in order to measure the x and y coordinate of the incoming particle. Lateral STKs are composed of three position-sensitive double planes of silicon detectors. Each plane has an active area of 95 cm × 66.5 cm and is equipped with 10 Type A ladders (Fig. 4 panel 3) or 14 Type B ladders (Fig. 4 panel 4). Each Type B ladder is assembled with 5 single sided SSDs. The material between the active parts will be minimized to ensure the physics performance while keeping the stability of the overall mechanics. The SSD is the key component in STK and the HERD baseline is to use the same SSD as in DAMPE [4] (Fig. 4

panel 1). With a strip pitch of 121 μm and a readout pitch of 242 μm we expect a resolution of about 40 μm after proper alignment. The readout of one every other strip (corresponding to 384 channels per ladder) is a balance among a good performance in terms of spatial resolution and a limited power consumption.

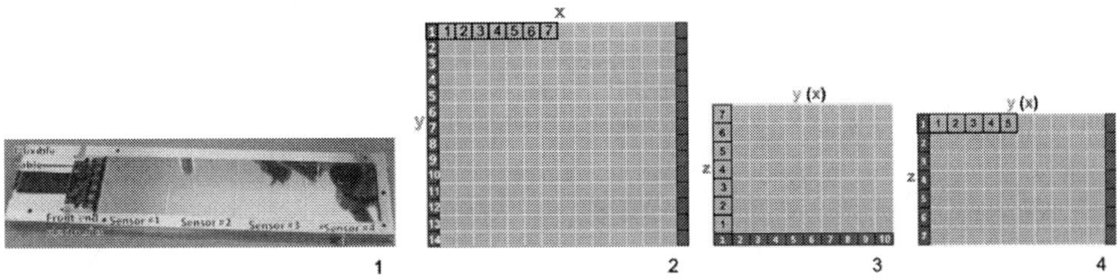

Figure 4. DAMPE ladder equipped with 4 SSDs made by Hamamatsu with an area of 95 mm × 95 mm (1). Sketch of a top STK plane equipped with 28 Type A ladders (2). Sketch of a lateral STK plane equipped with 10 Type A ladders (3). Sketch of a lateral STK plane equipped with 14 Type B ladders (4).

3.2.2 FIT

For the lateral STK, the requirement of a full acceptance puts very stringent constraints on the placement of front-end and readout electronics. For this reason a scintillating fiber tracker provides a very interesting alternative to a silicon strip tracker since fiber mats can be easily adapted to the needed geometry. There are several motivations to choose a scintillating fiber tracker: it would be simpler to build such a long detector (O(1 m)), the dead area would be reduced (no gap between sensors, no dead area on the silicon detector, front-end and readout electronics placed outside of support trays), it would be less expensive to add more layers, thus improving the point spread function of low energy photons and more robust because wire bonds on the detectors are not needed.

Figure 5. Sketch of the 4-side FIT detector (left). Sketch of one x-y plane (center). Sketch of a FIT module (right).

According to the current design, the FIT will be composed of four identical sectors (Fig. 5 left) equipped with 5 x-y planes (Fig. 5 center). A x plane is made of 7 x-modules, a y plane of 10 y-modules. The module is the elementary unit of the FIT detector (Fig. 5 right). It includes a fiber mat and 3 silicon photomultiplier (SiPM) arrays to read-out the scintillation light. The overall mass will be ~ 250 kg, the overall dimensions $(1.4 \times 1.4 \times 0.9)$ m^3. The fiber mat has two possible lengths: 1.06 m for x-modules and 77 cm for y-modules, and a width of 97.8 mm to match 3 SiPM arrays. The mat is made of 6 layers of fibers. The used fibers are made by Kuraray (SCSF-78MJ type), they have a round section with a diameter of 250 μm and the peak emission is at 450 nm wavelength. Except for lenght and width, the fiber mat is the same used for LHCb fiber tracker upgrade [5]. Also the SiPM arrays so far used for tests are the same used for LHCb experiment: S133552-HRQ type, made by Hamamatsu with 2 chips/array and 64 channels/chip. Each channel contains 4×26 pixels of 57.5 μm × 62.5 μm area, with a gain of 10^6. A different SiPM array produced by Hamamatsu for the University of Geneva is under test. This is the S13552-10 type with 2 chips of 64 channels, customised with 23×163 pixels/channel of 10 μm × 10 μm. The latter would assure the measurement of charged particles up to at least $Z = 19$. A beam test of a FIT module prototype consisting of a 1 m fiber mat coupled to one S133552-HRQ SiPM array was conducted at the CERN SPS H2 beam line in 2017. Preliminary results for 100 GeV pions show a nominal light yield of 16 photoelectrons (Fig. 6 left), a good spatial resolution of 59.6 μm (Fig. 6 right, 70 μm would be sufficient for our application because it guarantees an angular resolution of 0.1°) and a hit efficiency of 99.7%.

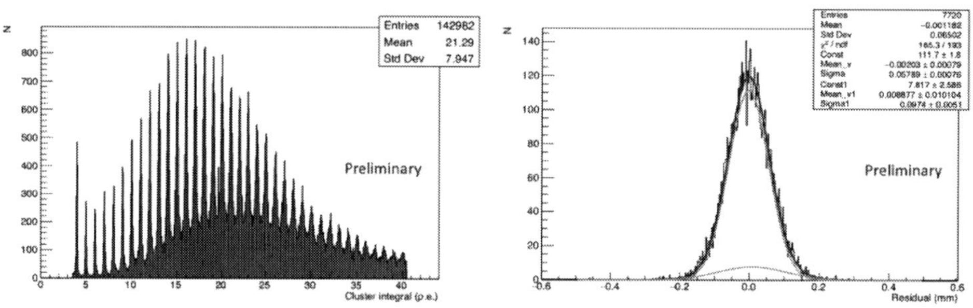

Figure 6. Beam test results for a FIT prototype module made of 1 m length fiber mat and one SiPM array: light yield (left) and hit residual distribution with double Gaussian fits (right).

An option to use 10 x-y planes for the fiber tracker has been proposed to provide unprecedented sensitivity to gamma-rays of 100 MeV to a few GeV.

4 The role of HERD in the multi-messenger astronomy

HERD with its large field of view and unique energy coverage will play a unique and complementary role in multi-wavelength studies across the electromagnetic spectrum with other space and ground telescopes involving radio, optical, X-ray, γ-ray, and in the search for the electromagnetic counterpart of gravitational waves and for contemporaneous flaring in photon data and neutrinos (IceCube, KM3Net). Thanks to the trio HERD + CTA + LHAASO there will be a simultaneous coverage of the same sources from a few GeV to a few PeV, in addition to an overlap of measured spectra. It will be possible to distinguish diffuse emission from localized contributions, to disentangle acceleration and propagation mechanism in SNRs, PWNe, pulsars, and in more extended objects as the Fermi bubbles, to study transient phenomena, which is crucial to analyse the properties of jets, and can help determining the extragalactic background light, intergalactic magnetic fields, and the validity of the Lorentz invariance. Moreover, HERD will produce ToO alerts for AGN, novae, binary systems.

5 Summary and outlook

HERD will operate on the CSS starting from 2025 for more than 10 years. It will be a calorimetric detector with unprecedented acceptance. It will have important and frontier scientific objectives in dark matter search, cosmic ray physics and gamma-ray astronomy. HERD could become the only space-borne high energy γ-ray detector, once the Fermi satellite will stop its operations. The baseline detector is defined and fulfils the requirements. To optimise the detector geometry a dedicated Monte Carlo simulation is being developed, based on the Geant4 framework. Several studies for further improvements and optimization are ongoing.

References

[1] M. Aguilar et al. (AMS), Phys. Rev. Lett. **113**, 221102 (2014)
[2] G. Ambrosi et al. (DAMPE), Nature **552**, 63 (2017), 1711.10981
[3] M. Ajello et al. (Fermi-LAT), Astrophys. J. **861**, 85 (2018)
[4] A. Tykhonov et al. (DAMPE), Nucl. Instrum. Meth. **A893**, 43 (2018), 1712.02739
[5] P. Hopchev (LHCb SciFi Tracker), *SciFi: A large Scintillating Fibre Tracker for LHCb*, in *5th Large Hadron Collider Physics Conference (LHCP 2017) Shanghai, China, May 15-20, 2017* (2017), 1710.08325

EPJ Web of Conferences **209**, 01041 (2019) https://doi.org/10.1051/epjconf/201920901041
RICAP-18

Helium flux and elemental composition of galactic Cosmic Rays with the DAMPE space mission

Margherita Di Santo[1,2,a], on behalf of the DAMPE collaboration

[1] *Dipartimento di Matematica e Fisica "E.De Giorgi", Università del Salento, 73100, Lecce, Italy*
[2] *Istituto Nazionale di Fisica Nucleare (INFN) - Sezione di Lecce, I-73100, Lecce, Italy*

Abstract. DAMPE (DArk Matter Particle Explorer) is a space mission project promoted by the Chinese Academy of Sciences (CAS), in collaboration with Universities and Institutes from China, Italy and Switzerland. The detector is collecting data in a stable sun-synchronous orbit lasting 95 minutes at an altitude of about 500 km. It has been launched in December 17th, 2015, from the Jiuquan Satellite Launch Center, in the Gobi Desert. The main goals of the mission are: indirect search for Dark Matter, looking for signatures in the electron and photon spectra with energies up to 10 TeV; analysis of the flux and composition of primary Cosmic Rays with energies up to hundreds of TeV; high energy gamma-ray astronomy. Preliminary results about the Helium flux and Cosmic Ray composition will be presented and discussed.

1 Introduction

Helium nuclei (He), together with Protons (p), are the most abundant components of Cosmic Rays (CRs). The study of their fluxes as a function of their kinetic energy is crucial to better understand where these particles come from and which are the accelaration and propagation mechanisms that rule their path all around the Galaxy. Several experiments (like AMS [1], CREAM [2], PAMELA [3] and ATIC [4]) found that the measured spectral shape of cosmic He nuclei cannot be well described by a single power law because data exhibit a spectral hardening at energies of hundreds of GeV. This feature of He flux (seen also in the p flux) gives rise to different and complex models which try to clarify the origin and the acceleration and diffusion processes of CRs. Here the DAMPE preliminary results about the elemental composition of CRs and the He flux will be presented and discussed.

2 The DAMPE detector

The DArk Matter Particle Explorer (DAMPE) [5] spacecraft hosts a powerful space telescope consisting of four sub-detectors, as schematically shown in Fig. 1, which are, from the top to the bottom of the satellite, the following:

1) *Plastic Scintillation array Detector* (*PSD*), composed by 82 bars of plastic scintillator assembled on two planes with a double layer configuration. It is used to measure the absolute value of the

[a] e-mail: margherita.disanto@le.infn.it

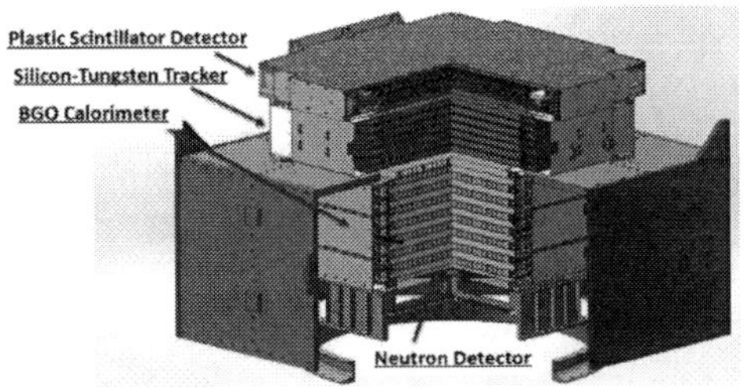

Figure 1. 3D view of the DAMPE detector.

charge of incident high-energy particles (with $Z \leq 28$) with a charge resolution $\leq 25\%$ for $Z = 1$, as well as to provide an anti-coincidence system for γ-rays.

2) *Silicon-Tungsten tracKer-converter* (*STK*), composed by twelve position-sensitive silicon detector planes and three thin layers of tungsten inserted in between the silicon planes (2, 3, 4 and 5) to increase the convertion probability of γ-rays in e^+/e^- pairs. The main purpose of the STK is the trajectory reconstruction of incoming particles and a further measurement of their charge.

3) *BGO calorimeter* (*BGO*), made up of 14 layers with a total area of $60\,\text{cm} \times 60\,\text{cm}$ and a depth of $\sim$ 32 radiation lengths and $\sim$ 1.6 interaction lengths ($\sim$ 40 cm). Each layer constists of 22 crystal bars of *Bismuth germanium oxide* ($Bi_3Ge_4O_{12}$, BGO). It provides the energy measurement, the electron and hadron identification and an additional contribution to the track reconstruction. It guarantees very good performances in the energy ranges of 1 GeV – 10 TeV for electrons and γ and 50 GeV – 100 TeV for nuclei.

4) *NeUtron Detector* (*NUD*) made of boron-doped plastic scintillator plates. It gives further information for electron/hadron discrimination.

3 DAMPE data taking and event selection criteria

The DAMPE satellite provides a good geometrical acceptance of $\sim 0.3\,\text{m}^2\text{sr}$ for galactic CRs events in the energy range 20 GeV–100 TeV. It is stably collecting $\sim$ 5M of events every day with a total amount of $\sim$ 100 GB/day of data transferred to the ground, with a trigger rate of $\sim$ 50 Hz. The dead time of the satellite is mainly due to three contributions: the lapse in which the detector crosses the South Atlantic Anomaly (SAA) region ($\sim$ 4.5% of the operation time), the on-orbit calibration data-taking ($\sim$ 1.5%) and the instrumental dead time ($\sim$ 18.5%).
In this work, every event recorded by DAMPE inside the SAA region is excluded. This is one of the so-called *pre-selection cuts*, together with the rejection of side track events and the requirement of total energy depositon $E_{dep} > 20$ GeV inside the BGO calorimeter. Once an event satisfies all the preselection requirements, the subsequent analysis goes through several steps made up of the following requests :

- High Energy Trigger (HET), which means an energy release higher than $\sim$ 10 MIP (1 MIP_{BGO} = 23 MeV) inside the first 4 layers of the BGO;
- at least one reconstructed track in the STK;
- compatibility between the STK track direction and the position of the hit on the PSD bars;

- match between the direction of the STK track and the direction of the BGO shower and, in case of more than one track, the closest to the BGO shower direction will be selected;
- match of the PSD charge measurement with the first STK layer charge measurement;
- discrimination between electromagnetic and hadronic induced event;
- choice of a proper PSD energy range to select the charge and so the incoming particles of interest.

The use of these event selection criteria is supported by the good agreement between the results of the analysis of flight-data and Monte-Carlo simulations.

4 Preliminary Charge Spectrum and Helium flux with DAMPE

A good estimate of the charge spectrum up to $Z = 28$ (Ni) has been obtained by using the charge measurement provided by the PSD. The top-most sub-detector of DAMPE is composed by 2 planes, each one with 41 bars of plastic scintillator. The bars on the top plane are parallel to the X-axis of the satellite coordinate system, while that of the second plane are parallel to the Y-axis. For this reason, the two planes can be called the YZ and the XZ views, respectively. The best estimate of the incoming particle charge can be provided by the arithmetic mean of the energy loss measurements in the two views XZ and YZ of the PSD. A preliminary result of the charge spectrum, shown in Fig. 2, has been obtained as the number of events recorded by DAMPE during 17 months as a function of the square root of the arithmetic mean of the two PSD views divided by 1 MeV in order to have a dimensionless number which is, according to the Bethe-Bloch formula, proportional to the charge of the incoming CR nucleus. Almost all the peaks are clearly visible up to Nickel.
Preliminary results about He acceptance and flux are shown in Fig. 3. The effettive acceptance, Fig. 3*a*, has been estimated by using a Monte-Carlo He sample and the following formula:

$$A_{EFF} = A_{GEN}\frac{N_{SURV}}{N_{GEN}} \tag{1}$$

where A_{GEN} is the Acceptance for the generated MC events, N_{GEN} is the number of the generated events and N_{SURV} is the number of all the generated events which survived to the same analysis cuts used for the on-orbit data.
The He flux measured by DAMPE as a function of the kinetic energy per nucleon, Fig. 3*b*, with a systematic uncertainty of $\sim 15\%$, shows a good agreement with other experiments. The statistical errors are plotted with bars while the systematics are represented by the grey band. Systematic errors have been evaluated by considering the differences between on-orbit data and MC data, taking into account the uncertainty due to the HET, the STK track selection and the PSD charge reconstruction in its two layers. Also the proton contamination in the Helium sample has been evaluated by using a mixed proton and Helium MC sample. The result is that this contamination is very small ($< 1.5\%$), while the Lithium pollution at low energies results negligible. A spectral hardening after ~ 100 GeV/n can be observed, confirming the previous results.

5 Conclusions

The DAMPE satellite is in a very stable data-taking since December 2015, providing satisfactory preliminary results about the All-Z charge spectrum and the He analysis, which have been presented

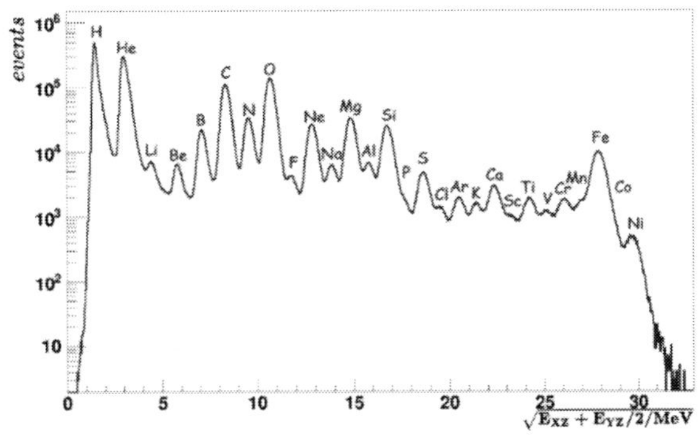

Figure 2. The All-Z charge spectrum reconstructed with the first 17 months of on-orbit data by using the arithmetic mean of the energy deposition measured in the XZ (E_{XZ}) and YZ (E_{YZ}) views of the PSD.

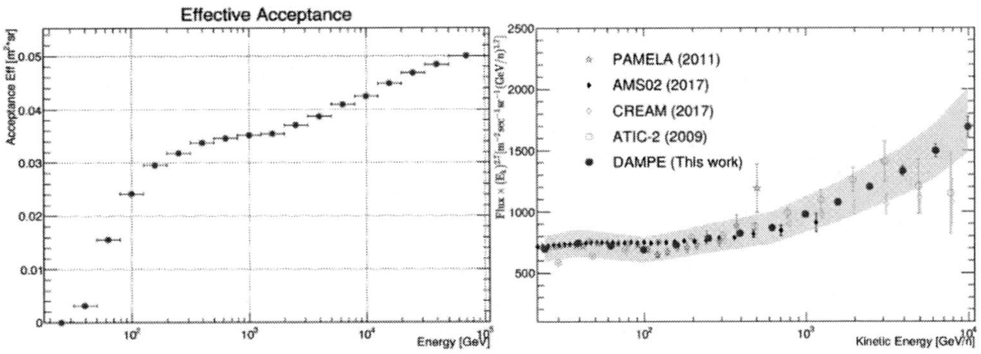

Figure 3. (a) Preliminary Cosmic Ray Helium effective acceptance. (b) Preliminary He flux as a function of the kinetic energy per nucleon measured by DAMPE compared with other experiments. The statistical errors are plotted with bars while the systematics are represented by the grey band.

here. By taking into account the charge measurement provided by the PSD, a very good charge spectrum has been obtained, with the peaks of almost all the nuclei from H to Ni ($Z = 1 - 28$) clearly visible. The preliminary He flux shows a good agreement with the previous experiments, confirming the hardening observed after $\sim$ 100 GeV/n. One of the next goals of the DAMPE mission is to go further with energies up to 100 TeV/n in order to provide an important contribution to the knowledge about the origin, the accelaration and propagation mechanisms of Cosmic Rays, as well as about their elemental composition.

References

[1] M. Anguilar et al. (AMS Collaboration), Phys. Rev. Lett. **115** (2015) 211101
[2] Yoon et al. (CREAM Collaboration), Astrophys.J. **839** (2017) 1
[3] O. Adriani et al. (PAMELA Collaboration), Science **332** (2011) 69
[4] A.D. Panov et al. (ATIC Collaboration), Bull. Russ. Acad. Sci.: Phys. **73** (2009) 564
[5] J. Chang et al. (DAMPE Collaboration), Astropart. Phys. **95** (2017) 6-24

EPJ Web of Conferences **209**, 01042 (2019) https://doi.org/10.1051/epjconf/201920901042
RICAP-18

Studies of the UHECR Mass Composition and Hadronic Interactions with the FD and SD of the Pierre Auger Observatory

J. M. Carceller[1,a], for the Pierre Auger Collaboration[2,b]

[1] *Departamento de Física Teórica y del Cosmos, University of Granada, Granada, Spain.*
[2] *Full author list and acknowledgments at: http://www.auger.org/archive/authors_2018_09.html*

Abstract.
With data on the depth of maximum $X_{\max}$ collected during more than a decade of operation of the Pierre Auger Observatory, we report on the inferences on the mass composition of UHECRs in the energy range $E = 10^{17.2} - 10^{19.6}$ eV and on the measurements of the proton-air cross section for energies up to $10^{18.5}$ eV. We also present the results on $X_{\max}$ obtained using the information on the particle arrival times recorded by the SD stations allowing us to extend the $X_{\max}$ measurements up to 10^{20} eV. The inferences on mass composition, in particular using the data of the SD, are subject to systematic uncertainties due to uncertainties in the description of hadronic interactions at ultra-high energies. We discuss this problem with respect to the properties of the muonic component of extensive air-showers as derived from the SD data.

1 Introduction

The Pierre Auger Observatory is the largest detector of ultra-high energy cosmic rays (UHECR) built so far. It comprises an area of 3000 km^2 and features two different types of detectors: The Fluorescence Detector (FD) and the Surface Detector (SD). With the FD, a direct measurement of the depth of maximum of shower profiles, $X_{\max}$, is performed using the ultraviolet light emitted by nitrogen as the shower develops in the atmosphere. While the FD operates only during moonless nights and has a duty cycle of around 15%, the SD has an almost 100% duty cycle. The SD measures the time of arrival of particles at the ground level and the signal produced by these particles by means of the Cherenkov light that particles emit when they pass through the water inside the stations of the SD. With its hybrid design, several techniques can be exploited to acquire information about mass composition. However, the interpretation of the results depends on the results of the simulations, and these depend on the choice of model for the hadronic interactions of high energy. For this reason, the study of hadronic interactions is of utmost importance for the correct interpretation of the results obtained in studies of mass composition.

[a] e-mail: jmcarcell@ugr.es
[b] e-mail: auger_spokespersons@fnal.gov

2 Mass composition

Mass composition is one of the fundamental questions that remain unsolved about ultra-high energy cosmic rays. Information about the mass composition of cosmic rays of such high energies can allow to constrain models about acceleration at the sources and to help to understand the propagation mechanisms to which these cosmic rays are exposed.

One of the most important observables for mass composition is the slant depth of maximum X_{max}. X_{max} is the depth where the number of particles (or the energy deposition) reaches its maximum and is measured with the Fluorescence Detector (FD) [1].

In Figure 1, it can be seen that the average value of X_{max} is different in simulations using proton and iron primaries. It scales linearly with the energy so that for a constant composition the evolution of X_{max} with the energy given by the elongation rate is:

$$D_{10} = \frac{\mathrm{d}\,X_{max}}{\mathrm{d}\,\mathrm{lg}(E/\mathrm{eV})} = 60\ \mathrm{g/cm^2/decade}. \tag{1}$$

However, from data it is found that $D_{10} = 79 \pm 1$ g/cm^2/decade between $10^{17.2}$ and $10^{18.33}$ eV. This result is an indication of a change in the primary mass from heavier to lighter because the elongation rate is significantly larger than for a pure composition. On the other hand, $D_{10} = 26 \pm 2$ g/cm^2/decade from $10^{18.33}$ eV onward, indicating a change in composition from lighter to heavier. The fluctuations of X_{max} also decrease above $10^{18.3}$ eV, indicating a change in composition towards a heavier primary [2]. These results can be directly compared to simulations because they are obtained once all the effects related to the detector, such as the resolution of the detector or the acceptance, are taken into account.

The study of mass composition using X_{max} can be extended beyond the average value and the fluctuations. For this, the distributions of X_{max} in data have been matched with those in simulations. Simulations of showers with different primaries (namely proton, helium, nitrogen and iron) have been made, and the fraction of each primary has been left free to find the distribution that best approximates the distribution of X_{max} in data. The result is that models can not find a combination of fractions that can reproduce the details of the observed X_{max} distributions [2, 3] because the fraction of p-values obtained for the fits that are small (< 0.1) is significantly above what is expected on statistical grounds.

Information for mass composition can also be extracted from the measurements of the SD. Using the risetime of the signal measured in each station, Δ is defined as an observable that measures how fast or slow the risetimes of an event are compared to an average benchmark. This benchmark describes the behaviour of the risetime as a function of the distance of each station to the core of the shower. This observable can be calibrated in hybrid events with X_{max}, and thus a value of X_{max} can be assigned to all the events, extending the current measurements of X_{max} up to 10^{20} eV. In Figure 2 the values of the X_{max} obtained with this technique (called the Delta method) are compared to those measured by the FD. These values have been transformed to $\langle \ln A \rangle$, the average logarithm of the mass number. The results obtained are compatible with what the FD measures, and at the highest energies the change in composition from lighter to heavier seems to be stopping, but uncertainties are large [4].

3 Hadronic interactions

Useful information can be extracted from cosmic-ray showers for the study of hadronic interactions. In particular, it can be proved that the tail of the distribution of X_{max} is sensitive to the proton-air cross section. This result, in combination with the Glauber model, has been successfully employed

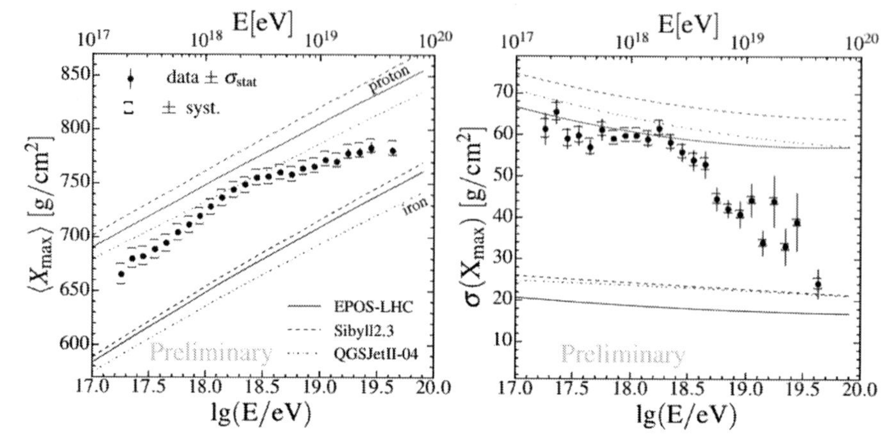

Figure 1. Evolution of the average value of X_{max} with the energy (left) and its standard deviation (right) for both data and simulations [2].

to obtain the proton-proton cross section at $\sqrt{s} = 57$ TeV, a center of mass energy that is higher than what current accelerators can achieve. The results for the inelastic and total cross section are [5]:

$$\begin{aligned} \sigma_{pp}^{\text{inel}} &= \left[92 \; \pm 7(\text{stat}) \; ^{+9}_{-11}(\text{sys}) \; \pm 7(\text{Glauber})\right] \text{mb}, \\ \sigma_{pp}^{\text{tot}} &= \left[133 \; \pm 13(\text{stat}) \; ^{+17}_{-20}(\text{sys}) \; \pm 16(\text{Glauber})\right] \text{mb}. \end{aligned} \tag{2}$$

Some discrepancies have been found between the number of muons measured in data and simulations. Inclined events (zenith $\theta > 60°$) that have been measured at the same time by the FD and SD have been studied. The interest in these events is that most of the signal comes from muons due to the absorption of the electromagnetic component of the shower in the longer path of atmosphere that they have to traverse to reach the detector. It is found that it should be necessary to increase the number of muons between 30% and 80% (depending on the model used for high-energy hadronic interactions) to match with simulations what is measured in the data [6]. In the left panel of Figure 3, the average logarithmic muon content is compared to simulations, and it is found to be significantly above what is obtained from simulations.

Another way of showing this discrepancy is to study simulations that have the same longitudinal profile (dominated by the electromagnetic component) as in data. For this, simulations are made until a longitudinal profile that is very similar to that in the data is obtained. Then, the signal at the ground measured by the SD is studied as the sum of the contribution of an electromagnetic and a hadronic component:

$$S_{\text{resc}}(R_E, R_{\text{had}})_{i,j} \equiv R_E \, S_{EM,i,j} + R_{\text{had}} \, R_E^{\alpha} \, S_{\text{had},i,j}, \tag{3}$$

where R_E is a factor that scales the electromagnetic component, R_E^{α} reflects the fact that the hadronic signal increases slower than linearly with energy, and R_{had} is the scaling of the hadronic component. Then, R_E and R_{had} are left free in a fit to match the signal measured in data. In the right panel of Figure 3, it can be seen that while the energy rescaling R_E is compatible with 1, meaning that no scaling is needed, the scaling of the hadronic component is significantly above 1 for the hadronic models used and compositions shown [7].

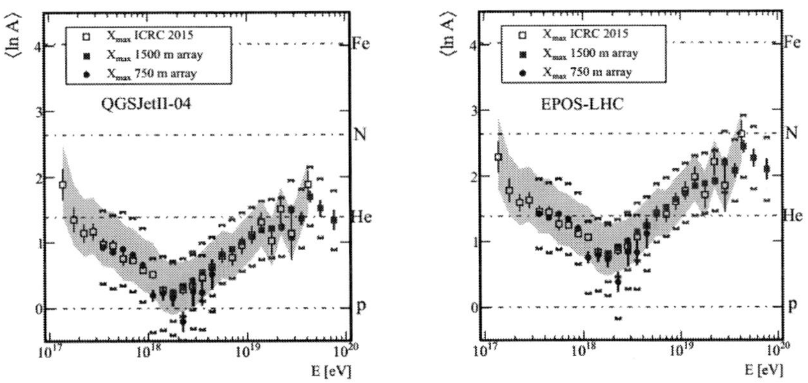

Figure 2. $\langle \ln A \rangle$ as a function of the energy. The results obtained with the Delta method (filled squares and filled circles) are compared to the values of $X_{\rm max}$ measured by the FD (open squares) [4].

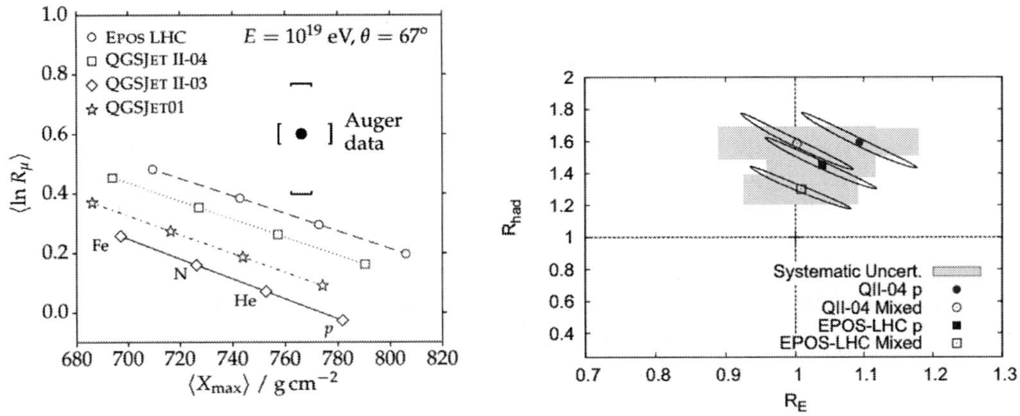

Figure 3. (Left) Average logarithmic muon content as a function of the average shower depth $\langle X_{\rm max} \rangle$. Model predictions are obtained from showers simulated for $\theta = 67°$ for proton and iron showers, and interpolated for helium and nitrogen showers [6]. (Right) Best-fit values of R_E and $R_{\rm had}$. The ellipses and grey boxes show the 1-σ statistical and systematic uncertainty [7].

References

[1] The Pierre Auger Collaboration, Phys. Rev. D **90**, 122005 (2014), 1409.4809
[2] Jose Bellido for The Pierre Auger Collaboration, PoS (ICRC2017) **506** (2017)
[3] The Pierre Auger Collaboration, Phys. Rev. D **90**, 122006 (2014), 1409.5083
[4] The Pierre Auger Collaboration, Phys. Rev. D **96**, 122003 (2017), 1710.07249
[5] The Pierre Auger Collaboration, Physical Review Letters **109**, 062002 (2012), 1208.1520
[6] The Pierre Auger Collaboration, Phys. Rev. D **91**, 032003 (2015), 1408.1421
[7] The Pierre Auger Collaboration, Physical Review Letters **117**, 192001 (2016), 1610.08509

EPJ Web of Conferences **209**, 01043 (2019) https://doi.org/10.1051/epjconf/201920901043
RICAP-18

Underground Nuclear Astrophysics: Present and future of the LUNA experiment

Carlo Gustavino [a]

[1] *INFN Sezione di Roma, I-00185 Roma, Italy*

Abstract. The evolution of celestial bodies is regulated by gravitation and thermonuclear reaction rates, while the Big Bang nucleosynthesis is the result of nuclear processes in a rapidly expanding Universe. The LUNA Collaboration has shown that, by exploiting the ultra low background achievable deep underground, it is possible to study the relevant nuclear processes down to the nucleosynthesis energy inside stars and during the first minutes of Universe. In this paper the main results of LUNA are overviewed, as well as the scientific program the forthcoming 3.5 MV underground accelerator.

1 Introduction

The synthesis of elements in stars are due to thermonuclear fusion reactions, which start from the most abundant and lightest element, the hydrogen, towards to synthesis of the heaviest elements. Therefore nuclear fusion reactions determine the energy production, the neutrino luminosity and the evolution of stars. In the stellar environment the thermal energy of nuclei is much lower than the Coulomb barrier. As an example, the core temperature of the Sun is about $15 \cdot 10^6\ K$, corresponding to a Maxwell-Boltzmann energy distribution peaked to about 1 keV, much lower than the 550 keV effective Coulomb barrier for the $p + p$ reaction. Indeed, the nuclear reactions take place via quantum tunneling effect and the cross section of charged particle reactions can be expressed as follows:

$$\sigma(E) = \frac{S(E)e^{-2\pi\eta}}{E} \tag{1}$$

In this formula, the exponential term takes into account the Coulomb barrier, while the astrophysical factor $S(E)$ contains all the nuclear effects. For non-resonant reactions, it is a smoothly varying function of energy. The Sommerfeld parameter η is given by $2\pi\eta = 31.29 Z_1 Z_2 (\mu/E)^{1/2}$. Z_1 and Z_2 are the nuclear charges of the interacting nuclei. μ is their reduced mass (in units of a.m.u.), and E is the center of mass energy (in units of keV) [1, 2].
Fusion reactions in stars occur in a well defined energy range, the so called Gamow peak, which arises from the convolution of the thermal energy distribution of nuclei and the tunneling probability through the Coulomb barrier between the interacting nuclei. For a given stellar temperature T the energy corresponding to the Gamow peak is larger than KT, typically in the order of tens of keV. Most of the reactions between light nuclei at such energies have cross sections of the order of $10^{-9} \div 10^{-12}$ barn.

[a]e-mail: carlo.gustavino@roma1.infn.it

Table 1. Selection of nuclear reaction already studied by LUNA (a), in progress (b) and in program (c).

Reaction	main motivation	accelerator	reference
$^{3}He(^{3}He,2p)^{4}He$ [a]	p-p chain	LUNA-50	[4]
$d(^{3}He,p)^{4}He$ [a]	electron screening	LUNA-50	[8 – –10]
$d(p,\gamma)^{3}He$ [a,b]	BBN	LUNA-50, LUNA-400	[11 – –13]
$3He(\alpha,\gamma)^{7}Be$ [a]	p-p chain, BBN	LUNA-400	[14, 15]
$^{14}N(p,\gamma)^{15}O$ [a]	CNO cycle I	LUNA-400	[16 – –20]
$^{25}Mg(p,\gamma)^{26}Al$ [a]	Mg-Al cycle	LUNA-400	[21 – –23]
$^{15}N(p,\gamma)^{16}O$ [a]	CNO cycle II	LUNA-400	[24, 25]
$^{17}O(p,\gamma)^{18}F$ [a]	CNO cycle III	LUNA-400	[26, 27]
$^{22}Ne(p,\gamma)^{23}Na$ [a]	Ne-Na cycle	LUNA-400	[28, 29]
$^{17}O(p,\alpha)^{14}N$ [a]	CNO cycle III	LUNA-400	[30, 31]
$d(\alpha,\gamma)^{6}Li$ [a]	BBN	LUNA-400	[32, 33]
$^{13}C(\alpha,n)^{16}O$ [b,c]	S-process	LUNA-400, LUNA-MV	[12, 34]
$^{22}Ne(\alpha,n)^{25}Mg$ [c]	S-process	LUNA-MV	[34]
$^{12}C+^{12}C$ reactions [c]	Carbon burning	LUNA-MV	[34]

These correspond, under typical laboratory conditions, to experimental counting rates ranging from a few events per day to a few events per month. This means that in realistic experimental conditions, the expected counting rate is prohibitively low with respect to the detector background induced by the cosmic rays. A drastic reduction of this background can be achieved by carrying out the measurement deep underground. The "Laboratori Nazionali del Gran Sasso" (LNGS) hosts the facility LUNA (Laboratory for Underground Nuclear Astrophysics), that is still the worldwide unique facility operating deep underground, in which cross sections at energies of astrophysical interest are measured. The average depth of 1400 m of rock leads to a reduction of the muon and neutron fluxes by a factor 10^6 and 10^3, respectively. The suppression of the cosmic ray induced background also allows an effective suppression of the γ-ray activity by a factor $10^3 \div 10^5$, depending on the photon energy.

2 The LUNA experiment

High beam current, long-term stability, and precise beam-energy determination are the requirements for an ion accelerator conceived for direct nuclear astrophysics measurements. The first feature maximizes the reaction rate, the second provides the long time typically needed for a cross-section measurement, and the third is important because of the exponential energy dependency of the cross section (see eq. 1).

The first accelerator used at LUNA was a 50 kV machine, consisting of a duoplasmatron ion source, an ion-beam extraction and acceleration system, and a double-focusing 90° analyzing magnet. This machine was designed to deliver beams of protons, ^3He and ^4He ions of about 400 μA at energies between 10 and 50 keV with an energy spread of less than 20 eV and a long-term stability [3]. With this accelerator have been measured, for the first time inside the solar Gamow peak, the cross section of the $^3He(^3He,2p)^4He$ reaction [4]. The achievement of measuring the $^3He(^3He,2p)^4He$ cross section inside the solar Gamow peak can be considered a milestone in the long way to the discovery of neutrino oscillation [5, 6]. In fact the existence of the Fowler resonance to explain the "Solar neutrino problem" was definitively rejected, pushing towards a new generation of experiments addressed to measure the neutrino oscillation parameters and the solar interior (Borexino, Kamland, SNO). It is

worth to point out that at the lowest energy of E_{cm} = 16.5 keV the measured cross section is 0.02 ± 0.02 pb, which corresponds to a rate of about 1 event/month, rather low even for the "silent" experiments of underground physics [4]. Even though the LUNA-50 accelerator has produced outstanding results, it should be considered as a pilot project toward the present facility, the LUNA-400 accelerator [7].
As it is summarised in table 1, the LUNA-400 accelerator is mainly employed to study reactions of hydrogen burning in the Sun and in other celestial bodies, e.g. massive stars, AGB stars, Novae. Also important is the study of several leading processes of Big Bang Nucleosynthesis (BBN). As an example, the ongoing study of the d(p,γ)^{3}He reaction is particularly important to derive the cosmological baryon density Ω_b and to constrain the possible existence of "dark radiation" at BBN epoch, i.e. the existence of light particles such as sterile neutrinos or axions, not foreseen in the standard model.
Although the LUNA-400 accelerator is presently full operative, the future of LUNA is addressed to study the successive phases of stars, towards the synthesis of heavier and heavier elements. In this concern the LUNA-MV project has been approved and funded, and will be operative in 2019. This machine has a terminal Voltage of 3.5 MV and can deliver beams of $^1H^+$, $^4He^+$, $^{12}C^+$ and $^{12}C^{++}$. The starting program will last about 4 years and foresees the study of the carbon burning through the cross section measurement of ^{12}C+^{12}C processes, such as ^{12}C(^{12}C,p)^{23}Na, ^{12}C(^{12}C,α)^{20}Ne and ^{12}C(^{12}C,n)^{23}Mg. Carbon Burning is of crucial importance to understand the evolution of stars. In fact, when He burning subsides, the lack of radiation pressure sends the core of the star into gravitational free fall, with the temperature increasing in response to the decrease in gravitational potential energy. This collapse is halted by one of two possible events: the temperature becomes sufficient to ignite the carbon in the star's core or electron degeneracy halts the contraction producing a carbon–oxygen white dwarf. Which of these paths a particular star follows is dependent on whether the ignition temperature for carbon fusion is reached ($\sim 5\times10^8$ K) or, analogously, whether the critical mass for carbon ignition ($\sim 9\ M_\odot$) is exceeded prior to electron degeneracy. Thus the end of helium burning marks a branching point in stellar evolution. Lower mass stars will become stable, electron-degenerate white dwarfs, while higher mass stars will enter the quiescent carbon burning phase of their evolution. As a consequence, this limit separates the progenitors of white dwarfs, novae and type Ia supernovae, from those of core-collapse supernovae, neutron stars, and stellar mass black holes. This mass limit also controls the estimations of the expected numbers of these objects in a given stellar population. The rate of the ^{12}C+^{12}C reaction, which is the trigger of the carbon burning, is a primary input to predict the fate of a star at this branching point. The larger the ^{12}C+^{12}C rate, the lower the temperature of the carbon burning, with a shorter duration of the C burning phase. Carbon burning also influences the nucleosynthesis of massive stars. As a matter of fact, the two main channels of this reaction release protons and α particles in a rather hot environment, thus allowing a complex chain of reactions involving nuclei from C to Si. Some of these reactions, e.g. the $^{13}C(\alpha,n)^{16}O$ and the $^{22}Ne(\alpha,n)^{25}Mg$, release neutrons and, in turn, activate the s-process which allows the production of heavy elements (see below). The ^{12}C+^{12}C rate also affects the outcomes of type Ia supernovae, that play a fundamental role in cosmology, allowing the measurements of distances and of the expansion rates of high redshift galaxies. The understanding of these different phenomena, of primary importance for astrophysics and cosmology, deserves a deeper experimental investigation of the ^{12}C+^{12}C cross section. As shown in table 1, the initial program with the LUNA-MV accelerator includes the study of the $^{13}C(\alpha,n)^{16}O$ and $^{22}Ne(\alpha,n)^{25}Mg$ reactions. These reactions release neutrons and, in turn, activate the s-process which allows the production of heavy elements. Finally, the measurement of the $^{12}C(\alpha,\gamma)^{16}O$ with LUNA-MV has been considered. This is a key reaction during the stellar phase of helium burning. However the existence of a beam induced background due to the $^{13}C(\alpha,n)^{16}O$ reaction suggests an inverse kinematic approach, with a ^{12}C ion beam impinging a ^{4}He gas target.

References

[1] C. Rolfs and W.S. Rodney: Cauldrons in the cosmos, University of Chicago Press (1988).
[2] C. Iliadis, Nuclear Physics of Stars (Weinheim:Wiley-VCH) (2007).
[3] U. Greife *et al*, NIM A 350 (1994). Volume 350, 327-337, (1994).
[4] R. Bonetti *et al*, Phys. Rev. Lett. **82**, 5205 (1999).
[5] J.N. Bahcall, M.H. Pinsonneault, Rev. Mod. Phys. **64**, 885 (1992).
[6] W.A. Fowler, Nature **238**, 24 (1972).
[7] A. Formicola *et al*, Nucl. Instr. and Meth. A **507** 609 (2003).
[8] P. Prati *et al*, Z. Phys. A 350, 171-176 (1994).
[9] H. Costantini *et al*, Phys. Lett. B **482** 43-49 (2000).
[10] M. Aliotta *et al*, Nucl. Phys. A **690** 790-800 (2001).
[11] C. Casella *et al*, Nuclear Physics A **706** 203-216. (2002).
[12] "LUNA proposal 20015-2018", submitted to the LNGS scientific committee.
[13] E. Di Valentino, C. Gustavino *et al.*, Phys. Rev. D **90**, 023543 (2014).
[14] D. Bemmerer *et al*, Phys. Rev. Lett. **97**, 122502 (2006).
[15] F. Confortola *et al*, Phys. Rev. C **75**, 065803 (2007).
[16] A. Formicola *et al*, Phys. Lett. B **591**, 61 (2004).
[17] G. Imbriani AA **420** 625Ð629 (2004).
[18] G. Imbriani *et al*, Eur. Phys. J. A, **25** 455-466 (2005).
[19] M. Marta *et al*, Phys. Rev. C 78, 022802 (2008).
[20] A. Lemut *et al*, Phys. Lett. B **634** 483 (2006).
[21] B.Limata *et al*, Phys. Rev. C **82**, 015801 (2010).
[22] F. Strieder *et al*, Phys. Lett. B **7070** 60-65 (2012).
[23] O.Straniero *et al*, The Astrophysical Journal, 763:100 (10pp) (2013).
[24] D.Bemmerer *et al*, J. Phys. G: Nucl. Part. Phys. **36** 045202 (2009).
[25] A.Caciolli *et al*, AA **533**, A66 (2011).
[26] D.A. Scott *et al*, Phys. Rev. Lett. **109**, 202501 (2012).
[27] A. Di Leva *et al*, Phys. Rev. C **89**, 015803 (2014).
[28] R. De Palo *et al*, Phys. Rev. C **94**, 055804 (2016).
[29] F. Ferraro *et al*, Eur. Phys. J. A **54** 44 (2018).
[30] C.G. Bruno *et al*, Phys. Rev. Lett. **117**, 142502 (2016).
[31] M.Lugaro *et al*, Nature Astronomy **1**, 0027 (2017).
[32] M. Anders *et al*, PRL **113**, 042501 (2014).
[33] D. Trezzi *et al*, Astroparticle Physics **89**, 57–65 (2017).
[34] The LUNA Collaboration, LUNA-MV Proposal 2018-2022. INFN/17-19/LNGS (2017).

Instruments optimizations for low energy Gamma-ray detection

Gonzalo Rodriguez[1,a] and Aldo Morselli[1]

[1] *INFN Roma Tor Vergata, Rome*

Abstract. There is an experimental gap in the study of the non-thermal universe in the photon energy range from 0.3 to 100 MeV. We have analyzed the performance of a detector with unprecedented sensitivity, angular and energy resolution and combined with polarimetric capability to study of the most powerful Galactic and extragalactic sources and with a line sensitivity in the MeV energy range two orders of magnitude better than previous generation of instruments that can determine the origin of key isotopes fundamental for the understanding of supernova explosion and the chemical evolution of our Galaxy.

1 Introduction

The energy domain between 0.3 and 100 MeV covers a crucial range of nuclear gamma-ray lines produced by radioactive decay, nuclear collision, positron annihilation, or neutron capture, which makes it as special for high-energy astronomy as optical spectroscopy is for phenomena related to atomic physics. Also some objects in the Universe have their peak emissivity at photon energies between 0.3 and 100 MeV (e.g. gamma-ray bursts, blazars, pulsars, etc.), meaning that it is in this energy band that their physical properties can be directly studied. This energy range is also known to feature a characteristic spectral turn-over associated to hadronic emission from pion decay, that can be use to distinguishing leptonic from hadronic processes.

e-ASTROGAM [1] is a gamma-ray space mission that have been proposed as the M5 Medium-size mission of the European Space Agency. It is a gamma-ray instrument that inherits from predecessors such as AGILE [2] and Fermi [3], as well as from the MEGA [4] prototype, but it takes full advantages of recent progresses in silicon detectors and readout microelectronics to achieve measurement of the energy and 3D position of each interaction within the detectors with an excellent spectral and spatial resolution. The main innovative feature of the e-ASTROGAM mission is the capability of joint detection in the Compton (0.3 - 30 MeV) and pair (> 10 MeV) energy ranges in a single integrated instrument. The mission aims at improving the sensitivity in the medium-energy gamma-ray domain by one to two orders of magnitude compared to previous missions.

2 The e-ASTROGAM telescope

The e-ASTROGAM telescope is made up of three detection system, see Figure 1, a silicon Tracker in which the cosmic gamma rays undergo a Compton scattering or a pair conversion, a Calorimeter

[a] e-mail: grferna@roma2.infn.it

to absorb and measure the energy of the secondary particles, and an anti-coincidence (AC) system to veto the prompt-reaction background induced by charged particles. The telescope has a size of $110x110x80\ cm^3$ and a mass of 820 kg.

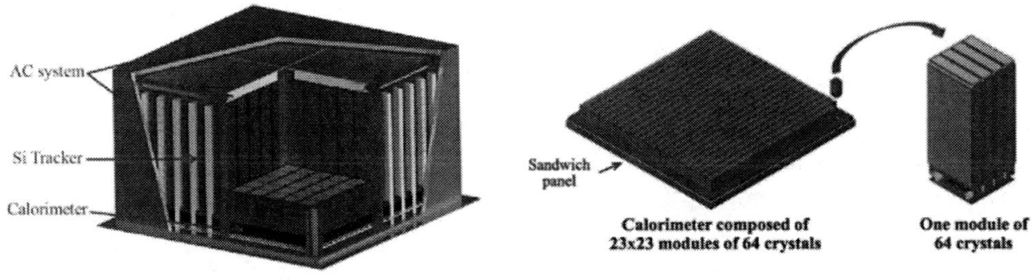

Figure 1: Overview of the e-ASTROGAM payload showing the silicon Tracker, the Calorimeter and the Anti-coincidence system.

The e-ASTROGAM Tracker is a double-sided strip detectors (DSSD) that comprises 5600 DSSDs arranged in 56 layers. It is divided in four units of 5x5 DSSDs, the detectors being wire bonded strip to strip to form 2-D ladders. The interlayer distance is 10 mm. Each DSSD has a geometric area of $9.5x9.5\ cm^2$, a thickness of 500 μm, and a strip pitch of 240 μm. The total detection area amounts to $9025\ cm^2$ and the total Si thickness to 2.8 cm, which corresponds to 0.3 radiation length on axis.

The Calorimeter is a pixelated detector made of a high-Z scintillation material ? Thallium activated Cesium Iodide. It consists of an array of 33,856 parallelepiped bars of CsI(Tl) of 8 cm length and $5x5\ mm^2$ cross section, read out by silicon drift detectors (SDDs) at both ends, arranged in an array of 529 (= 23x23) elementary modules comprising each 64 crystals. The Calorimeter thickness, 8 cm of CsI(Tl), makes it a 4.3 radiation length detector having an absorption probability of a 1-MeV photon on axis of 88%.

The third main detector of the e-ASTROGAM payload consists of an Anti-coincidence (AC) system made of segmented panels of plastic scintillators covering the top and four lateral sides of the instrument, requiring a total active area of about $4.7\ m^2$. The AC detector is segmented in 33 plastic tiles (6 tiles per lateral side and 9 tiles for the top). All scintillator tiles are coupled to silicon photomultipliers (SiPM) by optical fibers.

3 Silicon Tracker: Geometry optimization in the Compton regime

We have focus our optimization studies on the silicon tracker geometry. We have perform the simulations, the reconstruction and analysis of the events using the MEGAlib framework [5]. MEGAlib was develop to simulate tracker detectors in the Compton regime, and recently was update to include also the pair production regime. The detector is simulated using a wrapped of the well know Geant4 framework [6]. The output of the simulations is reconstructed using the tools *revan* and *mimrec*.

In the energy range from 0.3 to 5 MeV we measured the properties of the incident photon using the Compton effect. From the interaction we have two products, a photon and an electron. If the electron have enough energy will escape the silicon layer and will be detect as it hits other layers. Otherwise we will only have the hits from the photon. We have to distinguish between events with electron tracking and without electron tracking. This will have implications in the angular resolution.

https://doi.org/10.1051/epjconf/201920901044

3.1 Energy Resolution - Photopeak events

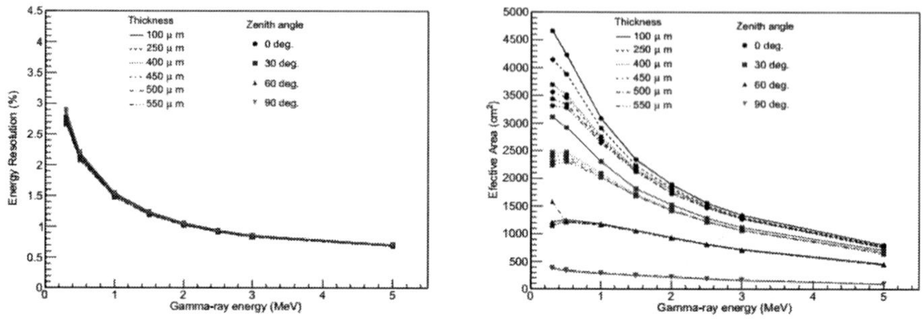

Figure 2: Energy resolution and Effective area from the photopeak area

We have to use the photopeak events to calculate the energy resolution in the Compton regime. In the Figure 2 we show the results, in the left panel we see that the energy resolution does not depends on any geometrical parameters use in the simulation. We have observe an energy resolution of 3% at low energies that decrease to about 1% at 5 MeV. In the right panel we show the dependence of the effective area as a function of the energy for different thickness and zenith angles.

3.2 Angular Resolution - Compton events

For the angular resolution we have to take into account the events reconstructed w/o electron tracking. For simplicity we show the results for events simulated with zenith angle at 0 degrees.

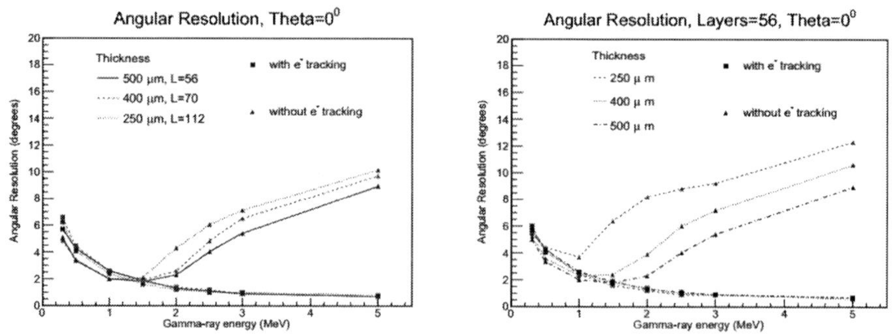

Figure 3: Angular resolution from the reconstruction of Compton events, Left panel: Fix the radiation length. Right panel: Fix the number of layers.

In the Figure 3 we show, on the left panel, the results for 3 different thickness when we have fixed the number of total radiation lengths to $0.3X_0$. For events without electron tracking, triangles points, we observe that at approximately 1.5 MeV the resolutions start to get worst for all thickness. This is not the case for the reconstructed events with electron tracking, square points, the angular resolutions gets better as the energy increase, going from $6°$ at 0.3 MeV to $1°$ at 5 MeV. For energies below 1.5 MeV both reconstruction give approximately the same angular reconstruction.

In the right panel, we show the results for geometries with different thickness but the same number of layer, in this case 56. For events with electron tracking we observe the same results when the total radiation length was fixed. We have found a different situation when we have reconstructed events with no electron tracking. In this case depending on the thickness the angular resolutions start getting worst at different energies. We can see that for 250 μm at energies less than 1 MeV the angular resolutions start decreasing, but for 400 and 500 μm this behaviour start at 1.5 MeV or above.

It is clear that we have to make an event selection to obtain the best angular resolution as a function of the energy and for each simulated geometry.

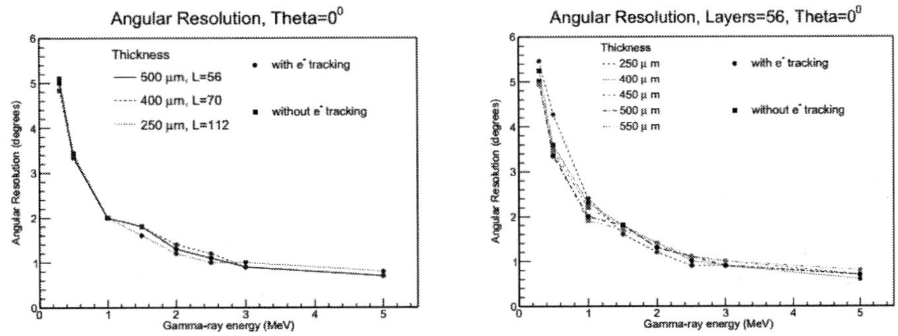

Figure 4: Angular resolution, after event selection, from the reconstruction of Compton events, Left panel: Fix the radiation length. Right panel: Fix the number of layers.

In the Figure 4 we show the results of the angular resolution after the event selection. As we can observe regardless of the geometry we have obtain the same angular resolution within the uncertainties. The angular resolution goes from $5°$ at 0.3 MeV to $1°$ at 5 MeV.

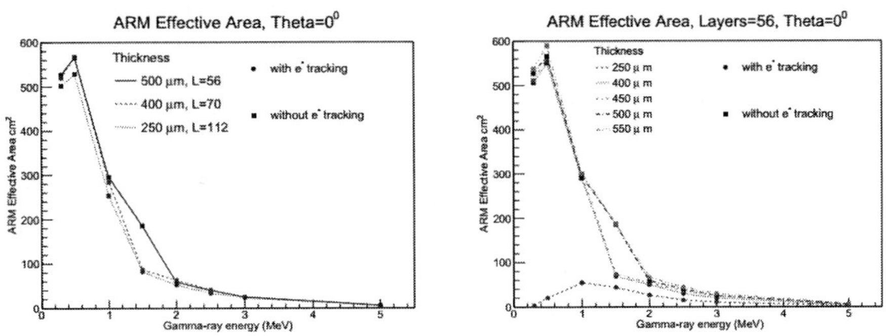

Figure 5: Effective area from the reconstruction of Compton events, Left panel: Fix the radiation length. Right panel: Fix the number of layers.

Using the event selection that minimizes the angular resolution we show in the Figure 5 the effective area. In the left panel, when the number of total radiation lengths is fixed we have obtain the same effective area independent of the simulated geometry. In the right panel, we have kept the number of layer fix to 56, in this case we observe that, for low energies, for thickness of 250 μm the effective area drop dramatically in comparison with 400 and 500 μm.

4 Conclusions

We have simulated different geometries for the eASTROGAM silicon tracker using MEGAlib framework. We have studied the Compton and Pair production regime by simulates different energies range from 0.3 to 5 MeV.

In the Compton regime we have found that the energy resolution does not depend on the simulate geometry.

For the angular resolution we have to make an event selection between reconstructed events w/o electron tracking to obtain the best possible resolution. After the selection the angular resolution is independent of the geometry. We have found difference in the effective area when the number of interactions lengths is smaller than $0.3X_0$, we have observed that, for low energies, for thickness of 250 μm the effective area drop dramatically in comparison with 400 and 500 μm. The similar study in the pair production regime is on-going

5 Acknowledgements

The research leading to these results has received funding from the European Union's Horizon 2020 Programme under the AHEAD project (grant agreement n. 654215)

References

[1] A. De Angelis et al., Science with e-ASTROGAM (A space mission for MeV-GeV gamma-ray astrophysics) Journal of High Energy Astrophysics Volume 19, August 2018, Pages 1-106 https://doi.org/10.1016/j.jheap.2018.07.001 [arXiv:1711.01265]

[2] Tavani, M., Barbiellini, G., Argan, A., et al., The AGILE Mission, Astronomy and Astrophysics 502, 995-1013 (Aug. 2009).

[3] Atwood, W. B., Abdo, A. A., Ackermann, M., et al., The Large Area Telescope on the Fermi Gamma-Ray Space Telescope Mission, Astrophysical Journal 697, 1071-1102 (June 2009).

[4] Kanbach, G., Andritschke, R., Zoglauer, A., et al., Development and calibration of the tracking Compton/Pair telescope MEGA, Nuclear Instruments and Methods in Physics Research A 541, 310-322 (Apr. 2005).

[5] A.Zoglauer et al. MEGAlib - The Medium Energy Gamma-ray Astronomy Library, New Astronomy Reviews, Volume 50, Issues 7-8, October 2006

[6] S.Agostinelli et al, Geant4 - a simulation toolkit, Nuclear Instruments and Methods in Physics Research A 506 (2003) 250-303

ESAIM: PROCEEDINGS AND SURVEYS, Vol. ?, 2019, 1-10
Editors: Will be set by the publisher

GRAVITATIONAL WAVES: WHERE WE ARE, WHERE WE GO

Fulvio Ricci[1]

Abstract. We review the present status of the Gravitational wave detectors on the Earth, focusing the attention on the present innovations and the longer term perspectives to improve their sensitivity. Then we conclude mentioning few potential searches of new Physics phenomena to be performed with these detectors and those of the third generation.

The detectors status

In 2019-20120 the two Advanced gravitational-wave detectors, LIGO in USA, Virgo in Italy and KAGRA in Japan, will start to take data in coincidence. This worldwide network will be capable to improve the sky location of the transient GW events as neutron start - neutron star (BNS) or black-hole-neutron star (BHNS) coalescences, providing crucial information to the astronomy community and triggering the multi-messenger observations.

The last data taking of LIGO and Virgo ended at the end of August 2017. Since this time several improvements have been implemented on these detectors.

In the LIGO sites the main hardware changes invested the vacuum system. The main actions have been the replacement of the mirrors acting as test masses and the signal recycling, the installation of the squeezing light benches in both sites. In addition an extended campaign to fight against the stray light has been pursued and the preliminary results seem very promising.

Virgo changed is laser source and installed the squeezing bench built by the group of Albert Einstein Institute of the Max Planck in Hannover. The main change is marked by the re-introduction of the monolithic suspensions of the mirrors. These suspensions made of thin fused silica wires were used during the run of 2011 with the 22 kg mirrors. In 2017 the mirrors of 42 kg were suspended by means of metallic wires, more reliable but with a higher contribution of the suspension thermal noise limit.

At present the three instruments are in a commissioning phase and the start of the science run, O3, is planned for April 2019. The scientific target is to achieve sensitivities curves similar to those reported in the article [1] where the collaborations LIGO, Virgo and KAGRA predict the detector performances in the following years .

The KAGRA interferometer has the unique peculiarities to be installed in an underground area and the test mass mirrors are cooled at 20 K. The installation of the interferometer started in 2015 the first test was carried on a 3 km Michelson interferometer with room temperature mirrors. Then the installation progressed rapidly in the years 2017 and 2018 with the cryogenic payloads equipped with the end mirrors made of sapphire cooled successfully down to 18 K. The KAGRA international collaboration is working hard to join LIGO and VIRGO in the second part of the O3 data taking.

[1] Dipartimento di Fisica, Università di Roma *Sapienza* & INFN - Sezione di Roma,
Piazza A. Moro 2, 00185 Roma, Italy

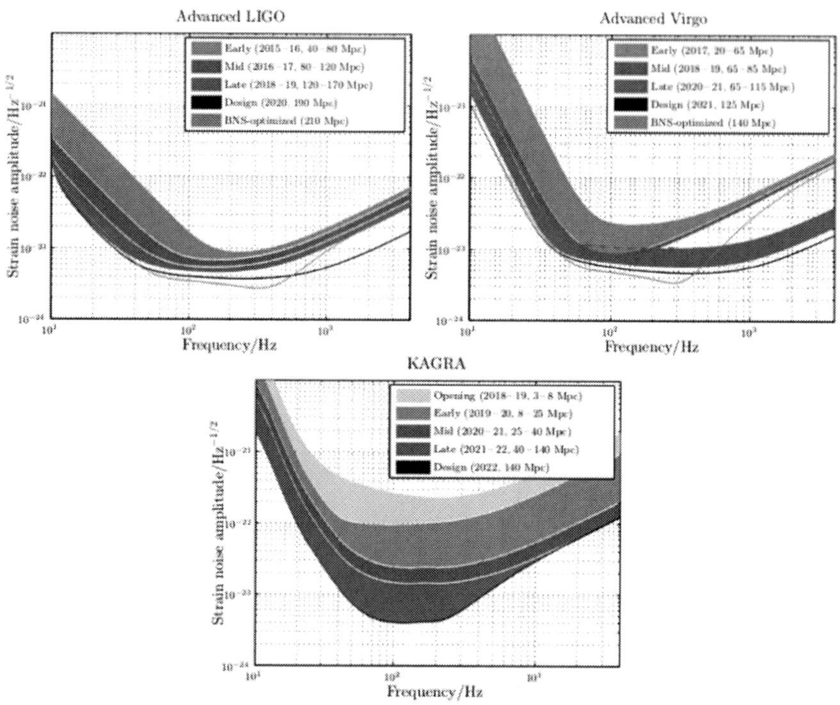

FIGURE 1. The target strain sensitivities of LIGO, Virgo and KAGRA as a function of frequency during the following data taking. The binary neutron star (BNS) range is the average distance in Mpc, to which these signals could be detected. The figure is extracted form the reference [1]

The next decade the gravitational community should converge toward the constructions of new detectors. The network of the advanced detectors, enriched by the inclusion of INDIGO, an other interferometer of the second generation whose construction is just started in India, should achieve the sensitivity values whose limitations are essentially related to the location and the space availability of the present facilities.

Within the international scientific community various initiatives are being carried out to study and implement third generation detectors (3G) able to expand by 1000 times the volume of the universe explored. To achieve this goal it is necessary a significant leap in the sensitivity of the instruments, which can be pursued on one hand by introducing new technological solutions, on the other by building a new infrastructure with different characteristics from the current ones.

European research agencies operating in the gravitational wave sector, brought together in the European APPEC Consortium, have started a preparatory work to present the application for a large research infrastructure for Earth based gravitational astronomy.

The present proposals, as for example the European project Einstein Telescope (ET) targets three crucial innovations, two of them studied already in KAGRA:

- the use of cryogenics to cool down the test mass

- the installation of the interferometer in an underground laboratory (or at lest the vacuum chambers hosting the test masses)
- the increase of the arm length of the interferometer

Italy has played a leading role in the study of gravitational waves and detector development thanks to the long-term support of INFN. This study produced significant scientific and technological achievements for Virgo, built in collaboration with of the French CNRS. Italy played a leading role also in the context of the ET design study, the project financed by the European Union in the Framework Program 7. On the base of this legacy INFN, with the full support of the local political and academic authorities, is presenting its candidature proposing to host the new detector at the Sos Enattos mine in Sardinia, an ideal place for the construction of the new infrastructure, compatible with the requirements of very low seismic and anthropic noise. The first stage has been accomplished already by Regional Government of Sardinia, which assigned to the University of Sassari a budget for the construction of the SAR-GRAV a research infrastructure in the Sos Enattos mine, a low noise laboratory devoted to host small scale gravity experiments.

New detectors imply also new technological solutions developed thanks to a robust R & D effort to be started from now days. . The R&D program paving the way for the realisation of the next generation of instruments, should be focused on some crucial technologies, as

- the improvement of the VIRGO seismic attenuation system, to improve the low frequency sensitivity and to maintain the INFN historic leadership in this field,
- the design, construction and test of a cryogenic payload,
- the development of innovative frequency dependent squeezing techniques for reducing quantum noise,
- the improvement of the (optical and mechanical) losses of the mirrors coatings, to reduce thermal noise.

1. Scientific targets of the Gravitational Wave Physics

The direct detection of GW signals, started on September 14th, 2015 with the observation of the coalescence of a binary black hole system (BBH) and then followed by several other observations confirmed the long-standing prediction of Albert Einsteins general theory of relativity and opened up a completely new way of observing Nature. The detected signals confirmed the existence of black holes with masses larger than 20 M_odot, but at the same time new questions arose: how many are the black holes in the Universe and what is their size? How are they formed? Since the stellar wind depends on metallicity, the question is how metallicity environment influences the formation?

Now days two models on the binary black hole formation are dominant. The first one hypotheses that we are dealing with two object formed and exploded at the same time from two stars with similar spins oriented in the same direction. The second one assumes that the black holes in a stellar cluster sink to the center of the cluster and pair up spin randomly oriented. The observation of several BBH events should permit to disentangle the two hypothesis on the base of the observed spin distribution In addition the detailed reconstruction of the BH mass function distribution can tell us about the Universe formation. Other questions are open. Do miniature black holes exist? In the case of a positive observation we can conclude that they have formed immediately after the Big Bang. Rapidly expanding space may have squeezed some regions into tiny, dense black holes less massive than the sun. Under a simplified hypothesis of a uniform distribution of BBH creation on the universe history, with a 3G detector we expect to observe up to 10^5 y^{-1} BBH with an impressive signal to noise ratio (SNR) of the order of 10^4 for rare events. The population study will be influenced by the detector sensitivity and the achievable SNR. GW signal amplitude depends on $\mathcal{M}^{5/2}$, where $\mathcal{M}$ = the chirp mass defined as $\mathcal{M} = (m_1 m_2)^{3/5}/(m_1 + m_2)^{1/5}$ Higher $\mathcal{M}$ should imply easier detection of the GW signal, but on the other hand the signal duration decreases with $M_{tot} = m_1 + m_2$ tends to decrease the SNR. In addition we note that too massive systems emit GW signals at frequency out of the bandwidth of the Earth detectors.

GW is the new tool not just for funding a new chapter of Astronomy, we will rely on this new probe to expand our comprehension of the fundamental laws of the Physics. It is widely spread the believe that General Relativity (GR) is incomplete because it is at odds with quantum mechanics. For this reason the GR tests in the strong field regime are crucial and the detector network can perform tests also via the observation of the polarisation states of GW.

BH is the main example of and object dominated by extreme gravity and GW seems to be the best way to probe its properties. The spectrum of Quasi Normal Modes (QNM) is characterised only by the BH mass and angular momentum. The detection of a few modes from the ringdown signal can allow for precision measurements of the BH mass and spin [2]. In addition the detection of higher multipole moments can be used to perform null-hypothesis tests of the no-hair theorems of general relativity. 3G and space detectors as

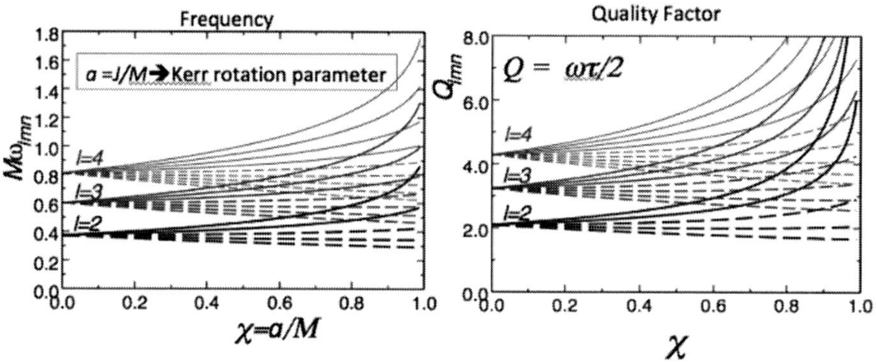

FIGURE 2. Frequencies and quality factors for the fundamental modes with $l = 2, 3, 4$ and different values of m. Solid lines refer to $m = l, .., l$ (from top to bottom), the dotted line to m = 0, and dashed lines refer to $m = 1, .., l$ (from top to bottom). Quality factors for the higher overtones are lower than the ones we display here.

LISA are ideal instruments for the ring-down observation of the BBH post merger phase. High-SNR events will give individual tests of GR, while the combination of many events collected with both advanced and 3G interferometers provide stronger constraints on GR.

The Einstein theory of Gravitation predicts that whenever crosses the BH event horizon the information it contains is lost, in contradiction with the expectations of quantum mechanics for which information can neither be created nor destroyed. The existence of Hawking radiation is directly related to this issue, because if in the evaporation the BH disappears all of the information into them flow. The detection of GW due to the excitation of the BH quasi normal modes , and in general the detection of the part of the BBH signal after the merger phase at high SNR, can tell us more about the BH nature and the properties of its horizon. The existence of echoed signals could suggest that BH are not bounded by a GR horizon but instead by a quantum-mechanical Planck-scale structure. The time delay between the echoes will be connected to the exotic compact object resembling a black hole and it should depend on its compactness, while the decay and shape of each pulse should encodes its reflective properties [4], [3].

In the future GW can probe the existence of the most exotic objects emerging from the complex mathematical framework of the GR equations. Again looking at the BH quasi normal modes we can try to disentangle the existence of classical BH from a more exotic traversing wormhole. In fact, it has been shown, under very specific assumptions [3], that a point particle, which plunges radially into a wormhole, when it crosses each of the light rings curves, it excites QNM characteristic modes trapped between the light-ring potential wells. In this very specific case the point particle plunges radially and emerges in another universe. When the particle crosses each of the light rings curves, it excites QNM characteristic modes trapped between the light-ring potential wells. The wormhole GW signal is dephased to the respect of that of a classical BH, due to the light travel time from

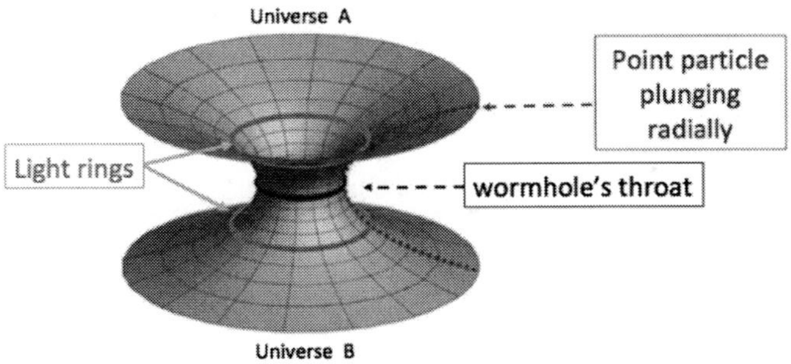

FIGURE 3

the throat to the light ring. To get evidence of the waveform difference will be a challenge, but if we succeed, it will represent a giant step forward in the exploration of our universe.

In conclusion, the future of the GW physics will be rich of new surprises and conundrums to be solved.

REFERENCES

[1] Abbott B P. et al., Living Reviews in Relativity December, 21:3 (2018)
[2] Berti E.,Cardoso V., Starinets O., Classical and Quantum Gravity, 26 (16) (2009)
[3] Cardoso V. , Franzin E., Pani P., PRL 116, 171101 (2016)
[4] Mark Z. , Zimmerman A. , Du S. M., Chen Y., PRD 96, 084002 (2017)

Balloon-borne Cosmic Microwave Background experiments

Silvia Masi[1,2,a], A. Coppolecchia[1,2], E. Battistelli[1,2], P. de Bernardis[1,2], F. Columbro[1,2], G. D'Alessandro[1,2], L. Lamagna[1,2], A. Paiella[1,2], F. Piacentini[1,2], and G. Presta[1,2]

[1] *Dipartimento di Fisica, Sapienza Universitá di Roma, Italy*
[2] *INFN Roma, Italy*

Abstract. Stratospheric balloon experiments play a unique role in current Cosmic Microwave Background (CMB) studies. CMB research has entered a precision phase, harvesting the detailed properties of its anisotropy, polarization and spectrum, at incredible precision levels. These measurements, however, require careful monitoring and subtraction of local backgrounds, produced by the earth atmosphere and the interstellar medium. High frequencies (larger than 180 GHz) are crucial for the measurements of interstellar dust contamination, but are degraded by atmospheric emission and its fluctuations, even in the best (cold and dry) sites on earth. For this reason, new balloon-borne missions, exploiting long-duration and ultra-long duration stratospheric flights, are being developed in several laboratories worldwide. These experiments have the double purpose of qualifying instrumentation and validating methods to be used on satellite missions, and produce CMB science at a relatively fast pace, synergically to ground-based CMB observatories.

1 Introduction

The Cosmic Microwave Background (CMB) is a faint background of thermal mm-wave photons, filling the entire Universe. These photons where generated a few milliseconds after the big bang, when matter and antimatter annihilated. They were thermalized by repeated Thomson scatterings in the primeval ionized fireball, during the first 380000 years. Later, neutral hydrogen atoms could form, and the universe became neutral (last scattering). At that point CMB photons formed a dazzling 3000K blackbody: the universe was everywhere like in the atmosphere of a red supergiant star. Since then, CMB photons did not interact with matter anymore, and were diluted, and redshifted, due to the expansion of the universe. Today they form a 2.725K blackbody. Having traveled in the Universe for most of its history, CMB photons carry important information on all phases of its evolution, through three main observables: the spectrum, the anisotropy, the polarization.

The power spectra of intrinsic (*primary*) CMB anisotropy is perfectly well fit by an Inflationary, Λ-CDM universe (see e.g. [1–9]).

The CMB is slightly polarized, due to anisotropic Thomson scattering at recombination. The component due to density fluctuations (irrotational E-modes) has been measured and is consistent with expectations (see e.g. [10–15]). The component due to tensor fluctuations (the gravitational waves generated by the hypothetical inflation phase in the very early universe), called inflationary B-modes due to their rotational nature, is still to be measured, and there are many on going and planned

[a] e-mail: silvia.masi@roma1.infn.it

missions focusing on this challenging measurement, from the ground, from stratospheric balloons, and from space (see e.g. [16]).

During their long travel, crossing half of the observable cosmos, CMB photons interact with the large scale structure of the universe. These *secondary* interactions provide additional precision observables of the CMB. Gravitational lensing of CMB photons provide a way to study the development of dark matter structures in the earliest phases of the structure formation process, exploiting the resulting distortion of the CMB anisotropy map (see e.g. [17]). This same effect produces B-modes of the CMB polarization field at small angular scales (see e.g. [18] and references therein). In addition, ionized structures in the universe (clusters of galaxies, filaments) produce inverse Compton scattering in CMB photons (the Sunyaev-Zeldovich effect, [19, 20]), which can be isolated exploiting its well defined and characteristic spectral distortion.

The specific brightness of a 2.725K blackbody peaks at a frequency of 159 GHz. In this high frequency microwaves band, the Earth atmosphere is only partially transparent, and only in the best sites (extremely cold and dry locations). As a result, the long-wavelength side of the CMB spectrum can be accessed quite efficiently from the ground, but the short-wavelength side cannot. Moreover, the emission of interstellar dust, a polarized foreground contaminating precision measurements of the CMB, rises steeply with frequency, and must be monitored and subtracted to obtain clean CMB measurements (see e.g. [21]). These facts drove the development of balloon-borne measurements of the CMB, carrying high-frequency CMB telescopes (with custom photometers, spectrometers and polarimeters), above the bulk of atmospheric emission and fluctuations, using stratospheric balloons.

2 Stratospheric Balloons

2.1 Generalities

Stratospheric balloons are near-space carriers able to reach altitudes of $\sim$ 40 km (corresponding to an environmental pressure of $\sim$ 3 mbar). Conventional stratospheric balloons are not sealed, and the gas pressure inside the balloon is the same as the surrounding atmospheric pressure. These balloons provide flight durations reaching 2 months (long duration stratospheric balloons, LDB), if the solar illumination is stable. Recently operated *super-pressure balloons* are sealed, and provide much longer flight durations (ultra long-duration flights, ULDBs). The lifted payload masses range from a few kg to more than two tons, depending on the balloons size, ranging from a few thousands cubic meters volume at float, to more than 1 million cubic meters. The heaviest payloads (2 tons) are larger and heavier than what can be reasonably carried by a satellite, at a mission cost which is roughly 100 times less than for a satellite mission. Once launched, operated and terminated, a stratospheric balloon payload is recovered and can be flown again, for improved measurements, after a relatively short amount of time since the first flight. This allows a staged development of the instrumentation, which is not possible in the case of satellite missions. Stratospheric balloons are thus important as precursors of satellite missions. They allow the community to test and qualify innovative space instrumentation, and improve it before using it in space. Last but not least, stratospheric missions are a very good way to educate and prepare students and young researchers. They can participate directly in all phases of the development: from the idea, to optimization, planning, construction, test, calibration, flight, data analysis of the mission.

2.2 Flight Options

We focus on LDBs and ULDBs, since CMB surveys require in general long integrations and the repetition of the measurements in different days, to test their robustness against systematic effects (sun illumination, ground spillover, etc.). LDB flights are carried out very successfully by NASA-CSBF in the summer season in Antarctica (see [22] and references therein), where a stable vortex allows the payload to circumnavigate the south pole. This is a consolidated facility, flying every summer season a few stratospheric payloads for several weeks flights, thanks to NSF and NASA-CSBF.

The Italian Space Agency has supported a pioneering program to fly LDBs from the Arctic airport of Longyearbyen (Svalbard, 78°N) [23–25]. This allowed to fly two heavy payloads (SORA in 2009 and OLIMPO in 2018) and several lighter payloads (PEGASO, DUSTER), even during the Arctic night season, with a stratospheric environment potentially very appealing for CMB survey experiments [26]. In figure 1 we show the launch of a stratospheric balloon from Svalbard, during the Arctic winter.

Figure 1. Launch of a stratospheric balloon from Svalbard during the Arctic winter.

Successful ULDB flights have been carried out (https://blogs.nasa.gov/superpressureballoon/) keeping a 700 kg payload in the stratosphere for $\sim$ 46 days, and look very promising for future CMB missions.

3 CMB missions on stratospheric balloons

The CMB community exploits the stratospheric environment to carry out sensitive observations at high frequency, high resolution, and at the largest angular scales. This is allowed by the strongly reduced atmospheric emission and noise, compared to the best CMB observation sites on Earth (Atacama desert, Antarctica, ..).

In figure 2 we plot the brightness, and the brightness fluctuations due to quantum fluctuations of the photon background, at different altitudes in the mm-wave band. The advantage of the operation of mm-wave instruments for the CMB from the stratosphere, without even considering atmospheric turbulence, is evident. In terms of achievable photon-noise-limited sensitivity, the improvement at 140 and 220 GHz is more than one order of magnitude, implying a reduction of more than two orders of magnitude of the integration time required for a given measurement. The presence of turbulence in atmospheric emission enhances the balloon environment with respect to ground even more. The high opacity of the Earth atmosphere makes balloon-borne measurements the only sub-orbital option for CMB and interstellar dust measurements at frequencies above 320 GHz.

After a pioneering period at the end of the seventies, a series of very successful balloon-borne CMB missions (see e.g. [1–4]) has been crucial for starting the so-called *precision cosmology* era. The LDB flight opportunities have been used by the BOOMERanG experiment [1], the TopHAT experiment [27], and more recently by a second generation of array-based CMB polarimeters: EBEX [28–31] and SPIDER [32–34]. Important missions to survey the CMB and polarized dust at high frequencies have been proposed [35].

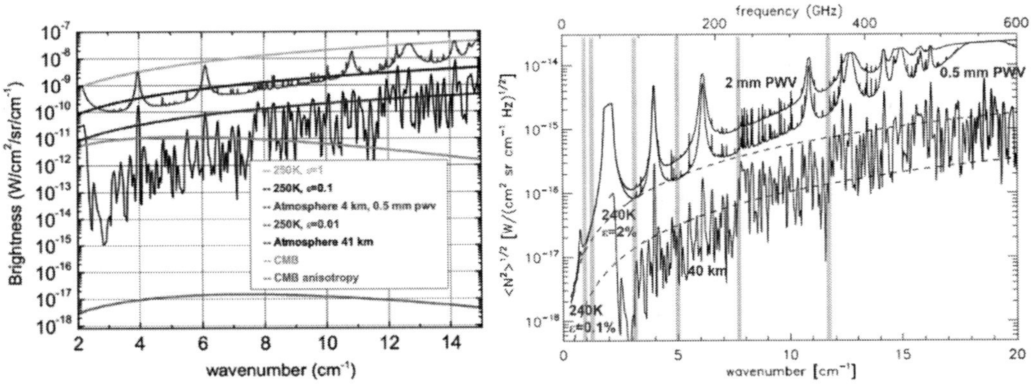

Figure 2. Left: atmospheric brightness in the mm range. Right: quantum fluctuations of the brightness. Both are shown for different altitudes. The dashed lines in the right panel represent the brightness fluctuations of room temperature optical systems, for low and high extreme values of their emissivity (2 % and 0.1 %).

The OLIMPO experiment [36], a large (2.6m aperture) CMB telescope aimed at spectral measurements of the Sunyaev-Zeldovich effect in the range 120 to 480 GHz, has been launched from Longyearbyen on July 7th, 2018 (see figure 3). This experiment has several innovative features, including the use - for the first time in the stratosphere - of arrays of Kinetic Inductance Detectors [37, 38] and a differential Fourier transform spectrometer [39–41]. See *http://olimpo.roma1.infn.it* .

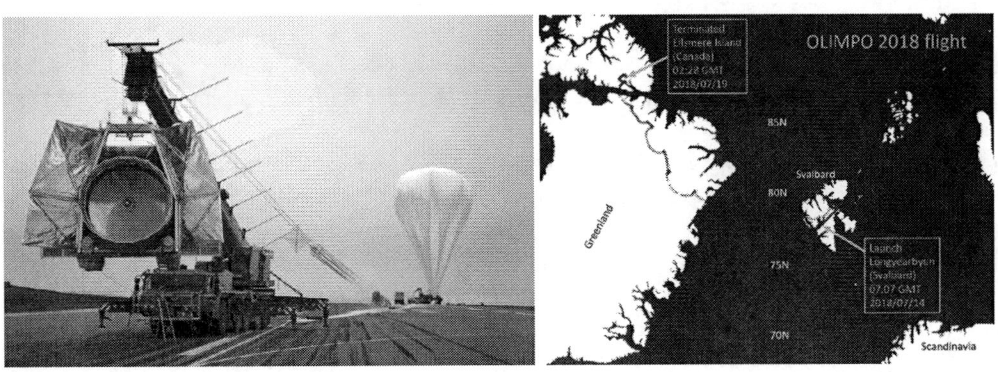

Figure 3. Launch (left) and ground path (right) of the OLIMPO mission.

The Short Wavelength Instrument for the Polarization Explorer (SWIPE) is a coarse-resolution balloon-borne CMB polarimeter, part of the Large Scale Polarization Explorer (LSPE) [42] mission, to be flown during the Arctic night. The instrument uses multi-moded TES bolometers [43, 44], sensitive to a total of 8800 radiation modes in the 150, 220, 240 GHz bands, and a cryogenic rotating HWP to modulate CMB polarization [45] and mitigate systematic effects. Its flight is planned for the 2020/2021 season. See *http://lspe.roma1.infn.it* .

4 Conclusions

Stratospheric balloons offer great opportunities for CMB research. They provide high-frequency measurements, which cannot be obtained from the ground, and are mandatory to remove the contaminating polarized signal from interstellar dust. They allow to validate in near-space the performance of key components for the final CMB polarization mission. They also provide an opportunity for measurements of spectral distortions of the CMB, either isotropic or anisotropic (SZ effect). Their development and availability to the scientific community should be continuously supported.

References

[1] P. de Bernardis, P.A.R. Ade, J.J. Bock, J.R. Bond, J. Borrill, A. Boscaleri, K. Coble, B.P. Crill, G. De Gasperis, P.C. Farese et al., Nature **404**, 955 (2000), astro-ph/0004404

[2] S. Hanany, P. Ade, A. Balbi, J. Bock, J. Borrill, A. Boscaleri, P. de Bernardis, P.G. Ferreira, V.V. Hristov, A.H. Jaffe et al., The Astrophysical Journal Letters **545**, L5 (2000), astro-ph/0005123

[3] J.E. Ruhl, P.A.R. Ade, J.J. Bock, J.R. Bond, J. Borrill, A. Boscaleri, C.R. Contaldi, B.P. Crill, P. de Bernardis, G. De Troia et al., The Astrophysical Journal **599**, 786 (2003), astro-ph/0212229

[4] A. Benoît, P. Ade, A. Amblard, R. Ansari, É. Aubourg, S. Bargot, J.G. Bartlett, J.P. Bernard, R.S. Bhatia, A. Blanchard et al., Astronomy and Astrophysics **399**, L19 (2003), astro-ph/0210305

[5] C.L. Bennett, M. Halpern, G. Hinshaw, N. Jarosik, A. Kogut, M. Limon, S.S. Meyer, L. Page, D.N. Spergel, G.S. Tucker et al., The Astrophysical Journal Supplements **148**, 1 (2003), astro-ph/0302207

[6] Planck Collaboration, P.A.R. Ade, N. Aghanim, M. Arnaud, M. Ashdown, J. Aumont, C. Baccigalupi, M. Baker, A. Balbi, A.J. Banday et al., Astronomy and Astrophysics **536**, A1 (2011), 1101.2022

[7] C.L. Bennett, D. Larson, J.L. Weiland, N. Jarosik, G. Hinshaw, N. Odegard, K.M. Smith, R.S. Hill, B. Gold, M. Halpern et al., The Astrophysical Journal Supplements **208**, 20 (2013), 1212.5225

[8] Planck Collaboration, P.A.R. Ade, N. Aghanim, M.I.R. Alves, C. Armitage-Caplan, M. Arnaud, M. Ashdown, F. Atrio-Barandela, J. Aumont, H. Aussel et al., Astronomy and Astrophysics **571**, A1 (2014), 1303.5062

[9] Planck Collaboration, Y. Akrami, F. Arroja, M. Ashdown, J. Aumont, C. Baccigalupi, M. Ballardini, A.J. Banday, R.B. Barreiro, N. Bartolo et al., arXiv e-prints (2018), 1807.06205

[10] J.M. Kovac, E.M. Leitch, C. Pryke, J.E. Carlstrom, N.W. Halverson, W.L. Holzapfel, Nature **420**, 772 (2002), astro-ph/0209478

[11] E.M. Leitch, J.M. Kovac, C. Pryke, J.E. Carlstrom, N.W. Halverson, W.L. Holzapfel, M. Dragovan, B. Reddall, E.S. Sandberg, Nature **420**, 763 (2002), astro-ph/0209476

[12] A.C.S. Readhead, S.T. Myers, T.J. Pearson, J.L. Sievers, B.S. Mason, C.R. Contaldi, J.R. Bond, R. Bustos, P. Altamirano, C. Achermann et al., Science **306**, 836 (2004), astro-ph/0409569

[13] D. Barkats, C. Bischoff, P. Farese, L. Fitzpatrick, T. Gaier, J.O. Gundersen, M.M. Hedman, L. Hyatt, J.J. McMahon, D. Samtleben et al., The Astrophysical Journal Letters **619**, L127 (2005), astro-ph/0409380

[14] F. Piacentini, P.A.R. Ade, J.J. Bock, J.R. Bond, J. Borrill, A. Boscaleri, P. Cabella, C.R. Contaldi, B.P. Crill, P. de Bernardis et al., The Astrophysical Journal **647**, 833 (2006), astro-ph/0507507

[15] T.E. Montroy, P.A.R. Ade, J.J. Bock, J.R. Bond, J. Borrill, A. Boscaleri, P. Cabella, C.R. Contaldi, B.P. Crill, P. de Bernardis et al., The Astrophysical Journal **647**, 813 (2006), astro-ph/0507514

[16] J. Delabrouille, P. de Bernardis, F.R. Bouchet, A. Achúcarro, P.A.R. Ade, R. Allison, F. Arroja, E. Artal, M. Ashdown, C. Baccigalupi et al., JCAP **4**, 014 (2018), 1706.04516

[17] A. Challinor, R. Allison, J. Carron, J. Errard, S. Feeney, T. Kitching, J. Lesgourgues, A. Lewis, I. Zubeldia, A. Achucarro et al., JCAP **4**, 018 (2018), 1707.02259

[18] Planck Collaboration, P.A.R. Ade, N. Aghanim, M. Ashdown, J. Aumont, C. Baccigalupi, A.J. Banday, R.B. Barreiro, N. Bartolo, S. Basak et al., Astronomy and Astrophysics **596**, A102 (2016), 1512.02882

[19] R.A. Sunyaev, Y.B. Zel'dovich, Astrophysics and Space Science **7**, 3 (1970)

[20] J.E. Carlstrom, G.P. Holder, E.D. Reese, Annual Review of Astronomy and Astrophysics **40**, 643 (2002), astro-ph/0208192

[21] Planck Collaboration, R. Adam, P.A.R. Ade, N. Aghanim, M. Arnaud, J. Aumont, C. Baccigalupi, A.J. Banday, R.B. Barreiro, J.G. Bartlett et al., Astronomy and Astrophysics **586**, A133 (2016), 1409.5738

[22] B.D. Stilwell, *The Latest Developments in NASA's Long Duration Balloon Systems*, in *40th COSPAR Scientific Assembly* (2014), Vol. 40 of *COSPAR Meeting*, pp. PSB.1–48–14

[23] S.E. Peterzen, R. Ibba, K. Bøen, P. Dragoy, S. Masi, A. Cardillo, I. Musso, D. Spoto, *Ultralight to heavy long duration balloon development from Svalbard, Norway*, in *17th ESA Symposium on European Rocket and Balloon Programmes and Related Research*, edited by B. Warmbein (2005), Vol. 590 of *ESA Special Publication*, pp. 403–406

[24] S. Peterzen, S. Masi, P. Dragoy, R. Ibba, D. Spoto, Memorie della Societá Astronomica Italiana **79**, 792 (2008)

[25] P. de Bernardis, S. Masi, OLIMPO and LSPE Teams, *Precision CMB measurements with long-duration stratospheric balloons: activities in the Arctic*, in *Astrophysics from Antarctica*, edited by M.G. Burton, X. Cui, N.F.H. Tothill (2013), Vol. 288 of *IAU Symposium*, pp. 208–213

[26] F. Piacentini, A. Coppolecchia, P. de Bernardis, G. Di Stefano, A. Iarocci, L. Lamagna, S. Masi, S. Peterzen, R. Giovanni, arXiv e-prints (2018), 1810.05565

[27] R.F. Silverberg, E.S. Cheng, J.E. Aguirre, J.J. Bezaire, T.M. Crawford, S.S. Meyer, A. Bier, B. Campano, T.C. Chen, D.A. Cottingham et al., The Astrophysical Journal Supplements **160**, 59 (2005)

[28] P. Oxley, P.A. Ade, C. Baccigalupi, P. deBernardis, H.M. Cho, M.J. Devlin, S. Hanany, B.R. Johnson, T. Jones, A.T. Lee et al., *The EBEX experiment*, in *Infrared Spaceborne Remote Sensing XII*, edited by M. Strojnik (2004), Vol. 5543 of *Proceedings of the SPIE*, pp. 320–331, astro-ph/0501111

[29] EBEX Collaboration, A.M. Aboobaker, P. Ade, D. Araujo, F. Aubin, C. Baccigalupi, C. Bao, D. Chapman, J. Didier, M. Dobbs et al., The Astrophysical Journal Supplements **239**, 7 (2018), 1703.03847

[30] EBEX Collaboration, M. Abitbol, A.M. Aboobaker, P. Ade, D. Araujo, F. Aubin, C. Baccigalupi, C. Bao, D. Chapman, J. Didier et al., The Astrophysical Journal Supplements **239**, 8 (2018), 1803.01018

[31] EBEX Collaboration, A. Aboobaker, P. Ade, D. Araujo, F. Aubin, C. Baccigalupi, C. Bao, D. Chapman, J. Didier, M. Dobbs et al., The Astrophysical Journal Supplements **239**, 9 (2018), 1702.07020

[32] T.E. Montroy, P.A.R. Ade, R. Bihary, J.J. Bock, J.R. Bond, J. Brevick, C.R. Contaldi, B.P. Crill, A. Crites, O. Doré et al., *SPIDER: a new balloon-borne experiment to measure CMB polarization on large angular scales*, in *Society of Photo-Optical Instrumentation Engineers (SPIE) Conference Series* (2006), Vol. 6267 of *Proceedings of the SPIE*, p. 62670R

[33] A.S. Rahlin, P.A.R. Ade, M. Amiri, S.J. Benton, J.J. Bock, J.R. Bond, S.A. Bryan, H.C. Chiang, C.R. Contaldi, B.P. Crill et al., *Pre-flight integration and characterization of the SPIDER balloon-borne telescope*, in *Millimeter, Submillimeter, and Far-Infrared Detectors and Instrumentation for Astronomy VII* (2014), Vol. 9153 of *Proceedings of the SPIE*, p. 915313, 1407.2906

[34] J.M. Nagy, P.A.R. Ade, M. Amiri, S.J. Benton, A.S. Bergman, R. Bihary, J.J. Bock, J.R. Bond, S.A. Bryan, H.C. Chiang et al., The Astrophysical Journal **844**, 151 (2017), 1704.00215

[35] S. Bryan, P. Ade, J.R. Bond, F. Boulanger, M. Devlin, S. Doyle, J. Filippini, L. Fissel, C. Groppi, G. Holder et al., *BFORE: a CMB balloon payload to measure reionization, neutrino mass, and cosmic inflation*, in *Millimeter, Submillimeter, and Far-Infrared Detectors and Instrumentation for Astronomy IX* (2018), Vol. 10708 of *Society of Photo-Optical Instrumentation Engineers (SPIE) Conference Series*, p. 1070805, 1807.05215

[36] S. Masi, E. Battistelli, D. Brienza, L. Conversi, A. Cruciani, P. de Bernardis, M. De Petris, P. Fiadino, A. Iacoangeli, L. Lamagna et al., Memorie della Societá Astronomica Italiana **79**, 887 (2008)

[37] A. Paiella, E.S. Battistelli, M.G. Castellano, I. Colantoni, F. Columbro, A. Coppolecchia, G. D'Alessandro, P. de Bernardis, S. Gordon, L. Lamagna et al., *Design and Electrical Performance of the Kinetic Inductance Detectors of the OLIMPO Experiment*, in *2017 16th International Superconductive Electronics Conference (ISEC)* (2017), pp. 1–3

[38] A. Paiella, A. Coppolecchia, L. Lamagna, P. Ade, E. Battistelli, M.G. Castellano, I. Colantoni, F. Columbro, G. D'Alessandro, P. de Bernardis et al., Journal of Cosmology and Astroparticle Physics **2019**, 039 (2019)

[39] P. de Bernardis, S. Colafrancesco, G. D'Alessandro, L. Lamagna, P. Marchegiani, S. Masi, A. Schillaci, Astronomy and Astrophysics **538**, A86 (2012), 1111.4588

[40] A. Schillaci, G. D'Alessandro, P. de Bernardis, S. Masi, C. Paiva Novaes, M. Gervasi, M. Zannoni, Astronomy and Astrophysics **565**, A125 (2014), 1402.4091

[41] G. D'Alessandro, P. de Bernardis, S. Masi, A. Schillaci, Applied Optics **54**, 9269 (2015), 1509.09221

[42] S. Aiola, G. Amico, P. Battaglia, E. Battistelli, A. Baó, P. de Bernardis, M. Bersanelli, A. Boscaleri, F. Cavaliere, A. Coppolecchia et al., *The Large-Scale Polarization Explorer (LSPE)*, in *Ground-based and Airborne Instrumentation for Astronomy IV* (2012), Vol. 8446 of *Proceedings of the SPIE*, p. 84467A

[43] M. Biasotti, D. Bagliani, V. Ceriale, D. Corsini, P. De Bernardis, F. Gatti, R. Gualtieri, L. Lamagna, S. Masi, G. Pizzigoni et al., *Large area TES spiderweb bolometer for multi-mode cavity microwave detect*, in *Millimeter, Submillimeter, and Far-Infrared Detectors and Instrumentation for Astronomy VII* (2014), Vol. 9153 of *Proceedings of the SPIEE*, p. 915308

[44] R. Gualtieri, E.S. Battistelli, A. Cruciani, P. de Bernardis, M. Biasotti, D. Corsini, F. Gatti, L. Lamagna, S. Masi, Journal of Low Temperature Physics **184**, 527 (2016), 1602.07744

[45] A. Buzzelli, P. de Bernardis, S. Masi, N. Vittorio, G. de Gasperis, Astronomy and Astrophysics **609**, A52 (2018), 1709.04842

Integration and testing of the Mini-EUSO telescope

G. Cambiè[1] and L. Marcelli[1]
for the JEM-EUSO collaboration
e-mail: giorgio.cambie@roma2.infn.it

[1] *INFN and University of Rome Tor Vergata, Physics Department, via della Ricerca Scientifica 1, 00133, Roma (RM).*

Abstract. The Mini-EUSO telescope, approved by ASI (Italian Space Agency) and Roscosmos (Russian Space Agency) as long-term program of space experiments to be launched on the Progress spacecraft during the Expedition 60/61 in 2019, will be placed on a nadir-facing, UV transparent window on the ISS Russian Zvezda module looking down on Earth. As one of the JEM-EUSO collaboration missions, its first objective is to observe UV light produced by UHECR (Ultra High Energy Cosmic Ray) events from upper atmosphere ($\sim$ 400 km altitude), moreover, from this point of view, Mini-EUSO will perform, for the first time, high-resolution mapping of night-earth UV (300 - 400 nm) emission, focusing on terrestrial background sources, TLEs (Transient Luminous Events), bioluminescence, meteoroids burning in atmosphere, search for SQM (Strange Quark Matter) signals, as well as detecting and mapping space debris. The integration of the instrument, currently underway at the University of Rome Tor Vergata, is at an advanced stage in order to be compliant with a launch opportunity in June 2019. The on ground test results will be presented.

1 Introduction

Along with TA-EUSO, EUSO-Balloon and EUSO-SPB, Mini-EUSO belongs to the JEM-EUSO collaboration missions [1], which involves more than 16 countries, 93 institutes and 351 people all over the world. The JEM-EUSO collaboration aims to detect UV light produced by UHECR-induced extensive air showers (EAS) due to the interaction in atmosphere with N_2 molecules from the vantage point of low Earth orbit, thereby largely increasing the effective detector volume [2]. Mini-EUSO instrument comprises a compact telescope of dimensions $37 \times 37 \times 62\ cm^3$ with a large field of view ($\pm 22°$), based on an optical system employing a focusing optics, a sensor array of MAPMTs (Multi Anode PhotoMultiplier Tubes) with its read-out electronics and different ancillary sensors for complementary measurements in the near infrared (NIR) (1500 - 1600 nm) and visible (VIS) (400 - 780 nm) range.

2 Instrument overview

Mini-EUSO instrument comprises two double sided PMMA Fresnel lenses, 25 cm in diameter, which will focus light onto a focal surface, called PDM (Photo Detector Module), consisting of an array

 https://doi.org/10.1051/epjconf/201920901047

of 36 Hamamatsu Photonics (model R11265-M64) MAPMTs. Details are shown in table 2a. Each dinodes of MAPMTs is powered by a Cockroft-Walton high voltage power supply placed inside a ceramic pad. All MAPMTs are covered by a 2 mm, anti-reflective, BG3 UV filter placed on the entry window. In addition to the main detector sensible in the UV range, Mini-EUSO contains two ancillary cameras, NIR and VIS, provided by PointGrey. All subsystems are shown in fig. 1.

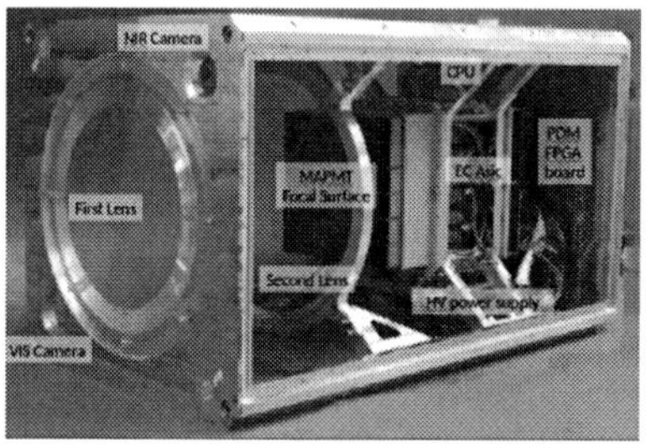

(a) Mini-EUSO telescope body

(b) Mini-EUSO PDM

Figure 1: Mini-EUSO apparatus

Furthermore, for the first time in space, Mini-EUSO telescope will take advantage of taking data through a 8 × 8 channels SiPM-MPPC (Silicon PhotoMultiplier-Multi Pixel Proportional Counter) C14047 Hamamatsu array. The data processing chain comprises three dedicated SPACIROC3 Asic boards, a Xilinx Zynq XC7Z030 system on chip (Kintex7 FPGA), a PCIe/104 form factor CPU and an ATMegaS128 rad hard based board for analog sensors read out with a sampling rate of 2.5 μs called Gate Time Units (GTUs). The LVPS (Low Voltage Power Supply) feed the entire Mini-EUSO subsystem which filters and stabilize 27 V coming from ISS, thus preserving the entire instrumentation from spike or polarization inversion through several Vicor DC/DC Converters. A scheme of power supply is shown in fig. 2b.

Lens Dimension	$25 \times 11\ mm^2$
Focal Lenght	$30\ cm$
N° of Pixels	2304
FoV/PDM	$44/2.6 \times 10^2\ km$
FoV/pixel	$0.8/6.11\ km$
Mass	$\sim 27\ kg$
Power Consumption	≤ 60

(a) Mini-EUSO key parameters

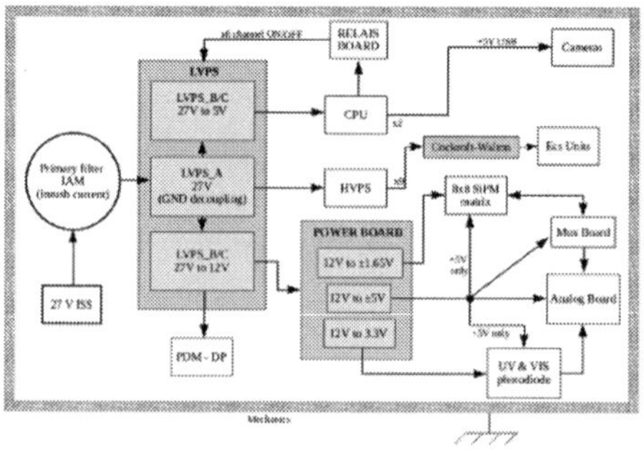

(b) Mini-EUSO subsystem power distribution scheme

Figure 2

3 Integration and test

The JEM-EUSO collaboration worked on the integration of the apparatus during whole 2017 and 2018 focusing on: mechanics and supports assembly, operational tests on electronic boards, subsystem interfaces, instrumentation response and efficiency. In fig. 3 are shown some of the apparatus integration basic steps and the test bench setup done at the hosting institutes: Université Paris Diderot - APC laboratoire (Astroparticule et Cosmologie), Riken research institute (Wako, Tokyo) and University of Rome Tor Vergata. We focused our attention on the qualification and efficiency test on the Photo Detector Module comprising the MAPMTs array and the front-end electronics. Several aspects have been taken into account.

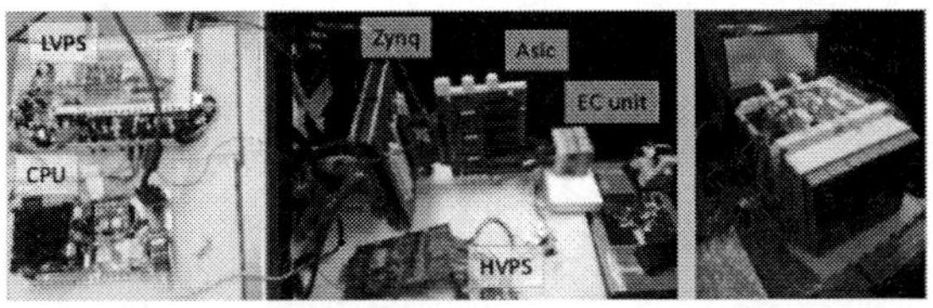

(a) Test bench inside a black box for electronics operational tests and MAPMTs response

(b) Test on fresnel lenses point spread function through a laser shot

Figure 3: Mini-EUSO Integration and test bench procedure

Tests on the correct operativity, power supply and response of the whole detector were made, as well as the PMTs efficiency analysis to provide the correct single photoelectron counting mode working point, through Asic ADC graphical analysis response (S-curve). This can be achieved from fig. 5b, where in the top we have a sample of S-curve which shows counts as a function of ADC channels while on bottom its slope analysis allows to determine the right ADC channel working point. Simulation of PDM response to the signal produced by interested events such as UHECR (fig. 5c), meteoroids and SQM efficiency detection by magnitude (fig. 5d), satellite and space debris passage on the field of view, star classification and so on [4] were made. Most of the work was done through the ETOS (Euso TO Screen) scientific software which was expressly generated for all experiments belonging to the JEM-EUSO program. It is the main quick-view and quick-analysis software written both in Python and C++ which operates on the ROOT files generated with ETOT (Euso TO Tree) from the raw CPU data. A demonstration on how ETOS works is shown in fig. 4.

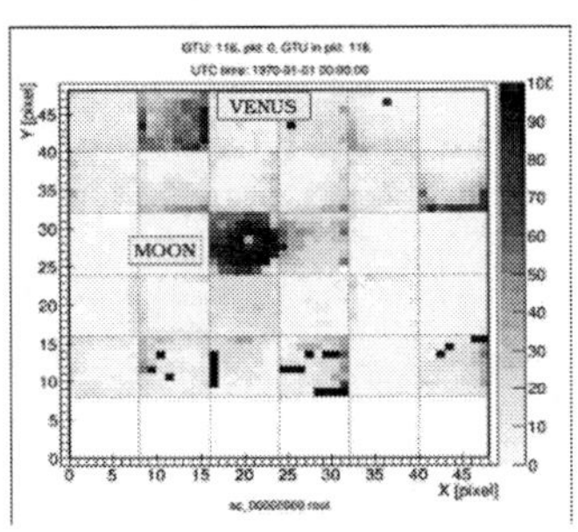

Figure 4: Tokyo night time PDM reconstruction with ETOS software.

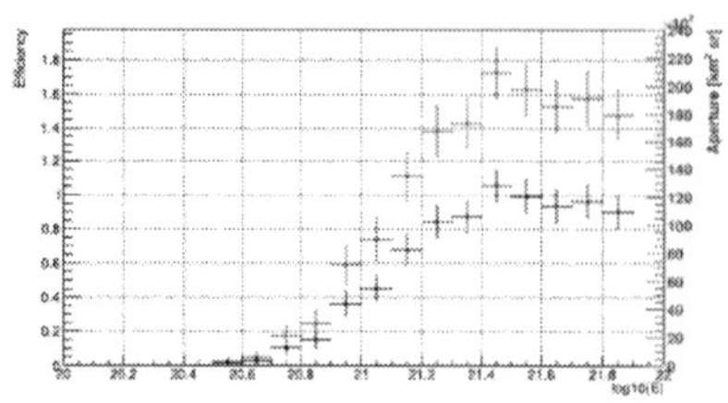

(a) The detection efficiency (on the left axis, in black) and geometric aperture (on the right axis, in red) as a function of the EAS energy in eV [3]

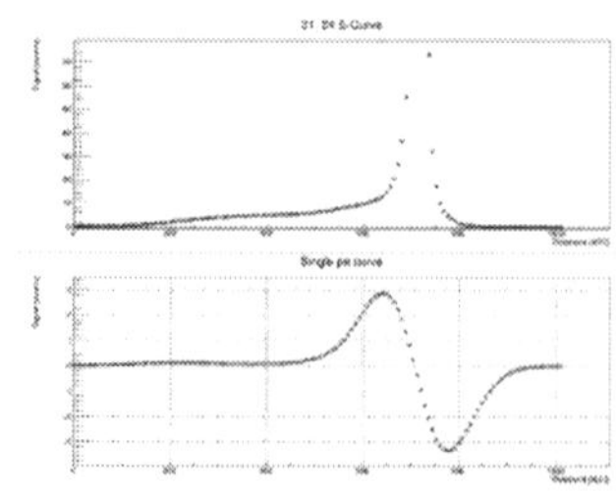

(b) Single pixel S-curve and its derivative in order to understand the single photoelectron work point

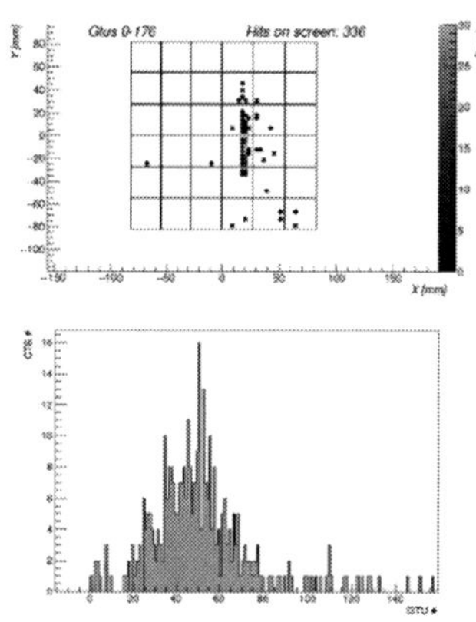

(c) Top: Photon counts observed in the Mini-EUSO focal surface for a simulation of an $E = 1 \times 10^{21}$ eV event with an inclination of 80° to the nadir. Bottom: Light curve for the same event [4]

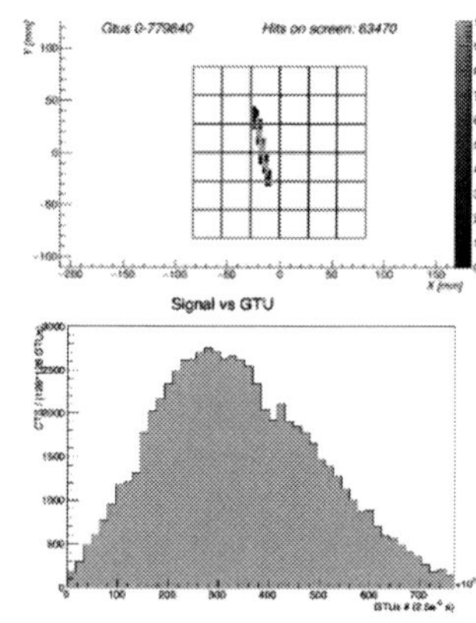

(d) Top: Expected light track of a meteor of absolute magnitude M = +5. Bottom: Expected light profile. Each time bin on x-axis corresponds to an integration time of 40.96 ms, corresponding to Mini-EUSO data trigger level 3 [4]

Figure 5: Mini-EUSO PDM efficiency analysis and simulations

References

[1] M. Casolino, A. Belov, M. Bertaina, G. Cambie', F. Capel, T. Ebisuzaki, P. Klimov, M.I. Panasyuk, P. Picozza, M. Ricci, Proceedings of Science, 301, 369 (2017). PoS(ICRC2017)369

[2] A. V. Olinto, E. Parizot, M. Bertaina, and G. Medina-Tanco, Proceedings of the 34th International Cosmic Ray Conference **236**, (2015), PoS(ICRC2015)623,.

[3] F. Capel, A. Belov, M. Casolino, P. Klimov, Advances in Space Research **62(10)**, 2954-2965, (2018), 1729.00405.

[4] M. Bertaina, S. Biktemerova, K. Bittermann, et al., Advances in Space Research, **53(10)**, 1515-1535, (2014).

Main scientific results of the DAMPE mission

Paolo Bernardini[1,2,a], on behalf of the DAMPE Collaboration

[1] *Dipartimento di Matematica e Fisica, Università del Salento, Lecce, Italy*
[2] *Istituto Nazionale di Fisica Nucleare, Sezione di Lecce, Lecce, Italy*

Abstract. DAMPE (DArk Matter Particle Explorer) is a satellite-born experiment, resulting from the collaboration of Chinese, Italian, and Swiss institutions. Since December 2015, DAMPE flights at the altitude of 500 km and collects data smoothly. The detector is made of four sub-detectors: top layers of plastic scintillators, a silicon-tungsten tracker, a BGO calorimeter (32 radiation lengths), and a bottom boron-doped scintillator to detect delayed neutrons. The main goal of the experiment is the search for indirect signals of Dark Matter in the electron and photon spectra with energies up to 10 TeV. Furthermore DAMPE studies cosmic charged and gamma radiation. The calorimeter depth and the large acceptance allow to measure cosmic ray fluxes in the range from 20 GeV up to hundreds of TeV. An overview of the latest results about light component (p+He) of charged cosmic rays, gamma astronomy and electron and positron spectrum will be presented.

1 The detector

The DArk Matter Particle Explorer (DAMPE) is a satellite-born experiment and it is funded by the strategic space projects of the Chinese Academy of Sciences with the contribution of Italian and Swiss agencies [1, 2]. It is devoted to measure the fluxes of charged cosmic rays (electrons, protons and heavier nuclei), to study the high energy gamma signal of astrophysical origin and to search for indirect dark matter signatures.

The detector design was performed looking for the proper compromise between research goals and limitations on geometry, power consumption and weight. In Fig. 1 the four sub-detectors of the DAMPE experiment are shown:

- PSD, in place of Plastic Scintillator strip Detector, used to measure the charge (Z) of cosmic rays and to select gamma events (events in anti-coincidence with the PSD signal),
- STK, in place of Silicon-Tungsten tracKer-converter, to reconstruct the trajectory of incident particles, to repeat the charge measurement and to convert photons in electron-positron pairs,
- BGO, in place of bismuth germanate imaging calorimeter (14 layers) of about 32 radiation lengths, to reach high resolution in energy measurement and to distinguish between hadronic and electromagnetic showers,
- NUD, in place of NeUtron Detector, to further increase the hadronic shower rejection power.

[a] e-mail: paolo.bernardini@le.infn.it

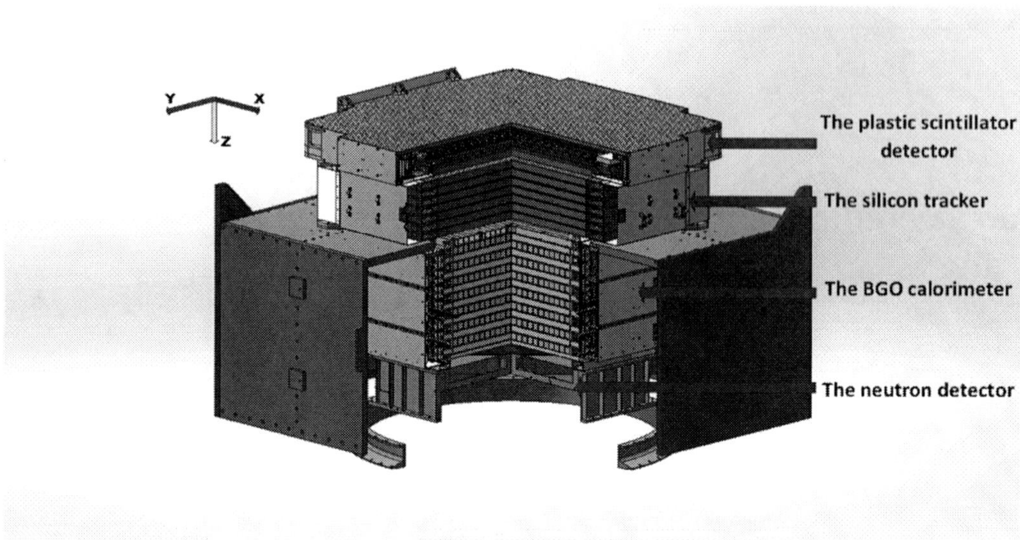

Figure 1. The DAMPE detector

DAMPE performances were verified by a series of beam tests at CERN [1, 3, 4]. The wide energy range of electron beam and the high purity of proton beam allowed to check energy resolution, linearity (Fig. 2) and e/p separation (see Fig. 8 in the following). Also beams of argon and lead fragments were used to check the PSD capability to measure Z for heavy ions (Fig. 3 for the argon beam). Furthermore sea-level muon test have been performed during different stages of the DAMPE assembly. Then the energy response to Minimum Ionizing Particles (MIPs), the efficiency and the detector alignment have been checked with a large sample of cosmic ray muon events.

The satellite has been successfully put on orbit on December 17, 2015. After few days all detectors were powered on and the data acquistion began very soon. Presently DAMPE flies at the altitude of 500 km on a Sun-synchronous orbit.

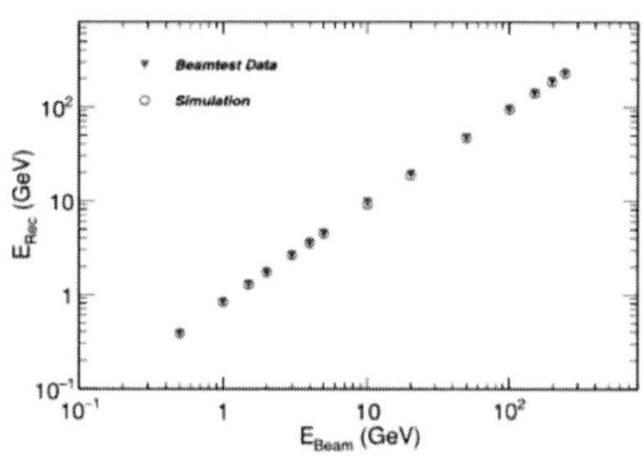

Figure 2. Beam test - Reconstructed energy versus electron beam energy.

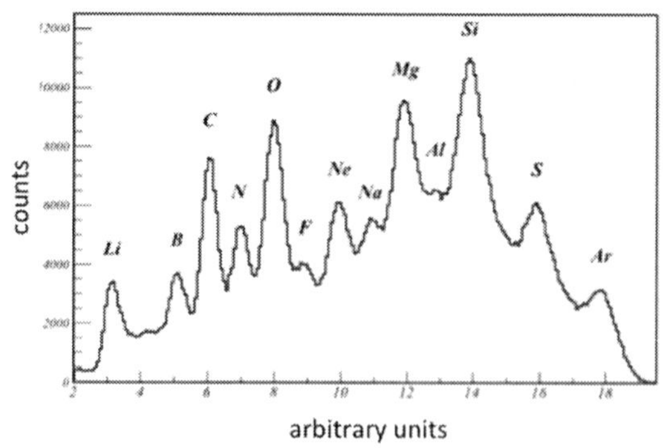

Figure 3. Beam test - PSD charge spectrum of fragments of an argon beam (helium peak has been removed).

 https://doi.org/10.1051/epjconf/201920901048

On orbit the detector is calibrated by using the standard scientific data and dedicated calibration data. The data acquisition system runs continuously in observation mode for most of the time and switches to calibration mode for 40 seconds twice per orbit. All the calibration results (pedestal values, dynode ratios, energy gains, trigger thresholds and so on) are consistent with the ground ones and very stable after temperature correction. The STK alignment allows a significant improvement of the tracking resolution [5]. In Fig. 4 (left panel) the PSD signal due to minimum ionizing protons is plotted as a function of time in the period December 2016 - March 2017. A similar plot is shown in the right panel of Fig. 4 for BGO signal due to helium nuclei not showering in the BGO calorimeter. In both cases the detector response is very stable.

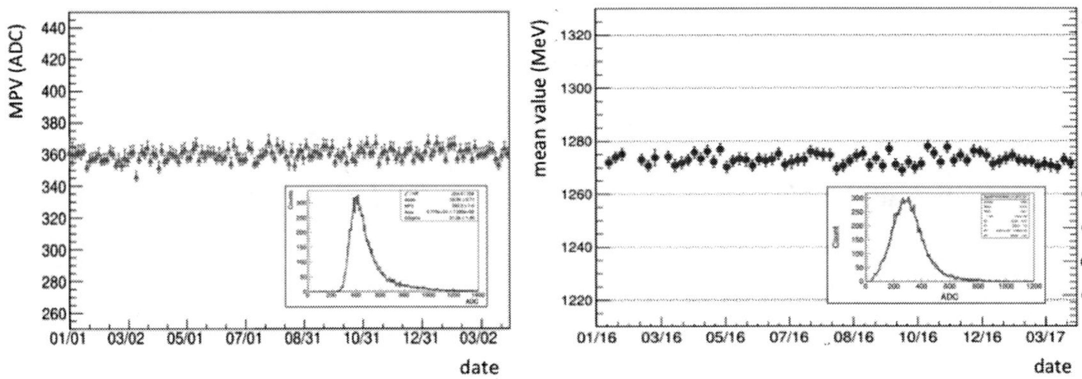

Figure 4. Left panel: the deposited energy in PSD for minimum ionizing protons is fitted with the convolution of a Landau function with a Gaussian one (insert). The Most Probable Value (MPV) s shown as a function of time. Right panel: similar plot for minimum ionizing helium nuclei in BGO (the mean value of the deposited energy is displayed versus time instead of MPV).

2 Cosmic rays

The cosmic ray (CR) studies are based on the charge measurement (square root of the energy release in PSD is proportional to Z), the STK tracking and the BGO imaging. The event shape in the calorimeter allows to distinguish electromagnetic showers from hadronic ones. Furthermore the unprecedented thickness of the calorimeter allows to push the direct measurement of CR spectra to highest energies with excellent energy resolution. Indeed DAMPE CR measurements are going to reach energies of some hundreds of TeV.

The template fit for protons and helium in Fig. 5 shows an excellent data-Montecarlo agreement and allows to estimate the contamination in each peak. The preliminary measurement of proton and helium fluxes are presented in Fig. 6.

3 Gamma astronomy

Another aim of the DAMPE mission is the gamma astronomy by collecting events without hits on the PSD layers. In less than 1 year DAMPE is able to scan the whole sky by looking for bright gamma sources, detecting their fluxes and possible variability. The gamma periodicity of two brightest pulsars (Geminga and Vela) has been measured in the energy range from 1 GeV to 100 GeV [6]. In Fig. 7 the flux of the variable high energy gamma emission detected by DAMPE from two active galactic nuclei (CTA102, 3C454.3) is shown [7].

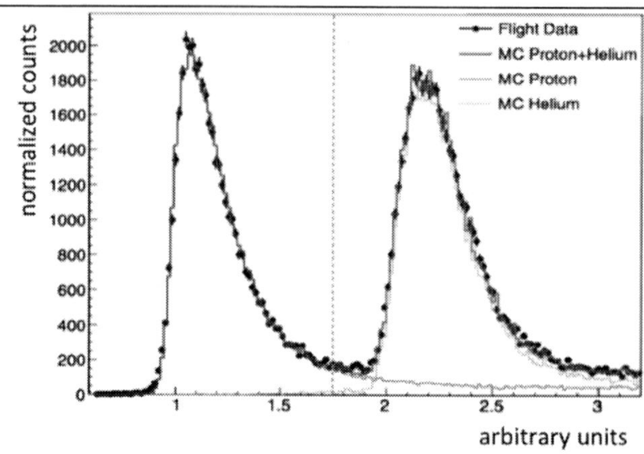

Figure 5. Combined PSD signal for protons and helium nuclei, for BGO energies in the range from 447 GeV to 562 GeV. The color codes to distinguish on-orbit and simulated data are shown in the plot. The cut to select proton candidates is represented by the vertical dotted line.

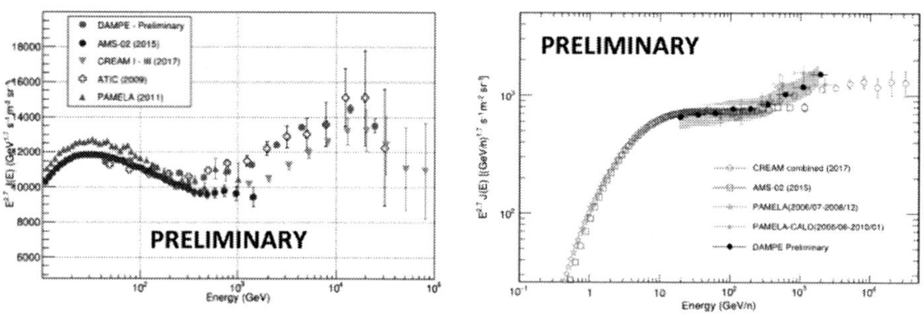

Figure 6. Spectra of protons (left panel) and helium nuclei (right panel).

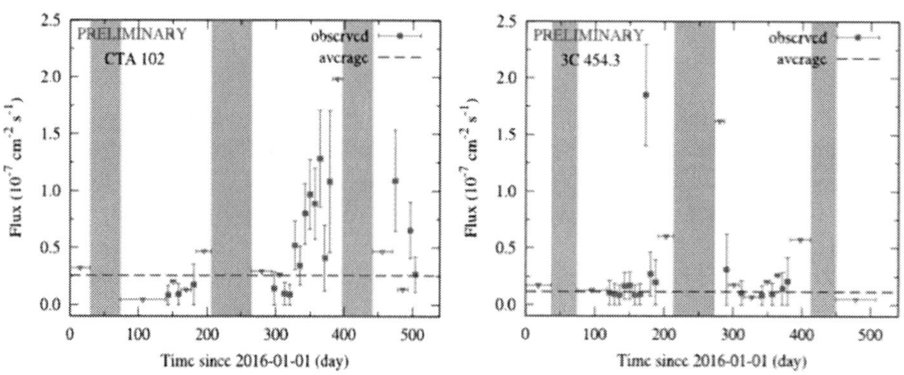

Figure 7. The light curves for two active galactic nuclei (left: CTA 102, right: 3C454.3).

4 Measurement of electron and positron spectrum

In the last years some measurements [8–14] reported an unexpected shape of the flux for CR electrons and positrons (CRE). Furthermore evidence for a spectral break in the TeV range of the CRE flux has been provided by indirect measurements [15, 16], although the results were affected by significant uncertainties.

In around two years of data taking DAMPE made a direct measurement of CREs in the range from 25 GeV to 4.6 TeV [17] with unprecedented accuracy (high energy resolution and low background). Indeed the depth and the imaging properties of the BGO calorimeter are exploited to recognise the showers initiated by CRE. The identification algorithm is based on the spread of the BGO hits (*spread*) and on the fraction (F_{last}) of the energy on the lower layer with respect to the sum of energies released on all the BGO layers. The quantities *spread* and F_{last} are combined in the parameter

$$\zeta = F_{\mathrm{last}} \frac{(spread/\mathrm{mm})^4}{8 \times 10^6}. \tag{1}$$

The efficiency of the ζ-parameter in order to distinguish electrons and protons has been verified in the beam test (Fig. 8). Then the CRE sample is selected by requiring $\zeta < 8.5$ (see Fig. 9, left panel). The CRE flux measured by DAMPE as a function of energy is shown in right panel of Fig. 9 with measurements by others experiments. A smoothly broken power law is preferred to fit the largest part of the spectrum instead of a single power law model. More details are available in [17].

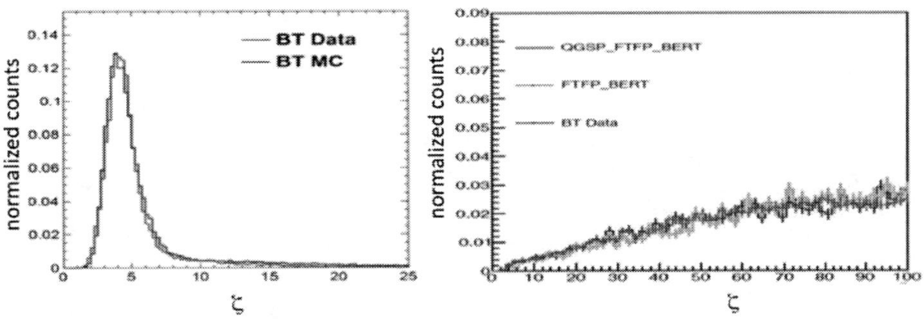

Figure 8. Beam test - The ζ-distributions for electrons (left panel) and for protons (right panel). In the proton plot the real data are compared with simulations based on different models of the hadronic interactions.

5 Conclusions

The DAMPE detector works extremely well since its launch more than 2.5 years ago and it is expected to operate stably in the next few years. Many analyses based on DAMPE data are going on successfully on cosmic rays and gamma astronomy. The measurement of CRE spectrum is particularly remarkable. Indeed the direct detection of a spectral break at ~0.9 TeV confirms what found by previous indirect measurements, clarifies the shape of the CRE spectrum at energies above 1 TeV and sheds light on the physical origin of the sub-TeV CREs.

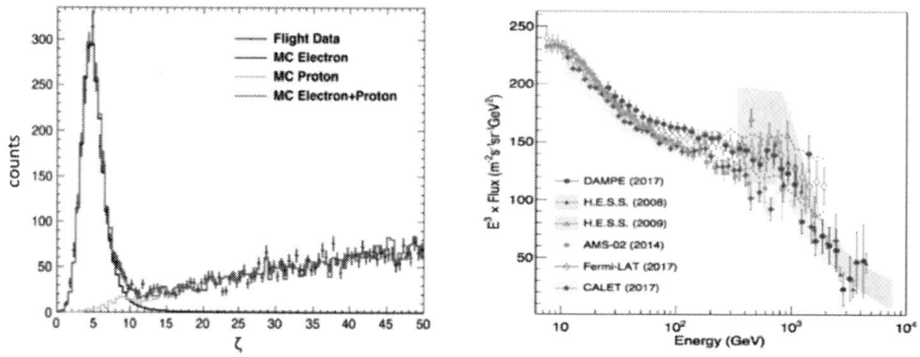

Figure 9. Left: Distribution of the ζ parameter (the cut $\zeta < 8.5$ is applied to select the CRE sample). Right: electron and positron spectrum measured by different experiments (DAMPE: red dots, HESS systematic errors: grey band).

References

[1] Chang J., Chinese J. Space Science **34**, 550 (2014)
[2] Chang J. *et al.*, Astropart. Physics **95**, 6 (2017)
[3] Azzarello P. *et al.*, Nuclear Instruments and Methods A **831**, 378 (2016)
[4] Zhang Z. *et al.*, Nuclear Instruments and Methods A **836**, 98 (2016)
[5] Tykhonov A. *et al.*, Nuclear Instruments and Methods A **893**, 43 (2018)
[6] Salinas M.M. *et al.*, PoS(ICRC2017) 709 (2017)
[7] Yuan Q. *et al.*, PoS(ICRC2017) 617 (2017)
[8] Chang J. *et al.* (The ATIC Collaboration), Nature **456**, 362 (2008)
[9] Abdo A.A. *et al.* (The Fermi-LAT Collaboration), Phys. Rev. Lett. **102**, 181101 (2009)
[10] Abdollahi S. *et al.* (The Fermi-LAT Collaboration), Phys. Rev. D **95**, 082007 (2017)
[11] Aguilar M. *et al.* (The AMS-02 Collaboration), Phys. Rev. Lett. **113**, 221102 (2014)
[12] Adriani O. *et al.* (The PAMELA collaboration), Phys. Rev. Lett. **106**, 201101 (2011)
[13] Adriani O. *et al.* (The CALET Collaboration), Phys. Rev. Lett. **119**, 181101 (2017)
[14] Adriani O. *et al.* (The CALET Collaboration), Phys. Rev. Lett. **120**, 261102 (2018)
[15] Aharonian F. *et al.* (The H.E.S.S. Collaboration), Phys. Rev. Lett. **101**, 261104 (2008)
[16] Aharonian F. *et al.* (The H.E.S.S. Collaboration), Astron. Astrophys. **508**, 561 (2009)
[17] Ambrosi G. *et al.* (The DAMPE Collaboration), Nature **552**, 63 (2017)

SiPM optical modules for the Schwarzschild-Couder Medium Size Telescopes proposed for the CTA observatory

G. Ambrosi[1,2], M. Ambrosio[3], C. Aramo[3], B. Bertucci[1,2], E. Bissaldi[3], M. Bitossi[3], C. Boiano[3], C. Bonavolontà[3,7], M. Caprai[1], L. Consiglio[3], L. Di Venere[4], E. Fiandrini[1,2,a], N. Giglietto[4,5], F. Giordano[4,5], M. Ionica[1], F. Liciulli[4], S. Loporchio[4,5], V. Masone[3], R. Paoletti[6,8], A. Rugliancich[6,8], L. Tosti[1,2], V. Vagelli[1,2], M. Valentino[9], and for the CTA SCT Project

[1] *INFN Perugia, Italy*
[2] *Dipartimento di Fisica e Geologia, Università degli Studi di Perugia, Italy*
[3] *INFN Napoli, Italy*
[4] *INFN Bari, Italy*
[5] *Dipartimento Interateneo di Fisica dell'Università e del Politecnico di Bari, Italy*
[6] *INFN Pisa, Italy*
[7] *Dipartimento di Fisica, Universitá di Napoli, Italy*
[8] *Sezione di Fisica, Dipartimento SFTA dell'Università di Siena, Italy*
[9] *CNR-Spin, Napoli, Italy*

Abstract. Silicon Photomultipliers (SiPMs) are excellent devices to detect the faint and short Cherenkov light emitted in high energy atmospheric showers, and therefore suitable for use in imaging air Cherenkov Telescopes. The high density Near Ultraviolet Violet SiPMs (NUV-HD3) produced by Fondazione Bruno Kessler (FBK) in collaboration with INFN were used to equip optical modules for a possible upgrade of the Schwarzschild-Couder Telescope camera prototype, in the framework of the Cherenkov Telescope Array project. SiPMs are 6×6 mm^2 devices based on 40×40 μm^2 microcells optimized for photo-detection at the NUV wavelengths. More than 40 optical modules, each composed by a 4×4 array of SiPMs, were assembled. In this contribution we report on the development and on the assembly of the optical modules, their validation and integration in the camera.

1 The SCT telescope proposed for the CTA observatory

The Cherenkov Telescope Array (CTA) Consortium is an international collaboration to build the next generation of Imaging Air Cherenkov Telescopes (IACTs) for the detection of γ rays in the 20 GeV-300 TeV range [1]. Two arrays of IACTs of different dimensions and mirror optics technologies will be deployed, one in the Southern and one in the Northern hemisphere, to achieve the full sky coverage. Medium Size Telescopes (MST) based on a Davies-Cotton optics and photomultiplier tube (PMT) readout provide the core sensitivity of the observatory in the 100 GeV-10 TeV energy range. A possible upgrade of the MST telescope based on a dual-mirror optics, the Schwarzschild-Couder

[a] e-mail: emanuele.fiandrini@pg.infn.it

Telescope (SCT), is proposed as an alternative for medium telescopes [2]. The SCT optical system is designed to provide compensation of optical aberrations and focus the light on a compact high resolution imaging camera. This results in a smaller point spread function (PSF), improved angular resolution, a wider field of view (8°) and improved background rejection compared to single mirror solution with a similar cost per telescope. The SCT camera can be consequently equipped with

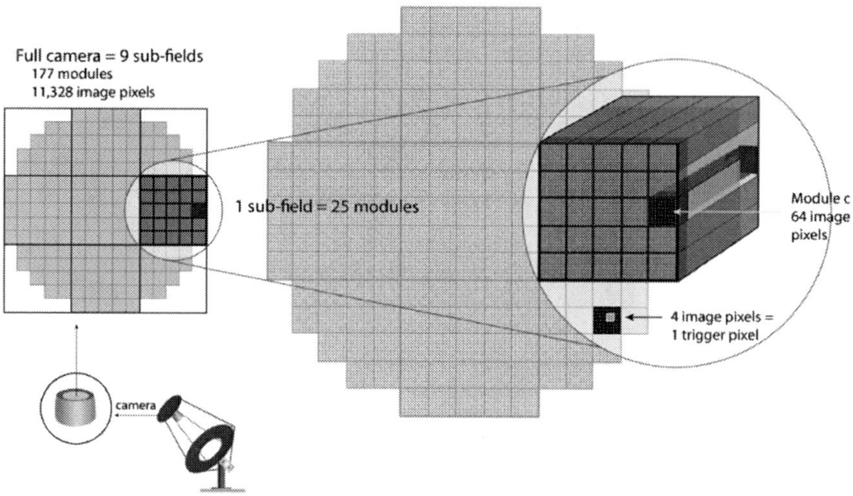

Figure 1. Top: scheme of the focal plane of the pSCT telescope. Bottom: Photo Detection Unit (PDU) made of 4×4 SiPMs.

imaging silicon photomultipliers (SiPM) pixels. The installation of a prototype for the SCT solution, the pSCT telescope [3], has been successfully completed at the Fred Lawrence Whipple Observatory (FLWO) in August 2018 and it is currenty being commissioned.

The pSCT camera [4] readout concept is based on SiPM detectors for photon detection and a TARGET-7 ASIC front-end electronics (see Sec. 3). The backplanes of the camera handle and manage the data acquisition for separate camera sectors. The fully equipped focal plane of the telescope, shown schematically in the top panel of Fig. 1, has a diameter of ~81 cm for a total active area of 0.4 m^2. It will host 177 modules in 9 sectors or sub-fields. Each module is made of 64 SiPMs arranged in 4 independent Photo Detection Units (PDU) equipped with 16 SiPMs, that constitute the single pixels of the pSCT camera. A PDU is shown in the bottom panel of Fig. 1. The pSCT camera will be equipped with a total of 11328 pixels. Each sensor corresponds to a 0.067° pixel in the sky, matching the PSF of the optical system. Hamamatsu S12642 SiPMs have been originally selected to equip the pSCT camera. A possible upgrade based on the third generation of Near Ultra Violet High Density (NUV-HD3) SiPMs produced by Fondazione Bruno Kessler (FBK) in collaboration with Istituto Nazionale di Fisica Nucleare (INFN) is being investigated [5]. The performances of the NUV-HD3 SiPM devices are described in [6]. The sensors have been optimized for the detection of UV Cherenkov photons. The active area of NUV-HD3 SiPMs is 6×6 mm^2, with a microcell area of 40×40 μm^2. They feature a breakdown voltage of ~27 V at room temperature with a temperature gradient of ~30 mV/°C, a Photo Detection Efficiency (PDE) peak of 60% at 350 nm dropping below 20% above 500 nm and a single photon dark count rate below 100 kHz/mm^2 at 20 °C.

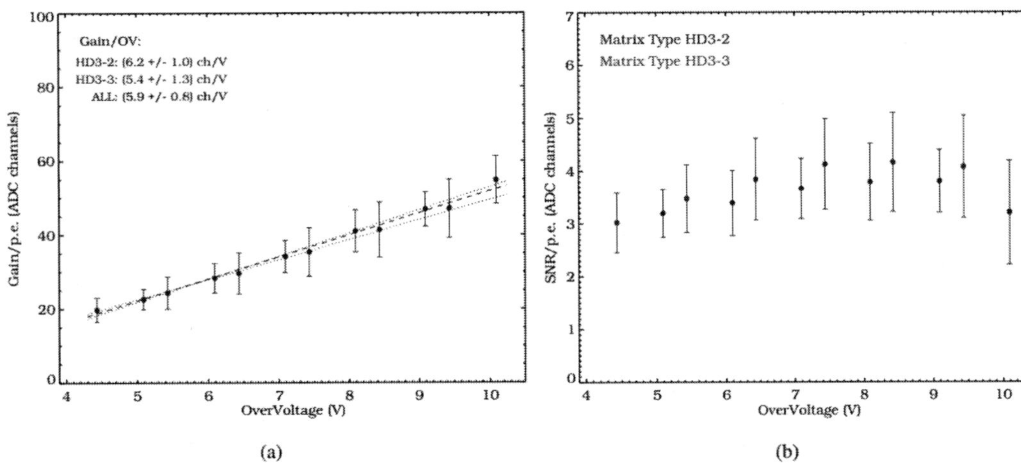

Figure 2. Average values for gain (a) and signal to noise ratio (SNR) (b) for more than 50 modules made of HD3-2 (blue) and HD3-3 (red) SiPMs.

2 Packaging and tests of the telescope optical units

The single SiPMs are designed and produced by FBK without any packaging on PCB. The assembly and packaging procedure of the optical modules was developed and made at the INFN laboratories. The basic mechanical unit is the PDU, made of 16 pixels. Four PDUs form a 64 pixels module that covers an area of 54×54 mm^2. INFN designed custom 27×27 mm^2 PCBs with 0.5 mm sensor-sensor distance and such that the same spacing is present between adjacent PDUs in a module. To precisely place the SiPMs on the PCBs and dispense the glue on the pads, a manual die-bonder machine was used. The alignment of the sensors has been checked with an optical metrology machine, resulting better than 40 µm. The maximum deviation from planarity across has been found to be ~80 µm, in good compliance with the optical requirements of the pSCT camera. Then, ultrasonic wire bonding from the top of the SiPMs to the PCB bonding pads to read out the current signal was made with a 20 µm Al/Si wire. The breakdown voltage of each sensor is measured analyzing the dark current as a function of the bias voltage to spot any defective sensor. Finally UV transparent resin was dispensed on top of the PDUs for mechanical protection.

After the assembly, each module was illuminated with a 380 nm pulsed laser. The charge signal of the 16 sensors has been read out signals from the 16 channels simultaneously using a DAQ system consisting of a charge to digital converter CAEN V792 QDC to obtain the charge of the signal. The integration time was set to 30 ns. The output signal of each SiPM was then amplified using a custom 16-channel front end board to match the dynamic range of the QDC [7]. The uniformity of the signal response for more than 50 modules have been investigated in the bias range from 31 V to 36 V. The average values for the gain and signal to noise ratio (SNR) shown in Figure 2 confirm the good level of uniformity for the modules performances.

3 Validation of the FEE readout chain

The pSCT digitization and readout electronics is based on the TARGET-7 ASIC [7], developed for the efficient signal readout of photosensors in high density cameras. The TARGET-7 ASIC handles first-level trigger -based on the signal of 4 adjacent sensors- and digitization for 16 channels. A switched capacitor ring buffer of 16384 units samples the analog signal at 1 GS/s with a maximum buffer depth of 16 μs. Extensive quality control tests have been performed on 9 TARGET modules to validate their functionalities. First, for each ASIC, the mean and the RMS of the electronics pedestals of the 16k capacitors have been evaluated in absence of light signal. Then, the SiPM modules coupled with the TARGET board have been illuminated with a NUV laser equipped with intensity filters. Figure 3(a) shows an example of a charge spectrum acquired with the TARGET board in conditions of low light intensity and 3(b) shows the average integrated amplitude as function of the light intensity for 16 channels of one PDU. The trend confirms the linearity of the digitizer. The operation of the TARGET-

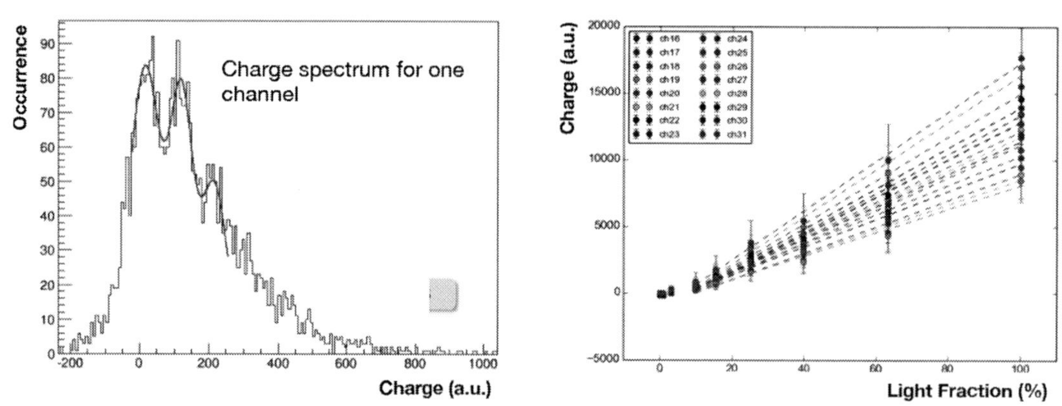

Figure 3. (a) Integrated signal over 16 ns of one channel with TARGET7. The red lines represent the gaussian fit used to find the peaks. (b) Average value of the charge distribution as function of the light intensity for 16 channels of one TARGET ASIC.

7 internal trigger has been verified by varying the intensity of light for a fixed trigger threshold and by varying the trigger threshold for a fixed light intensity. The trigger threshold is set by a DAC ranging from 0 to 3500 counts, where an increase of the DAC value corresponds to a lower physical threshold on the signal. The FEE quality control campaign has confirmed the performances of the

electronics modules before their integration on the pSCT camera, identifying defects for less than 1% of the channels (4 defective electronics channels out of 576).

4 State of the art and prospects

Six modules were installed on the pSCT telescope camera to populate a fraction of the central camera sector for validation and tests *in situ*. An additional batch of 9 modules (36 PDUs), for a total of 576 SiPMs, has been recently assembled using the latest production of NUV-HD3 SiPMs (HD3-4) and the modules are ready to be integrated in the camera to complete the central sector and measure the first lights. One additional sector of the pSCT telescope (25 modules, 1600 SiPMs) is planned to be integrated and operated in 2019. The experience gained from camera operations will be significant to confirm the possibility to operate a large, fast and compact SiPM camera for the medium size telescopes of the CTA observatory. An additional technological advance is represented by the development of new ASIC boards based on the TARGET paradigm. The new concept separates the sampling and digitization on a board (called TARGET C) and the trigger functions on the companion T5TEA ASIC to reduce the trigger noise induced by noise pickup from the analog sampling circuitry of the TARGET7 board [8]. An additional preamplifier ASIC, the SMART board intended to further decouple the analog pre-amplification from the digitization process, is currently being prototyped for the pSCT by INFN. The new FEE design is planned to be integrated with the pSCT backplanes together with the complete integration of the 177 sectors of the pSCT camera in the next future. The existing prototype will be successful proof of concept for the overall design of the SCT telescope, and the experience gained with the commissioning and operations will provide important feedbacks to open for the SCT telescope production for the CTA observatory.

Acknowledgements

This work was conducted in the context of the CTA Consortium for the pSCT project under the grant of the National Science Foundation "MRI Consortium: Development of a WideField-of-View Camera for the Schwarzschild-Couder Gamma Ray Telescope".

https://www.cta-observatory.org/consortium_acknowledgments/

References

[1] B. S. Acharya *et al*, in *Astrop. Phys.* 43 (2013) 3
[2] V. Vassiliev, S. Fegan, P. Brousseau, in *Astrop. Phys.* 28 (2007) (10-27)
[3] W. Benbow *et al*, in *AIP Conf.Proc.* 1792 (2017) no.1, 080005
[4] A. N. Otte *et al*, in PoS ICRC2015 (2016) 1023
[5] G. Ambrosi *et al*, in *Nuovo Cim. C* 41 (2018) no.1-2, 95
[6] G. Ambrosi *et al*, in *Nuovo Cim. C* 40 (2017) no.1, 78
[7] L. Tibaldo *et al*, in *PoS, ICRC2015* (2016) 932
[8] S. Funk *et al*, in *AIP Conference Proceedings* 1792, 080012 (2017)

Montecarlo simulation of the High Energy Particle Detector on board the satellite CSES

Luca Carfora for the CSES/HEPD collaboration[1,a]

[1]*University of Rome "Tor Vergata", Department of Physics, Via della Ricerca Scientifica 1 - 00133 Rome (Italy)*

Abstract. The High-Energy Particle Detector (HEPD) is an instrument devoted to the measurement of cosmic particles from few MeV up to hundreds of MeV. The HEPD will contribute to the China Seismo-Electromagnetic Satellite mission by measuring the precipitation of trapped particles and by studying the solar-terrestrial environment especially during impulsive events like coronal mass ejections and solar energetic particle emissions.
A Monte Carlo software was realized to study the performance of HEPD, such as its particle discrimination capability, the energy threshold for trigger and the maximum energy detectable in full containment.
This contribution reports the main features of the HEPD Monte Carlo simulation and some results of the detector performance based on it. A comparison with beam tests was carried out, showing a good agreement with the simulation.

1 HEPD: High Energy Particle Detector

The HEPD was developed by the Italian "Limadou" collaboration in the framework of the Chinese-Italian CSES mission.

The detector will range the low-energy component of cosmic rays, and in particular electrons in the range 3 - 100 MeV, protons in the range 30 - 300 MeV and light nuclei. It is composed by several sub-detectors, arranged as shown in Fig. 1, left: a tracker, with 2 planes of double-sided silicon microstrip detectors; a trigger plane, formed by 6 plastic scintillator paddles; a range-calorimeter, a tower of 16 plastic scintillator planes and a 3×3 matrix of inorganic scintillator (LYSO); with the tracker/trigger plane it is used for the particle identification and the energy reconstruction. A veto system based on a set of plastic scintillators rejects events that are not fully contained in the calorimeter.

HEPD is installed on board the CSES satellite, a scientific mission dedicated to monitor electromagnetic field and waves, plasma and particles perturbations of the atmosphere, ionosphere and magnetosphere induced by natural sources and anthropocentric emitters [1] and to study their correlations with the occurrence of seismic events [2][3]. Data collected by the mission will also allow studying solar-terrestrial interactions and cosmic ray solar modulation, with particular regard to particle spectra of the 24th and 25th solar cycle.

[a]e-mail: luca.carfora@roma2.infn.it

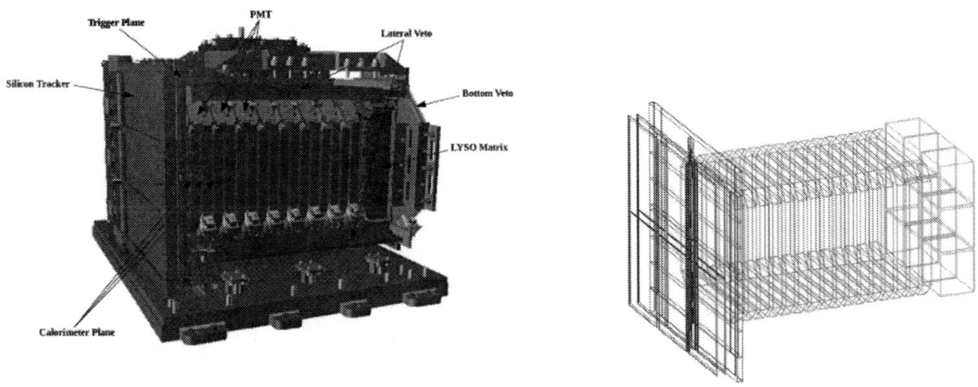

Figure 1. Left: An overview of HEPD. The trigger plane, below the silicon planes, is not visible. Right: example of the graphic transparent visualization of HEPD MC software; veto planes, HEPD box, satellite wall and passive materials are not shown.

2 HEPD MC simulation

The performance of HEPD using different particles and energies was studied with a Monte Carlo software, based on Geant4 package. Geant4 is a toolkit for the simulation of the passage of particles through matter. Its areas of application are in high energy, nuclear and accelerator physics and studies in medical and space science [4] [5] [6].

HEPD geometry is reported in Fig. 1, right. Different colours are used for different sub-detectors (blue for the silicon tracker, magenta for all the plastic scintillators and green for the LYSO crystals). The veto system and all mechanical structures have been hidden in the figure.

In the HEPD MC software there is the possibility of reproducing different environments, such as the beam test configurations, the muon configuration (atmospheric muons simulated with angular and energy distribution according to [7]) and the flight configuration. The production of optical photons inside the scintillators and the light collected by each PMT has also been added to the simulation.

2.1 Trigger energy threshold and maximum energy range

A simulation was carried out to study the particle energy thresholds with different trigger configurations and with maximum energy detectable, requesting in this case that a particle is fully contained in HEPD. The simulated data set consisted in 1 million of electrons and 1 million of protons with a uniform energy spectrum respectively between 1 and 100 MeV and 1 and 300 MeV and an angular spectrum uniform in $\cos^2\theta$.

Energy ranges have been studied with several selections on these events (shown in Figure 2):

- the minimum limit in Fig. 2 (a) and (b) is realized with different trigger configurations (some of which used in flight);
- the maximum range is evaluated as the energy of events fully contained in the calorimeter; the overall selection is one trigger paddle hit and no lateral veto hit (Fig. 2 (c) and (d)).

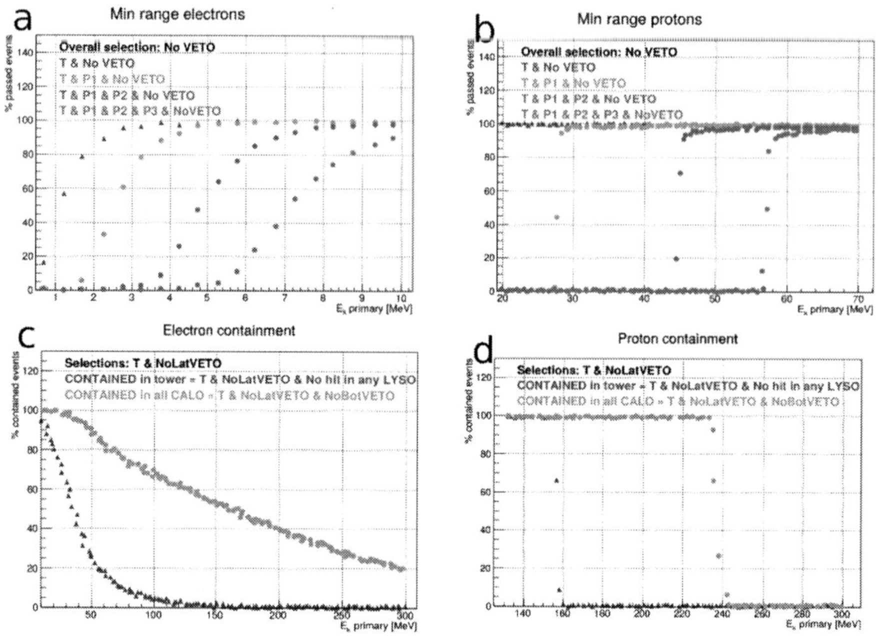

Figure 2. Left: Fraction of electrons (a) and of protons (b) that pass some trigger conditions as a function of primary energy; Right: fraction of electrons (c) and of protons (d) contained in the tower and in the whole calorimeter as a function of primary energy.

With a trigger configuration of one trigger paddle and the first two planes hit, electron minimum detectable energy is 6 MeV with 80% probability with respect to the overall events; for protons, it is 45 MeV. The maximum energy curve for electrons is not sharp because of Bremsstrahlung photons produced by electrons in scintillator material. Maximum energy for a proton contained in the calorimeter is 240 MeV.

2.2 Particle discrimination

Using the same data set described above, the particle identification was studied with the ΔE vs E method.

Fig. 3 shows the ΔE vs E method for particle discrimination, where ΔE is the energy deposit in one thin thickness and E is the total energy deposit measured by the detector. Two particle populations (protons on the top and electrons on the bottom) are clearly visible.

2.3 MC digitization

Digitization of the MC data consists in the conversion of the light collected by each PMT into a signal in ADC counts. The digitization factors were obtained comparing the peak and sigma values of ADC signals of PMTs recorded during proton test beams at different energies with distributions of light collected from PMTs in the corresponding HEPD simulations.

Both data and MC distributions were realized with the same selection:

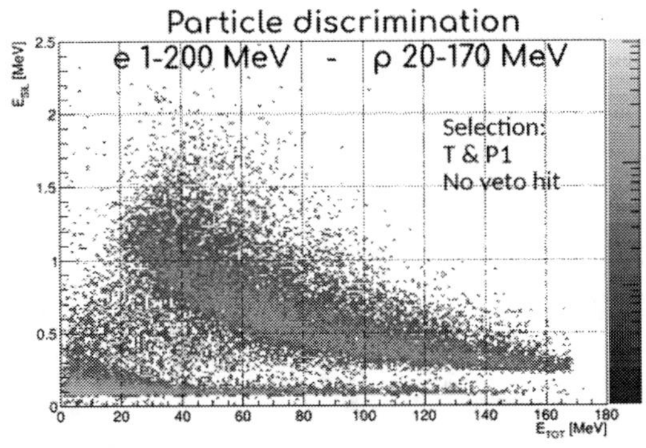

Figure 3. ΔE vs E method for particle identification with a sample of simulated electrons and protons; in the plot, E_{SIL} is the energy deposit in the 1st tracker plane, E_{TOT} is the energy deposit in the detector.

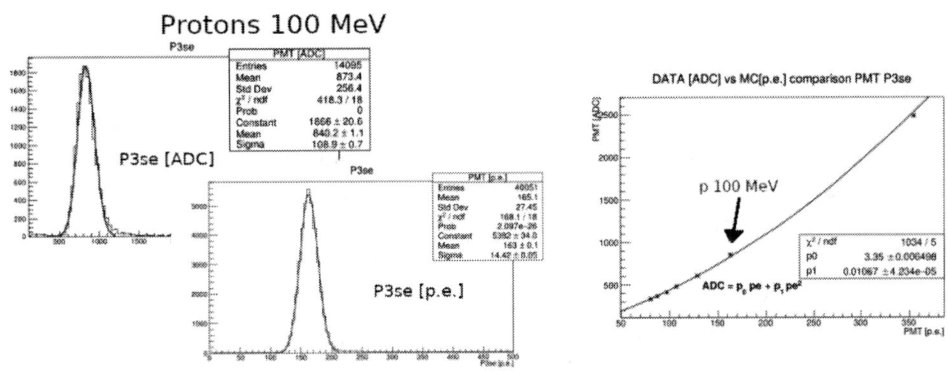

Figure 4. Left: distributions of ADC signal and p.e. (photo electrons) from PMT P3se in the 100 MeV proton run; both distributions were fitted with a gaussian; Right: peak of ADC signal as a function of light collected from PMT P3se in all proton runs; the curve was fitted with a parabola.

- most significant trigger paddle & P1 (both PMTs over 10σ);
- no hit in any veto (both PMTs under 5σ);
- trigger multiplicity equal to 1 (both PMTs under 5σ).

Values of the peak distributions were correlated in ADC-p.e. (photo electrons) plots (one for each PMT) and points were fitted with a function of the type: $ADC = p_0 \cdot PE + p_1 \cdot PE^2$. ADC is the signal pedestal-subtracted.

Digitizaton factors p_0 and p_1 were applied to MC light distributions to verify the correctness of the procedure; the comparison between real and reconstructed ADC distributions for some PMTs for protons and for a set of independent data (electrons beam test and muons) is shown in fig. 5.

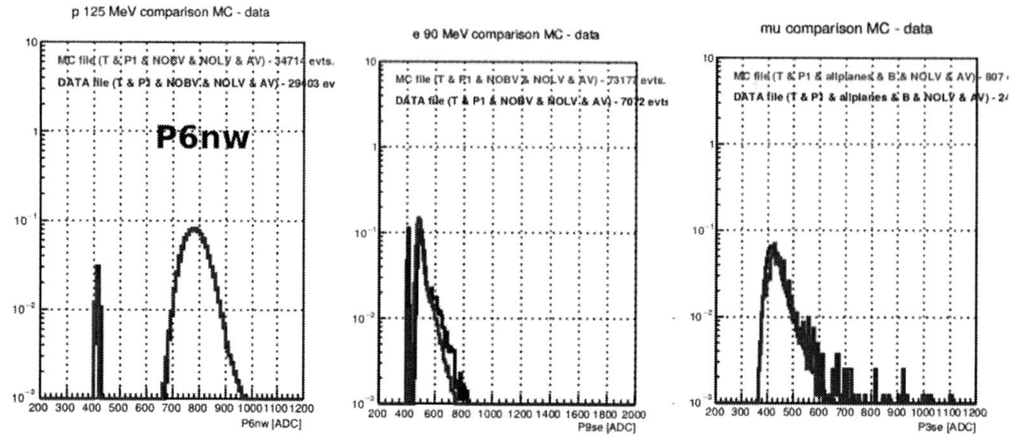

Figure 5. Comparison between distributions of real (blue) and reconstructed ADC signal from MC (red) for 3 PMTs for 125 MeV protons (left), 90 MeV electrons (center) and muons (right). Selections used are the same for the two distributions and are: most significant paddle & P1, no hit in any veto and trigger multiplicity equal to 1.

3 Conclusions

In this contribution some results of HEPD Monte Carlo simulation (energy range, particle discrimination) were presented; they show a good agreement with test beam and flight data analysis.

Furthermore, the digitization procedure for PMT scintillator signals was described. It will be used to support the equalization of PMTs and the calibration of the calorimeter.

References

[1] V. Vitale et al. (CSES/HEPD collaboration), EPJ Web of Conferences **136**, 4 (2017)
[2] X. Shen et al., Earthquake Science **28**, 303 (2015)
[3] A. De Santis et al., Physics and Chemisrty of the Earth **85-86**, 1 (2015)
[4] A. S. et al., Nucl. Instr. and Meth. in Phys. Res. A **506**, 250 (2003)
[5] J. Allison et al., IEEE Transactions on Nucl. Sc. **53**, 270 (2006)
[6] J. Allison et al., Nucl. Instr. and Meth. in Phys. Res. A **835**, 186 (2016)
[7] S. Cecchini, M. Spurio, Geoscientific Instrumentation Methods and Data System **1**, 185 (2012)

Status and perspectives of the radio detection of high-energy cosmic rays

Jörg R. Hörandel[1,2,3,*]

[1]Radboud Universiteit, Department of Astrophysics/IMAPP, P.O. Box 9010, 6500 GL Nijmegen, The Netherlands
[2]Nikhef, Science Park 105, 1098 XG Amsterdam, The Netherlands
[3]Vrije Universiteit Brussel, Department of Physics and Astronomy, B-1050 Brussels, Belgium

Abstract. Status and perspectives of the radio detection of high-energy cosmic rays are discussed with emphasis on recent developments at the LOFAR radio telescope and the Pierre Auger Observatory.

1 Introduction

High-energy cosmic rays impinging on the atmosphere of the Earth initiate cascades of secondary particles, the extensive air showers. Many secondary particles are electrons and positrons. They emit electromagnetic radiation with frequencies of tens of MHz during the shower development. The bulk of the radiation originates from the transverse separation of charges in the shower due to interactions with the Earth's magnetic field: the geomagnetic effect [1]. One expects the radiation intensity being proportional to $v \times B$, where v denotes the direction of the particles in the shower (shower axis) and B the Earth magnetic field. The geomagnetic radiation is linearly polarized.

A second, sub-dominant component arises through the longitudinal separation of charges in the shower (Askaryan effect [2]). This component is radially polarized. The direction of the electric field within the shower is measured with high resolution with LOFAR [3]. This allows one to quantitatively evaluate the contribution of the two emission processes to the overall shower emission. The results yield a dependence on the zenith angle and the distance to the shower axis for the ratio of the two radiation components. Values for the ratio range from about 5% to 20%, with an average contribution of the Askaryan effect of about 15%. A similar value has been obtained through polarization measurements with AERA [4].

Interference between the two radiation components with their different polarization behavior leads to a non-rotational symmetric distribution of the radiation intensity on the ground, as illustrated in Fig. 1 (left). This implies a non-trivial behavior of the lateral density distribution of the radio emission. Using the dense LOFAR core [8, 9] it became obvious that the lateral density distribution is not rotationally symmetric and a simple description as e.g. proposed in [10] is not sufficient. Instead now, more complex, two-dimensional functions are being used to describe the azimuthal dependence of the radiation strength [5, 8]. As an example, a parameterization introduced in [8] is depicted in Fig. 1 (right). The colors represent different intervals in azimuthal angles. The non-rotational symmetry is clearly visible.

*e-mail: j.horandel@astro.ru.nl

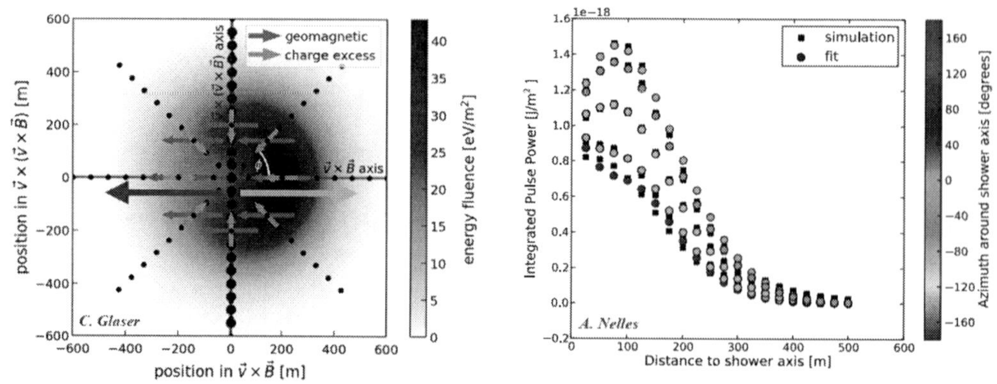

Figure 1. Left: Distribution of the energy fluence (in the 30 – 80 MHz band) of an air shower with 60° zenith angle [5]. Superimposed is the polarization direction of the geomagnetic and charge-excess emission processes at different positions in form of arrows. Simulated with CORSIKA/CoREAS [6, 7]. Right: Lateral distribution of the radiation power on the ground for bins in azimuthal angle, following a parameterization described in [8].

2 LOFAR and AERA

We focus here on the description of recent results obtained by the LOFAR radio telescope and the Auger Engineering Radio Array (AERA) at the Pierre Auger observatory.

LOFAR [11, 12] is a distributed radio telescope. Its antennas are distributed over northern Europe with the densest concentration in the north of the Netherlands. The antennas of LOFAR are grouped into stations. A station consist of 96 low-band antennas (10 – 90 MHz) and 48 high-band antennas (110 – 240 MHz). The 23 stations within the ~ 5 km^2 core are distributed in an irregular pattern that maximizes uv-coverage, or spatial frequencies for standard interferometric observations. At the center of the LOFAR core six stations are located in a roughly 320 m diameter area, called the Superterp, providing both, the shortest baselines for interferometric observations and the densest population of antennas, ideal for cosmic-ray observations. An array of scintillation detectors installed on the Superterp complements the set-up [13, 14]. The particle detector array is used to measure the properties of air showers and to issue a trigger signal to read out the radio antennas.

The AERA project [15, 16] comprises more than 150 antennas installed on an area covering about 17 km^2 inside the Auger observatory in the field of view of the High Elevation Telescopes (HEAT). HEAT are fluorescence light detectors aimed to measure air showers with energies above 10^{17} eV. The radio antennas are installed on sub-grids with different spacings, ranging from 144 m to 375 m in order to compensate for the change of the size of the radio footprint as a function of cosmic-ray energy. Two types of antennas are installed [17]: 24 stations are equipped with two logarithmic-periodic dipole antennas, oriented in magnetic North-South and East-West directions. The signals are amplified and filtered before they are introduced into a filter-amplifier and digitization chain. There, they are band pass filtered between 30 and 80 MHz and digitized. The remaining antennas are an active bowtie antenna, called "butterfly antennas". This antenna is highly sensitive towards the ground which enhances the antenna gain. All radio detector stations operate autonomously employing solar power systems.

Both installations are calibrated *in situ* with high precision, both for the signal strength [18, 19] and the signal arrival time [20, 21].

3 Properties of cosmic rays

The ultimate goal of the radio measurements of air showers is to derive the properties of the incoming cosmic rays, namely their arrival direction, energy and particle type (nuclear mass).

The **direction** of the cosmic ray is obtained through a precise measurement of the arrival time of the wavefront at the individual antennas of an array. LOFAR, with its high antenna density was valuable to clearly understand the precise shape of the radio wavefront in air showers [22] and it could be shown that a hyperboloid is an appropriate analytical description. This confirms earlier work by LOPES [23]. From measurements of the arrival time of the shower front the direction of the cosmic rays can be determined with a resolution of better than 1°.

The **energy** of the air shower is derived from the measurement of the radio energy density on the ground [24, 25]. A function [8] is fitted to the measured values in order to calculate the integral of the total energy in the frequency band 30 – 80 MHz transported in an air shower to the ground. This quantity is proportional to the shower energy. For a set of high-quality showers an energy resolution of better than 25% has been achieved [24, 25].

At LOFAR, using results from simulations and a comparison to measurements from a particle detector array an energy resolution around 30% is obtained [9, 26].

A cosmic ray with an energy of 1 EeV delivers about 15.8 MeV of energy to the ground in the frequency range from 30 to 80 MHz [24, 25]. The radio emission measured on the ground $E_{30-80\,\mathrm{MHz}}$ can be used to establish an absolute calibration of the energy scale E_{CR}, using the universal formula

$$E_{30-80\,\mathrm{MHz}} = [15.8 \pm 0.7(\mathrm{stat}) \pm 6.7(\mathrm{syst})]\ \mathrm{MeV} \times \left(\sin\alpha \frac{E_{\mathrm{CR}}}{10^{18}\,\mathrm{eV}} \frac{B_{\mathrm{Earth}}}{0.24\,\mathrm{G}}\right)^2, \qquad (1)$$

knowing the magnetic field B_{Earth} at the location of the detector and the angle α between the direction of the Earth magnetic field and the direction of the air shower.

Experimentally, the most challenging subject is the determination of the **particle type** or its nuclear mass. The basic observable in air shower physics is the depth of the shower maximum X_{max}, being proportional to the nuclear mass A of the cosmic ray $X_{\mathrm{max}} \propto \ln A$. The radio signals are in principle emitted on the Čerenkov cone. Thus, the size of the footprint of the radio emission on the ground scales well with the distance from the emission region (basically X_{max}) to the antennas (on the ground) [9]. Knowing the vertical profile of the atmosphere, the depth of the shower maximum X_{max} can be derived. For LOFAR a high-precision method has been developed to measure X_{max} [27, 28]. For each measured cosmic ray, dedicated simulations are conducted, taking into account the measured energy and direction of the cosmic ray. The predicted signals in the particle and radio detectors are compared to the measured values on a statistical basis and a best fit value is obtained for X_{max}. A resolution fo X_{max} of better than 20 g/cm^2 has been obtained.

This method is also applied to AERA data. The obtained X_{max} values from the radio measurements have been compared to X_{max} obseravtions from the fluorescence light telescopes. In the first analyses a resolution of the order of 40 g/cm^2 has been achieved. The resolution depends on the number of antennas participating in a measurement and thus, on the quality selection cuts applied [29].

Combining the various efforts around the world one can state that the radio detection technique is now mature and the properties of cosmic rays are now being measured on a regular basis with such devices with state-of-the-art accuracies of the order of 0.1° to 0.5° for the arrival direction, 20% to 30% for the energy, and 20 to 40 g/cm^2 for the depth of the shower maximum (used to study the cosmic-ray mass – particle type), with the accuracies mainly depending on the antenna density of the respective installation.

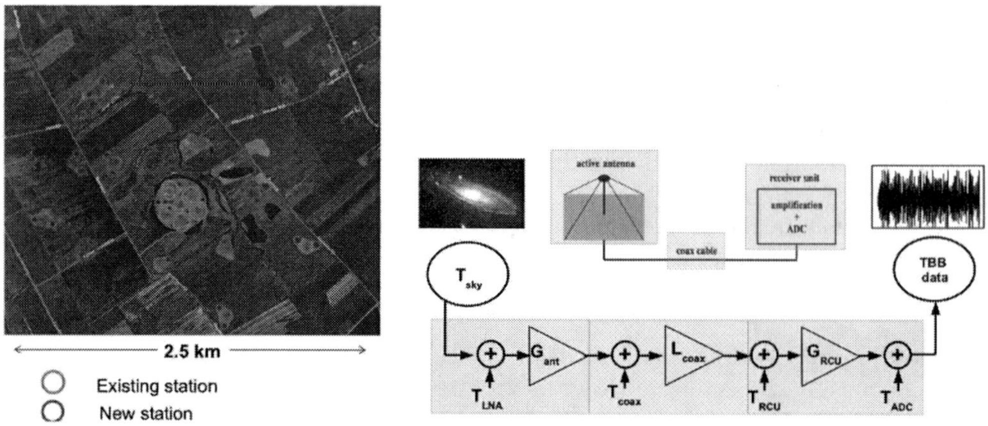

Figure 2. Left: Satellite image of the LOFAR dense core, illustrating the extension of the scintillator array. Blue circles indicate existing stations with particle detectors. In addition, particle detectors will be installed at the stations indicated with red circles. Right: LOFAR signal chain, illustrating the various steps from the observed signal on the sky to recorded ADC values, stored in ring buffers (transient buffer boards – TBB) [30].

4 Ongoing work and perspectives for the future

4.1 LOFAR

At present the particle detector array at the LOFAR observatory is being extended. Five LOFAR stations around the Superterp are being equipped with scintillator stations, as illustrated in Fig. 2 (left). At each station four scintillator detectors, 1 m^2 each, identical to the ones described in [13] will be installed. This will increase the effective area of the LORA particle detector array. At energies between 10^{16} and 10^{18} eV a $\approx 45\%$ increase of the number of air showers with a radio signal is expected [31].

At present also the calibration procedure for LOFAR is being improved [30], further reducing the systematic uncertainties with respect to previous work [18]. Radio emission from the Galactic center is used as a reference source. The complete signal chain, end to end, from the signal on the sky to the recorded ADC traces is evaluated, as illustrated in Fig. 2 (right). This method requires knowing the contributions of electronic noise to the system, and the original Galactic calibration is limited by the systematic uncertainty of the electronic noise. This is being improved through the ongoing efforts by modeling each step in the signal chain in order to understand the electronic noise introduced into the system. Using this method, a new absolute and frequency dependent calibration has been derived with a maximum of 14% systematic uncertainty for frequencies below 77 MHz. A comparison of the logarithmic slopes of calibrated, measured power spectra and simulated power spectra shows promising agreement.

4.2 The radio upgrade of the Pierre Auger observatory

At present, the Pierre Auger collaboration is working on an upgrade of the Observatory [33]. The physics case of the upgrade is outlined in [34]. The *key science questions* to be addressed are: What are the sources and acceleration mechanisms of ultra-high-energy cosmic rays (UHECRs)? Do we understand particle acceleration and physics at energies well beyond the

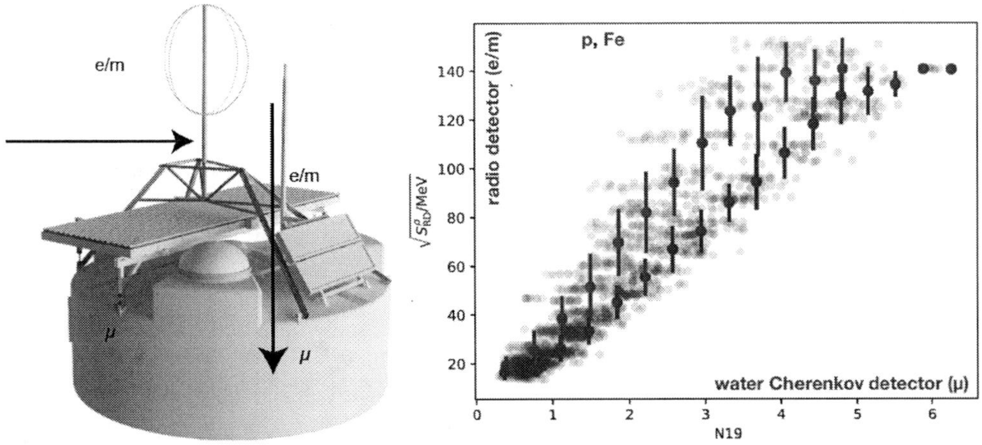

Figure 3. Left: Conceptional illustration of an upgraded station of the Surface Detector array of the Pierre Auger observatory, comprised of (from bottom to top) a water-Čerenkov detector, a layer of scintillators, and a radio antenna [32]. The arrows indicate the detection principles for vertical and horizontal air showers. In practice, the electromagnetic and muonic components are obtained by unfolding the signals from the Surface Scintillator Detector and the water-Čerenkov detector. Right: Electromagnetic energy in a shower as measured by the radio detector as a function of the muon content in a shower as measured by the water-Čerenkov detector [32]. Preliminary results from simulations are shown, see main text for more details.

LHC (Large Hadron Collider) scale? What is the fraction of protons, photons, and neutrinos in cosmic rays at the highest energies?

It is planned to install a radio antenna on each of the 1661 stations of the Surface Detector array of the observatory, forming a 3000 km^2 radio array, the largest radio array for cosmic-ray detection in the world [32, 35]. An artists impression of the planned set-up is given in Fig. 3 (left). It shows (from bottom to top) the water-Čerenkov detector with a layer of scintillators on top (Surface Scintillator Detector) and a radio antenna. With the combination of water-Čerenkov detector and Surface Scintillator Detector the electron-to-muon ratio (e/μ) is measured for vertical showers. In a similar way the combination of water-Čerenkov detector and Radio Detector will be used to measure the e/μ ratio for air showers with large zenith angles, in the following called horizontal air showers. In turn, the e/μ ratio will be used to derive the particle type of the incoming cosmic ray up to the highest energies. This is the main goal of the Auger upgrade, to measure the particle type of each incoming cosmic ray. Thus, the radio upgrade will increase the aperture of the observatory for mass-sensitive investigations, allowing to do e/μ separation for showers with a broad zenith angle range, from zenith with the Surface Scintillator Detector to the horizon with the Radio Detector.

There is an important difference related to the mass measurement of cosmic rays with AERA (for vertical showers) and the radio upgrade (for horizontal air showers). For vertical showers we use a geometrical method, correlating the size of the footprint on the ground to the distance from the observer to the shower maximum [15, 26, 36]. From this quantity, the depth of the shower maximum $X_{\max}$ is derived, which in turn is dependent on the logarithm of the number of nucleons (nuclear mass) A of the incoming cosmic ray $X_{\max} \propto \ln A$ [15]. For horizontal showers we aim to apply a different method: We will use a combination of radio antennas and the water-Čerenkov detectors to measure the e/μ ratio in air showers to determine $\ln A$.

In a first analysis the physics potential of the radio upgrade has been studied using CORSIKA[6]/CoREAS[7] simulations. 192 showers have been simulated, induced by half protons, have iron nuclei, with energies from 4 to 40 EeV and zenith angles between 60° to 80°. A full water-Čerenkov detector detector simulation and reconstruction has been included in the analysis, but only a simplified treatment of the radio signals was available. The normalized radiation energy has been calculated as described in [37]. The radiation energy has been "smeared out" in order to mimic reconstruction uncertainties. PRELIMINARY results are illustrated in Fig. 3 (right). The size of the e/m component (as measured with the Radio Detector) is plotted as a function of the muon number (as obtained with the water-Čerenkov detector). A clear separation between showers induced by protons and iron nuclei is visible. Please note: the flattening at the top-right is due to a technical flaw in the simulations, only showers up to a certain (too low) energy have been simulated.

The concept to measure horizontal air showers with an antenna array with km-scale spacing has been verified with the existing AERA at the Auger observatory [38, 39]. The size of the radio footprint of an air shower has been measured. The distance from the shower axis to the furthest away radio detector (in the shower plane, perpendicular to the shower axis) has been evaluated. For horizontal air showers (i.e. large zenith angles) the footprints reach sizes exceeding tens of km^2 and dozens of stations will have a signal above threshold. This is a very important result since it confirms experimentally that the radio emission from horizontal air showers can be measured with radio antennas on the standard Auger 1500 m grid.

References

[1] F. Kahn, I. Lerche, Royal Society of London Proceedings Series A **289**, 206 (1966)
[2] G. Askaryan, Journal of the Physical Society of Japan Supplement **17**, C257 (1962)
[3] P. Schellart, et al., JCAP **1410**, 014 (2014)
[4] A. Aab et al. (Pierre Auger Collaboration), Phys.Rev. **D89**, 052002 (2014)
[5] C. Glaser, S. de Jong, M. Erdmann, J.R. Hörandel, Astropart. Phys. **104**, 64 (2019)
[6] D. Heck et al., *Corsika: A monte carlo code to simulate extensive air showers*, Report FZKA 6019, Forschungszentrum Karlsruhe (1998)
[7] T. Huege, M. Ludwig, C.W. James, ARENA 2012, AIP Conf. Proc. 1535 pp. 128–132 (2013)
[8] A. Nelles, et al., Astropart.Phys. **60**, 13 (2015)
[9] A. Nelles, et al., JCAP **1505**, 018 (2015)
[10] H. Allan, *Progress in Elementary Particles and Cosmic Ray Physics* (J.G. Wilson and S.G. Wouthuysen eds., North Holland, 1971), p. 169
[11] M. van Haarlem, et al., Astronomy & Astrophysics **556**, A2 (2013)
[12] P. Schellart, et al., Astron.Astrophys. **560**, A98 (2013)
[13] S. Thoudam, S. Buitink, A. Corstanje, J. Enriquez, H. Falcke et al., Nucl.Instrum.Meth. **A767**, 339 (2014)
[14] S. Thoudam et al., Astropart. Phys. **73**, 34 (2016)
[15] J. Schulz, Pierre Auger Collaboration, Proceedings 34th International Cosmic Ray Conference, Den Haag p. PoS(ICRC2015)615 (2015)
[16] P. Abreu et al. (Pierre Auger Collaboration), Nucl.Instrum.Meth. **A635**, 92 (2011)
[17] P. Abreu et al. (Pierre Auger Collaboration), JINST **7**, P10011 (2012)
[18] A. Nelles et al., JINST **10**, P11005 (2015)
[19] A. Aab et al. (Pierre Auger), JINST **12**, T10005 (2017)
[20] A. Corstanje et al., Astron. Astrophys. **590**, A41 (2016)

[21] A. Aab et al. (Pierre Auger), JINST **11**, P01018 (2016)
[22] A. Corstanje, et al., Astropart.Phys. **61**, 22 (2015)
[23] W. Apel, J. Arteaga-Velázquez, L. Bähren, K. Bekk, M. Bertaina et al., JCAP **1409**, 025 (2014), 1404.3283
[24] A. Aab et al. (Pierre Auger), Phys. Rev. Lett. **116**, 241101 (2016)
[25] A. Aab et al. (Pierre Auger), Phys. Rev. **D93**, 122005 (2016)
[26] S. Buitink, et al., Phys.Rev. **D90**, 082003 (2014)
[27] S. Buitink et al., Proceedings 34th International Cosmic Ray Conference, Den Haag p. PoS(ICRC2015)368 (2015)
[28] S. Buitink et al., Nature **531**, 70 (2016)
[29] F. Canfora et al., Proceedings ARENA 2018 (2018)
[30] K. Mulrey et al., Proceedings ARENA 2018 (2018)
[31] S. Buitink et al., Proceedings ARENA 2018 (2018)
[32] J. Hörandel, Proceedings ARENA 2018 (2018)
[33] A. Castellina et al., Proceedings UHECR symposium, Paris (2018)
[34] A. Aab et al. (Pierre Auger), arXiv p. 1604.03637 (2016)
[35] J. Hörandel, Proceedings UHECR symposium, Paris (2018)
[36] T. Huege et al., Proceedings UHECR symposium, Paris (2018)
[37] C. Glaser, M. Erdmann, J.R. Hörandel, T. Huege, J. Schulz, JCAP **1609**, 024 (2016)
[38] A. Aab et al. (Pierre Auger), JCAP **1810**, 026 (2018)
[39] M. Gottowik et al., Proceedings ARENA 2018 (2018)

Precision Measurement of the monthly cosmic Ray fluxes e^-, e^+, p, He) with the Alpha Magnetic Spectrometer on the ISS

Maura Graziani

[1] *KIT, Karlsruhe Institute of Technology*

Abstract. The precision measurements of the monthly cosmic ray fluxes with Alpha Magnetic Spectrometer on the International Space Station are presented. Individual electron, positron, proton and helium spectra have been measured for each Bartel's rotation period (27 days) in the time range from May 2011 to May 2017. This period covers the ascending phase of solar cycle #24 together with the reversal of the Sun's magnetic field polarity through the minimum. The fluxes reveal a characteristic time dependence below 20 GeV. The data show a strong charge-sign dependent effects corresponding to the the polarity reversal of the solar magnetic field.

1 Introduction

The study of Cosmic Rays (CR) constitutes a unique instrument for understanding our universe. By means of the CR hadronic component, the knowledge about CR propagation through the galaxy and the Interstellar Medium can be improved. With the CR electromagnetic component (as well as the other rare components of CR like antiprotons) we can investigate the local CR sources, looking for indirect signs of Dark Matter. However, CR spectra, when measured near Earth, are significantly affected by the solar activity. The solar activity has a cycle of ~11 years, during which it increases reaching a maximum and then decreases again. The intensity of cosmic ray radiation is anti-correlated with the activity of the sun [1], and this is the so called Solar Modulation (SM) effect. In order to have a correct understanding of CR spectra out of the heliosphere, the SM should be well known and taken into account. A detailed study of the CR fluxes evolution with time is needed in order to develop and test different models of the SM effects based on the interaction of cosmic rays with the Heliosphere. The simultaneous measurements of e^- and e^+ (or p and $\bar{p}$) over a complete solar activity cycle can represent a sound test of the current charge-sign dependent modulation models. AMS-02 can provide the most accurate measurements of the time dependence of particle and anti-particle fluxes since 2011 thanks to its high acceptance and the excellent performance of the detector. In this letter, the time variation of CR electron (e^-), positron (e^+), proton (p) and Helium (He) fluxes during the first 6 years of data taking will be presented.

2 The AMS-02 detector

The AMS-02 is a large acceptance CR detector which has been installed during the STS-134 NASA Endeavour Space Shuttle mission in May 2011 on the International Space Station, where it will collect

CR until the end of the ISS operation, currently set to 2024. Thanks to the long exposure time combined with a large detector acceptance ($0.5\,m^2$ sr), AMS is able to study the primary CR fluxes in the energy range GeV-TeV with unprecedented precision and sensitivity. The core of the instrument is a spectrometer, composed of a permanent magnet (with a magnetic field with an intensity of 0.14 T), and of 9 layers of double-sided micro-strip silicon sensors that constitute the Tracker. Above and below the spectrometer two planes of Time of Flight counters are placed. A Transition Radiation Detector is located at the top of the instrument. The detector is completed with a Ring Imaging Cherenkov detector and an Electromagnetic Calorimeter. The central part of AMS-02 is surrounded by an Anti-Coincidence system. The AMS-02 detector is described in details in [2].

3 Flux measurements

The AMS data from May 2011 to May 2017 have been analysed for the measurement of e^-, e^+, p and He fluxes in time. The results have been published in [3, 4]. The applied analysis follows the formula used for the measurement of the time-averaged electron and positron fluxes [5], improving low-energy effective acceptance. The fluxes have been measured in 79 different time interval, each one corresponding to a different Bartel's rotation, in the rigidity range from 1 to 60 GV for p, from 1.9 to 60 GV for He, and in the energy range from 1 to 50 GeV for e^- and e^+.

3.1 Electron and positron fluxes in time

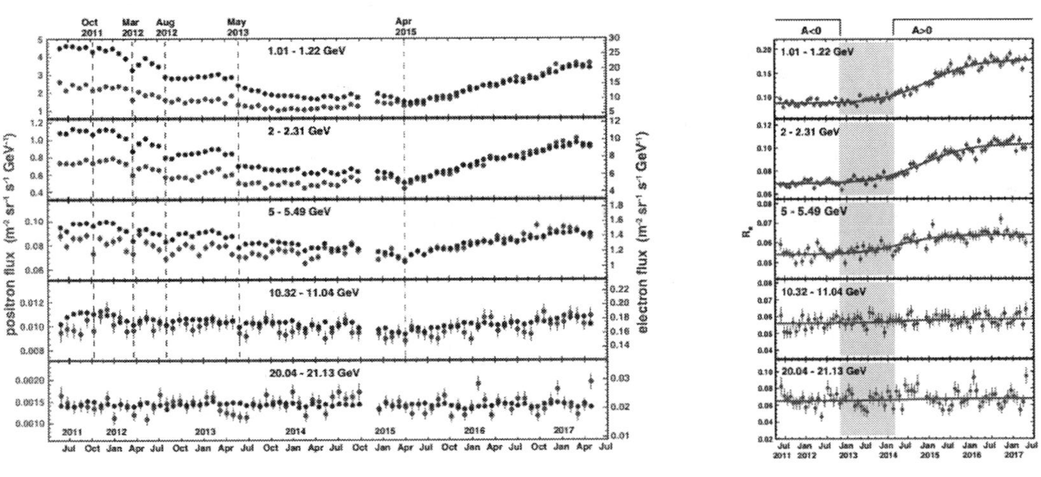

(a) (b)

Figure 1: (*a*) Fluxes of CRs e^+ (red, left axis) and e^- (blue, right axis) as functions of time, for 5 energy bins with the statistical uncertainties. Prominent and distinct time structures are marked by dashed vertical lines. (*b*) The e^+/e^- ratio, R_e, as a function of time with the statistical uncertainties. The red curves show a best-fit parametrisation obtained with an analytical function described in [3]. The polarity of the heliospheric magnetic field is denoted by $A < 0$ and $A > 0$. The period without well-defined polarity is marked by the shaded area.

In Figure 1a, the time dependence of the e^- and e^+ fluxes is reported for 5 characteristic energy bins (for more details see [6]). Both e^- flux and e^+ flux show a significant time dependence with a long-term time structure and short-term time structures (delimitated by the shaded lines). The time dependency of fluxes increases with the decrease of energy. Above 20 GeV, neither the electron flux nor the positron flux exhibits significant time dependence. For E<10 GeV, we observe that from May 2011 to May 2013, both for e^- flux and e^+ flux decrease with time. In July 2013 the solar magnetic field reversal occurred. From May 2013 to April 2015, the flux of electrons continues to decrease, but with reduced slope, while the positron flux begins to increase. Then, from April 2015 until May 2017, both fluxes rise steeply. The observed different behaviour between e^- flux and e^+ flux are related to the charge-sign dependent solar modulation effects. The main reason for this to occur is that when the solar magnetic field reverses its polarity, the galactic CRs of opposite charge will reach Earth from different heliospheric directions due to the magnetic drift. The charge-sign dependencies of solar modulation, can be clearly observed in Figure 1b in which the e^+/e^- ratio (R_e) is shown as a function of time for all energy bins up to 5 GeV. In R_e, the important, newly discovered short-term variations in the fluxes largely cancel, and a clear overall long-term trend appears. At low energies, R_e is flat at first, then smoothly increases after the time of the solar magnetic field reversal, to reach a plateau at a higher amplitude.

3.2 Helium and proton fluxes in time

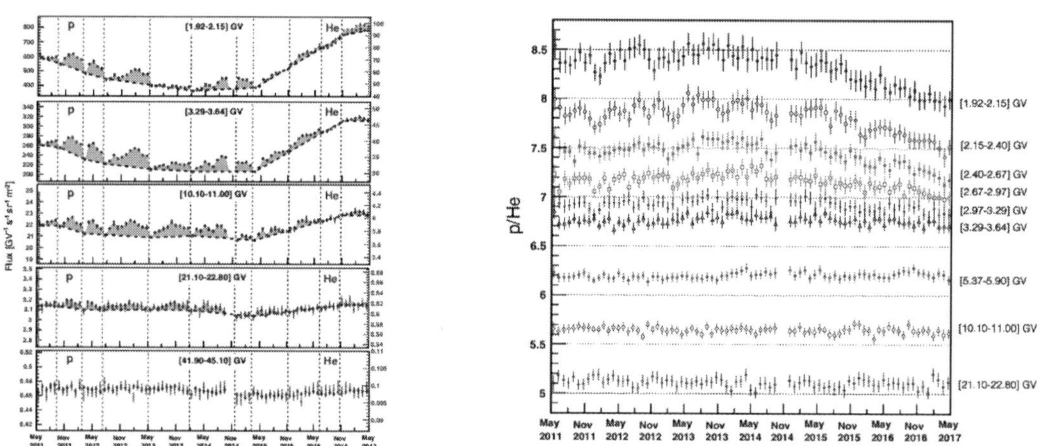

(a) (b)

Figure 2: (*a*) The AMS *p* (blue, left axis) and *He* (red, right axis) fluxes as function of time for 5 rigidity bins. The error bars are the quadratic sum of the statistical and time dependent systematic errors. Detailed structures (green shading and dashed lines to guide the eye) are clearly present below 40 GV. The red vertical dashed lines denote structures that have also been observed by AMS in the e^- flux and the e^+ flux. (*b*)The AMS p/He flux ratio as function of time for 9 characteristic rigidity bins. The errors are the quadratic sum of the statistical and time dependent systematic errors.

In Figure 2a, the time dependence of the *p* and *He* fluxes is reported for 5 characteristic rigidity bins (for more details see [7]). Both the *p* and *He* fluxes exhibit large variations with time at low rigidities which decrease with increasing rigidity. The structures in the *p* flux and *He* flux are nearly identical in both time and relative amplitude (indicated by the green shading) that decrease progressively with rigidity. The five red vertical dashed lines in the figure indicate the structure that have been also observed in the electron flux and the positron flux. After one year from the solar maximum (April 2014 for solar cycle 24), the amplitudes of the structures are considerably reduced and the proton and helium fluxes steadily increase at rigidities less than 40 GV. Figure 2b shows the AMS p/He flux ratio as a function of time for 9 rigidity bins. We can observe that, depending on the rigidity range, the p/He flux ratio shows two different behaviours in time. Above ~3 GV the ratio is time independent. Below ~3 GV the ratio has a long-term time dependence. An analytic fit on p/He ratio was performed and has shown that above 3.29 GV the p/He flux ratio is consistent with a constant value at the 95% confidence level. This shows the universality of the solar modulation of cosmic ray nuclei at relativistic rigidities. Below 3.29 GV, the observed p/He flux ratio is steadily decreasing with time.

4 Conclusion

The fluxes for p, He, e^+ and e^- as a function of time have been measured by AMS during the ascending phase of solar cycle 24 through its maximum and toward its minimum.

The unique performance of AMS-02 provides measurment of both e^+ and e^- fluxes as a function of time with an unprecedented high time granularity. Based on 23.5×10^6 events, we report the observation of short-term structures on the timescale of months coincident in both the e^- flux and the e^+ flux. These structures are not visible in the e^+/e^- flux ratio. The precision measurements across the solar polarity reversal show that the ratio exhibits a smooth transition over 830±30 days from one value to another. The midpoint of the transition shows an energy dependent delay relative to the reversal and changes by 260±30 days from 1 to 6 GeV.

The precision p flux and the He flux observed by AMS have fine time structures nearly identical in both time and relative amplitude. The amplitudes of the flux structures decrease with increasing rigidity and vanish above 40 GV. The amplitudes of the structures are reduced during the time period, which started one year after solar maximum, when the proton and helium fluxes steadily increase. In addition, above ~3 GV the p/He flux ratio is time independent. Below ~3 GV the ratio has a long-term decrease coinciding with the period during which the fluxes start to rise. Before AMS data, several effects had been proposed that lead to a time dependence of the p/He flux ratio at low rigidities, such as velocity dependence of the diffusion tensor, differences in the interstellar spectra of p and He, and the ^3He and ^4He isotopic composition [8–10].

AMS is measuring solar effects for all nuclei, particle and anti-particle fluxes in the present and next solar cycle providing information for the development of refined solar modulation models.

References

[1] M.S. Potgieter, Living Reviews in Solar Physics **10**, 3 (2013)
[2] Phys. Rev. Lett. **110**, 141102 (2013)
[3] Phys. Rev. Lett. **121**, 051102 (2018)
[4] Phys. Rev. Lett. **121**, 051101 (2018)
[5] Phys. Rev. Lett. **113**, 121102 (2014)
[6] See Supplemental Material at http://link.aps.org/supplemental/10.1103/PhysRevLett.121.051102
[7] See Supplemental Material at https://journals.aps.org/prl/supplemental/10.1103/PhysRevLett.121.051101/SM-p-He-vs-BR-REV.pdf
[8] The Astrophysical Journal Letters **849** (2017)
[9] G. Gloeckler, J.R. Jokipii, Phys. Rev. Lett. **17**, 203 (1966)
[10] I. Cholis, D. Hooper, T. Linden, Phys. Rev. D **93**, 043016 (2016)

Ultrahigh-Energy multi-messengers at the Pierre Auger Observatory

Francisco Pedreira[1,a] for the Pierre Auger Collaboration[2,b]

[1] *University of Santiago de Compostela, 15705,Santiago de Compostela, Spain*
[2] *Avda. San Martín Norte, Malargüe, Argentina*

Abstract. The study of correlations between observations of different messengers from extreme sources of the Universe has emerged as an outstanding way to make progress in astrophysics. The Pierre Auger Observatory is capable of significant contributions as an ultra-high energy particle detector, particularly through its capability to search for inclined showers produced by neutrinos. We describe the neutrino searches made with the Observatory with particular emphasis on the recent results following the detections of gravitational waves from binary mergers with Advanced LIGO and VIRGO, leading to competitive limits.

1 Introduction

The discovery of gravitational waves (GW) events by LIGO and in particular from the binary neutron-star (NS) merger GW170817 is one of the most important milestones in multi-messenger astronomy. The Pierre Auger Observatory plays a relevant role in the field of multi-messenger astronomy because with Auger we can detect Ultra-High Energy (UHE) neutrons, photons and neutrinos. In this contribution, we are going to focus on the latter, the neutrinos.

2 Search for neutrinos with Auger

2.1 Neutrino searching basics

The way to identify neutrinos is explained in the next lines. Protons, heavier nuclei, and photons interact shortly after entering the atmosphere, while neutrinos can initiate showers deep in the atmosphere. At large zenith angles, the atmosphere is thick enough so that the electromagnetic component of nucleonic cosmic rays gets absorbed, and the shower front at ground level is mainly composed of muons ("old" shower front). On the other hand, showers induced by neutrinos deep in the atmosphere have a considerable electromagnetic component at the ground ("young" shower front).

With the Pierre Auger Observatory we can detect neutrinos exploiting two different channels [1]. In the "downward-going" (DG) channel, showers with a large zenith angle (higher than 60°) are selected. In the "Earth-skimming" (ES) channel, we search for tau neutrinos that travel in a slightly

[a] e-mail: francisco.pedreira@rai.usc.es

[b] Full author list and acknowledgments at: http://auger.org/archive/authors_2018_09.html

upward direction with respect to the ground, that can interact close to the surface producing a tau lepton which escapes the Earth and decays in the atmosphere, close to the Auger surface detector (SD) of water-cherenkov detector (WCD).

The way to look for inclined showers is to use the shape of the pattern of triggered WCD on ground. These showers induce an elliptical elongated footprint on the ground where the apparent propagation speed of the signal is close to the speed of light. The second step is to look for "young" showers, meaning to look for signals with electromagnetic component. In the ~ 1600 WCD stations of the SD of the Pierre Auger Observatory, the signals produced by the shower particles are digitized with Flash Analog to Digital Converters with 25 ns resolution. This digitization allows us to distinguish signals induced by inclined showers initiated high in the atmosphere from the signals expected in inclined showers initiated close to the ground. The first of them appears as peaked signals narrow in time, but "young" showers induce signals more extended in time. We can distinguish them using the average value of the AreaOverPeak ($\langle AoP \rangle$). The AoP is the total signal recorded by each WCD divided by the peak value. In an "old" shower the $\langle AoP \rangle$ is close to 1, but in a "young" shower values of $\langle AoP \rangle > 1$ are expected according to MC simulations of UHEν-induced showers.

2.2 Data unblinding and limits

No candidates for neutrino-induced showers were found after imposing all the selection criteria (see Ref. [2] for full details) to the data collected between 1 January 2004 and 31 March 2017 in the Earth-skimming and downward-going channels.

Assuming a differential neutrino flux $dN(E_\nu)/dE_\nu = k \cdot E_\nu^{-2}$ and a $\nu_e : \nu_\mu : \nu_\tau = 1 : 1 : 1$ flavor ratio, an upper limit on the value of k can be obtained as:

$$k = \frac{N_{up}}{\int_{E_\nu} E_\nu^{-2} \varepsilon_{tot}(E_\nu) dE_\nu}, \tag{1}$$

where N_{up}, the actual value of the upper limit on the signal events depends on the number of observed events and expected background, as well as on the confidence level required (90 % C.L. in our case), and ε_{tot} is the total exposure as a function of the neutrino energy.

The result is plotted in Figure 1 (left), and the numerical value of the limit that we obtained is:

$$k \sim 5.0 \times 10^{-9}\ \text{GeV cm}^{-2}\text{s}^{-1}\text{sr}^{-1}. \tag{2}$$

As the Pierre Auger Observatory has a good angular resolution (less than 2.5°) [3], we can also obtain a limit to point-like sources of UHEν. Following a similar procedure, we calculated the upper limit to point-like sources, and the result is shown in Figure 1 (right). It is important to point out that with Auger SD we are most sensitive to neutrinos in the ES channel. For that reason, we are particularly sensitive to sources at declinations that are wiewed by Auger with zenith angles between 90° and 95°.

3 Gravitational wave events

3.1 First gravitational wave

In 2015 LIGO and Virgo detected two gravitational wave events caused by the merger of two black-holes [4]. The Pierre Auger Observatory searched for neutrinos in coincidence with these GW events. We applied the Earth-skimming and downward-going neutrino selection criteria to data collected ±500 s around GW events and one day after GW events [5]. No candidates for neutrino showers were found, and we placed constraints on the amount of energy radiated in the form of UHEν.

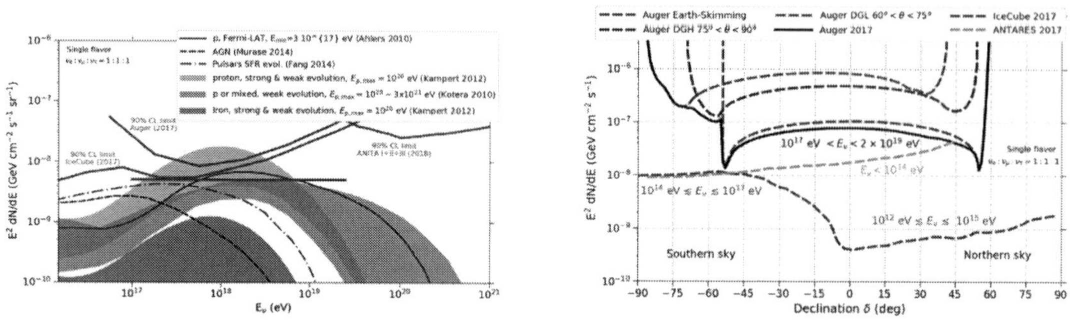

(a) Limit to the diffuse flux (b) Limit to point-like sources

Figure 1: Auger SD upper limits to fluxes of UHEν for the period 01 January 2004 to 31 March 2017 [2].

3.2 Follow-up of GW170817

The most important event in multi-messenger astronomy was the neutron star-neutron star merger GW170817. It was observed not only in gravitational waves, but also in radio, optical and X-rays. Moreover, Fermi detected a gamma-ray burst in coincidence with the GW detected by LIGO and Virgo [6].

ANTARES, IceCube and the Pierre Auger Collaboration searched for neutrinos in coincidence with GW170817 [7]. In Figure 2 we plot the skymap of the GW event at the moment of emission. The source of GW170817 was in an optimal position in the sky for the Pierre Auger Observatory, in the field of view of the ES channel, which is the most sensitive one in Auger.

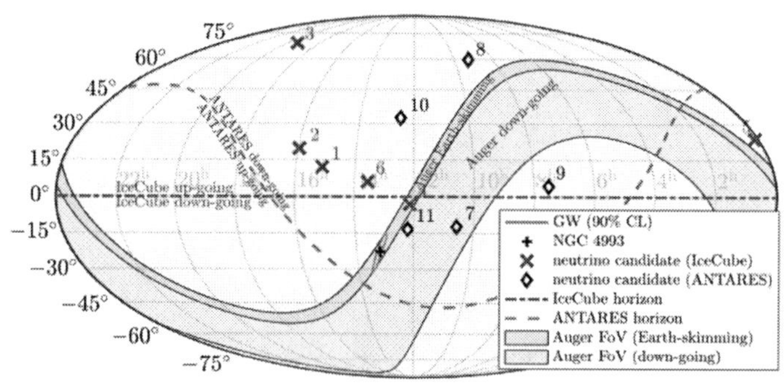

Figure 2: Localizations and sensitive sky areas for ANTARES, IceCube and Auger at the time of the GW170817 event in equatorial coordinates [7].

No candidates were found by any of the observatories. In Figure 3 we show the neutrino fluence limits for ±500 s and 14 days time-window where the searches were performed. The limit that the Pierre Auger Collaboration achieves is complementary to the Icecube and Antares limits. Despite the fact that Auger is not a dedicated ν experiment, produced a remarkably good limit at EeV energies. However, with Auger we cannot see the source all the time in the inclined directions where we search for UHEν due to the rotation of the Earth, and this is the reason why the limit in the 14 days time-window is one order of magnitude higher than the IceCube one.

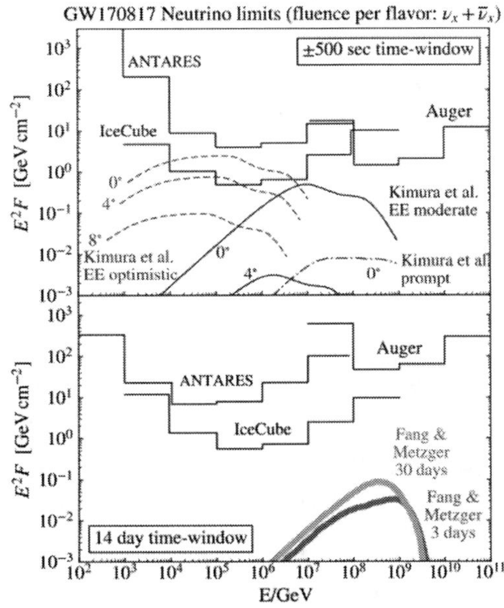

Figure 3: Upper limits (at 90 % C.L.) on the neutrino spectral fluence from GW170817 during a ±500 s window centered on the GW trigger time (top panel), and a 14-day window following the GW trigger (bottom pannel) [7].

4 Conclusions

Multi-messenger astronomy promises to give extraordinary insights into the working of the most extreme Universe, and all collaborations detecting particles and radiation should work together.

The Pierre Auger Observatory is a key detector in the multi-messenger astronomy at EeV energies due to the different kind of particles it can detect (photons, neutrons, and especially neutrinos). Indeed, with the Auger SD we have a very good sensitivity to UHE neutrinos, a large fraction of the sky can be probed with a good angular resolution.

References

[1] A. Aab et al. (Pierre Auger Collaboration), Phys. Rev. D **91**, 092008 (2015)
[2] E. Zas for the Pierre Auger Collaboration, PoS (ICRC 2017), 972 (2017)
[3] A. Aab et al. (Pierre Auger Collaboration), JCAP **08**, 019 (2014)
[4] B. P. Abbott et al. (LIGO Scientific and Virgo Collaborations), Phys. Rev. Lett. **116**, 061102 (2016)
[5] A. Aab et al. (Pierre Auger Collaboration), Phys. Rev. D. **94**, 122007 (2016)
[6] LIGO Scientific Collaboration et al., ApJL **848**, L12 (2017)
[7] A. Albert et al. (ANTARES, IceCube, Pierre Auger, LIGO Scientific and Virgo Collaborations), ApJL **850**, L35 (2017)

Astroparticle Physics with H.E.S.S.: recents results and near-future prospects

Emmanuel Moulin[1,a] for the H.E.S.S. collaboration

[1]*CEA, Irfu, Université Paris-Saclay, F-91191 Gif-Sur-Yvette Cedex, France*

Abstract. H.E.S.S. is an array of five Imaging Atmospheric Cherenkov Telescopes located in Namibia. It is designed for observations of astrophysical sources emitting very-high-energy (VHE) gamma rays in the energy range from a few ten GeVs to several ten TeVs. The H.E.S.S. instrument consists of four identical 12 m diameter telescopes and a 28 m diameter telescope placed at the center of the array. An ambitious Astroparticle Physics program is being carried out by the H.E.S.S. collaboration searching for New Physics in the VHE gamma-ray sky. The program includes the search for WIMP dark matter and axion-like particles, tests of Lorentz invariance, cosmic-ray electron measurements, and search for intergalactic magnetic fields. I will present the latest results on dark matter search from the observations of the Galactic Centre region, the search for Lorentz invariance violation with the 2014 flare observation of Markarian 501, and the first measurement of the cosmic-ray electron spectrum up to 20 TeV. The future of the H.E.S.S. Astroparticle Physics program will be discussed.

1 Introduction

Very-high-energy (VHE, $E \gtrsim 100$ GeV) gamma-ray astronomy has developed as a mature research field over the last few decades. Observations carried out by Imaging Atmospheric Cherenkov telescopes (IACT) enable to pursue rich astroparticle physics programmes to address contemporary fundamental topics such as the quest for dark matter in the form of weakly interacting massive or axion-like particles, tests of Lorentz invariance with the observations of bright and distant sources, the search for intergalactic magnetic fields with the observations of bright and hard spectum blazars, and measurements of the VHE cosmic-ray electrons to probe Galactic emission processes taking place in local astrophysical sources and/or the interstellar medium.

2 Search for heavy weakly interacting dark matter

A long-term observation programme is being carried out by the H.E.S.S. collaboration to search for signatures of dark matter (DM). Observations focus on close-by regions in the sky with expected high DM density and low gamma-ray background. The central region of the Milky Way is a prime target for DM detection due to its proximity and large DM density as expected at the minimum gravitational potential well of the Galaxy. However, the Galactic Centre (GC) environment is very complex and

[a]e-mail: emmanuel.moulin@cea.fr

numerous VHE gamma-ray emissions are detected that represent a challenging background for DM searches. Among them are HESS J1745-290 coincident with the supermassive black hole Sagittarius A* [1] and a diffuse emission [2] related to energetic protons accelerated to PeV energies [3]. Alternative targets are nearby dwarf spheroidal galaxies (dSph): they are composed of old stellar populations and harbor a low gas content, and thus are expected to be free from conventional background. This makes them targets for TeV DM that well complement GC region observations and could provide unambiguous detection of DM signals. However, the expected DM signals are a few orders of magnitude lower than that expected from the GC.

The H.E.S.S. DM strategy is based on the GC region and nearby dSph obsevations to look for VHE gamma rays expected from self-annihilation of heavy weakly interacting massive particles (WIMP). The energy-differential gamma-ray flux expected from the self-annihilation of DM particles of mass $m_{\rm DM}$ writes as:

$$\frac{d\Phi}{dE_\gamma}(E_\gamma,\Delta\Omega) = \frac{\langle\sigma v\rangle}{8\pi\, m_{\rm DM}^2}\frac{dN}{dE_\gamma}(E_\gamma)\times J(\Delta\Omega),\text{with } J(\Delta\Omega) = \int_{\Delta\Omega}\int_{\rm LOS} ds\, d\Omega\, \rho^2(r(s,\theta))\,. \quad (1)$$

$\langle\sigma v\rangle$ is the thermally-averaged velocity-weighted annihilation cross section and $dN/dE_\gamma(E_\gamma)$ the energy-differential gamma-ray spectrum per annihilation. The second term $J(\Delta\Omega)$ referred to as the *J-factor*, corresponds to the integral of the square of the DM density ρ along the line of sight (LOS) and in the solid angle $\Delta\Omega$. The coordinate r writes as $r = (r_\odot^2 + s^2 - 2r_\odot s\cos\theta)^{1/2}$, where s is the distance along the LOS, and θ is the angle between the direction of observation and the target. $r_\odot$ corresponds to the distance of the observer with respect to the target.

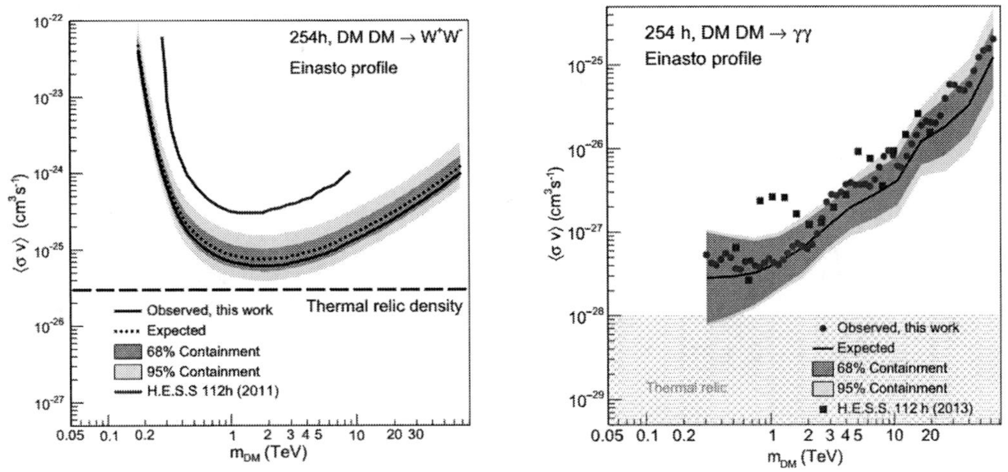

Figure 1. 95% C.L. upper limits on $\langle\sigma v\rangle$ as function of the DM mass from H.E.S.S. observations of the GC region assuming an Einasto profile for the DM distribution on the GC. The limits are shown for the W^+W^- (left) and the prompt $\gamma\gamma$ (right) channels, respectively. In each case, the natural scale expected for $\langle\sigma v\rangle$ is shown. The observed limits together with the expected limits and their associated 1 and 2σ containment bands are plotted. Figures extracted from Ref. [4] and Ref. [5], respectively.

The searches led so far in the GC region looked in a region of interest (ROI) of one degree in radius around the GC. The ROI has been defined according to the exposure map obtained from the

H.E.S.S.-I GC observation strategy and the background extraction technique. All the regions with conventional VHE gamma-ray emissions are excluded [6]. No significant excess signal with respect to background is found. Figure 1 shows the contraints obtained from 10-year H.E.S.S. observations of the GC region. The DM distribution assumes to follow an Einasto profile. The contraints are expressed in terms of 95% C.L. upper limits on $\langle\sigma v\rangle$ derived from a likelihood test ratio statistics analysis [6]. The observed limits together the expected limit with their 1 and 2σ containment bands are shown for DM self-annihilation in the W^+W^- (left) and the prompt $\gamma\gamma$ (right) channels, respectively. The limits reach $\langle\sigma v\rangle = 6\times10^{-26}$ cm^{3}s^{-1} for a DM mass $m_{\rm DM} = 1.5$ TeV in the W^+W^- channel [4], and $\langle\sigma v\rangle = 4\times10^{-28}$ cm^{3}s^{-1} for $m_{\rm DM} = 1$ TeV in the $\gamma\gamma$ channel [5]. They are the most constraining limits obtained so far for TeV DM.

In parallel, H.E.S.S. is leading a long-term observation program of nearby dwarf spheroidal galaxies (dSph) including Sagittarius, the classical dSphs Sculptor, Carina, Fornax, as well as the ultra-faint one Coma Berenices. No VHE gamma-ray emission has been detected so far. The limits reported by H.E.S.S. are summarized in Tab. 1.

Table 1. Upper limits on $\langle\sigma v\rangle$ towards classical and ultra-faint dSphs observed by H.E.S.S. The fourth column provides a round value value of the upper limits at 95% C.L. extracted from references given in the fifth column.

Target	Live time [hours]	Annihilation channels	Upper limits for $m_{\rm DM}$ = 1 TeV [cm^{3}s^{-1}]	Refs.
Sagittarius	85.5	$W^+W,^-ZZ$	2×10^{-23}	[7]
		$\gamma\gamma$	5×10^{-25}	[8]
Sculptor	11.8	$W^+W,^-ZZ$	10^{-22}	[7, 9]
		$\gamma\gamma$	8×10^{-25}	[8]
Carina	22.9	$W^+W,^-ZZ$	3×10^{-22}	[7, 9]
		$\gamma\gamma$	9×10^{-24}	[8]
Fornax	6.0	$W^+W,^-ZZ$	8×10^{-22}	[7]
		$\gamma\gamma$	3×10^{-24}	[8]
Coma Berenices	10.9	$W^+W,^-ZZ$	2×10^{-21}	[7]
		$\gamma\gamma$	8×10^{-25}	[8]
Combined	130	$\gamma\gamma$	5×10^{-25}	[8]

3 Test of Lorentz invariance

Lorentz invariance is a pillar of special relativity and is established as an exact symmetry up to the precision of current experiments. However, the Lorentz symmetry could be broken at a high-energy scale. In some quantum-gravity (QG) models the foamy structure of spacetime near the Planck scale ($\sim10^{19}$ GeV) could affect high-energy photon propagation in vacuum. One may thus expect a break of the Lorentz invariance at high energies [10]. A phenomenological way to account for possible Lorentz invariance violation (LIV) is to add extra terms in the dispersion relation of photons. The LIV terms are introduced as a series of powers n of the photon energy divided by the QG energy scale $E_{\rm QG}$. $n = 1$ corresponds to a linear LIV term, while $n = 2$ corresponds to a quadratic LIV. The LIV-modified dispersion relation of photons writes :

$$E \simeq p^2c^2\left[1 \pm \left(\frac{E}{E_{\rm QG}}\right)^n\right], \quad (2)$$

where E and p are the energy and momentum of the photon, and the signs + and - corresponds the superluminal and subluminal LIV scenarios, respectively.

The propagation of VHE gamma rays can be affected in two ways: *(i)* the propagation velocity of gamma rays would be dependent on their energy, which would translate into an energy-dependent time delay in the arrival time of gamma rays traveling over astrophysical distances [11]; and *(ii)* the reaction threshold of the pair-production process on ambient target photons would also be modified as a function of their energy [12, 13].

The strong flare of the blazar Markarian (Mrk) 501 in June 2014 has been observed by H.E.S.S. The data show a rapid variability of the flux with VHE gamma rays detected from $\sim$1 TeV up to 20 TeV. These exceptional features have been used to search for LIV-induced effects in the time-of-flight (ToF) and in the spectral measurement approaches. The ToF analysis looked for an energy dependent time delay in the light curve during the flare in two energy ranges: the low-energy one is defined with energies between 1.3 and 3.25 TeV, while the high energy one makes use of energies higher than 3.25 TeV. No significant energy-dependent lag in the arrival time of gamma rays is measured between the two regions [14]. The 95% C.L. lower limits on $E_{\rm QG}$ reaches 3.6×10^{17} GeV and 8.5×10^{10} GeV for the linear and quadratic terms, respectively. The H.E.S.S. spectrum of the Mrk 501 flare is well fitted by an intrinsic power law attenuated on the EBL using the optical depth as modeled in Ref. [15]. No significant deviation is shown as expected in presence of LIV. Table 2 summarizes the 95% C.L. constraints on $E_{\rm QG}$ for the ToF and spectral methods for subluminal LIV. For the spectral method, constraints on $E_{\rm QG}$ for the linear term reaches 2.6×10^{19} GeV, *i.e.* a factor of two higher than the Planck scale.

Table 2. Limits on Lorentz Invariance violation from the observations of the 2014 flare of Mrk 501. The constraints are expressed as 95% C.L. lower limits on E_{QG} for the linear (n=1) and quadratic cases (n=2) in case of subluminal LIV.

n	ToF	Spectrum
1	3.6×10^{17} GeV	2.6×10^{19} GeV
2	8.5×10^{10} GeV	7.8×10^{11} GeV

4 Cosmic-ray electron measurements

Cosmic-ray electrons (CRE) measurements with IACTs convey crucial information on the acceleration of electrons in nearby sources and their propagations in the local interstellar medium. However, only a limited number of nearby accelerators can contribute to the VHE cosmic-ray electron flux measured on Earth given that CREs suffer from strong energy losses during their propagation from inverse Compton scattering on interstellar radiation fields and synchrotron radiation in the μG Galactic magnetic field. CREs of TeV energies are expected to be produced within about one kpc from the Sun [16]. Among them are the pulsar wind nebulae Gemina, Monogem and Vela, and supernova remnants such as Vela and Vela Junior.

Since the discovery in 2008 by H.E.S.S. of an energy break in the CRE spectrum around 1 TeV [17], additional observations carried out in extragalactic fields up to 2010 provide now an increased photon statistics in the dataset by a factor of about 4. In addition to the standard data quality selection criteria, only extragalactic observations with Galactic latitudes $|b| > 8°$ are used to avoid contamination from Galactic diffuse emission and known VHE gamma-ray sources. Improvements in the data analysis technique based on goodness of fit yield a higher hadron rejection. The H.E.S.S. CRE spectrum is well fitted from 250 GeV up to 25 TeV by a broken power-law function with low and high-energy spectral indices of 3.04$\pm$ 0.01$_{\rm stat}$ and 3.78$\pm$ 0.02$_{\rm stat}$, respectively, and an energy break of $\sim$1 TeV [18]. For IACTs, this measurement suffers from systematic uncertainties coming from the event selection cuts involved in the analysis, the zenith angle and year-to-year dependency, as well as the atmospheric conditions of the data taking. A careful estimate of the contamination of misidentified protons in the CRE sample at the highest energies is crucial because of the steeper CRE spectrum compared to that of the protons, *i.e.* even a small fraction of misidentified protons could induce a spectral hardening that would mimic the contribution of a local source such as Vela or Cygnus loop. Given the present systematic uncertainties, the H.E.S.S. CRE measurement is consistent with both AMS02, CALET, Fermi and DAMPE measurements as well as the recently CRE spectrum measured by VERITAS [19]. The absence of significant spectral features detected in the high-energy end of the spectrum can be used to constrain local source models such as a nearby dark matter clump, local sources such as Vela or Cygnus Loop, and propagation models.

In addition to the spectral CRE measurements, another observable of prime interest to probe local CRE sources is the search for anisotropy in the arrival directions of the CREs. Predictions for the CRE dipole anisotropy in the single source scenario can reach several percents for Vela and Cygnus Loop [20] and a nearby dark matter clump [21]. While such a measurement is challenging for IACTs due to the dependency of the sensitivity to the telescope pointing pattern and the systematic uncertainties arising from the variation of the atmospheric conditions on the CRE flux measurement, no significant dipolar CRE anisotropy has been found by H.E.S.S. 95% C.L. upper limits for a dipolar anisotropy at the 5% level are derived for the whole sky [22].

5 Outlook

The search for dark matter with H.E.S.S. is crucial to complement the studies done at collider facilities, such as the LHC, or in direct detection experiments. With a sensitivity from the hundred GeV to the hundred TeV scale, H.E.S.S. is able to search for heavy weakly interacting DM particles at the TeV scale, a scale which would be difficult to probe at colliders. Since the chance of detection drives the observational strategy, DM searches with H.E.S.S. will focus first on deep observations of the GC region. A long-term and deep observation program of the inner several degrees of GC is being carried out since a few years in order to achieve the best possible sensitivity before CTA-South comes into operation. Secondly, observations of selected dwarf galaxies will complement those of the GC region. Some of the recently detected dSphs by the Dark Energy Survey (DES) are being observed, their location in the Southern-hemisphere sky guarantees the best sensitivity for H.E.S.S. under the best observational conditions.

LIV-induced effects will be searched both by looking for energy-dependent delay in the arrival time of VHE gamma rays, or by searching for deviations from the standard electron-positron pair production mechanism from the interaction of VHE gamma rays with EBL photons. For the former method, observations of transient or variable sources such as flaring AGNs, pulsars, or GRBs, are required, while for the latter variability is not a prerequisite. Bright and distant sources are a necessity and flaring AGN have been found to be prime objects so far. H.E.S.S. will continue to follow-up on alerts issued by other experiments for flaring AGN or GRBs provided observations are suitable for LIV analyses. Population studies are essential for the near-future LIV searches with astrophysical sources that will become a standard with CTA.

The search for axion-like particles (ALP) that could be detected via their coupling to photons in external magnetic fields will be pursued via observations of distant sources like AGN, where the ALP-photon conversion may occur in the magnetic field of the host galaxy cluster, in the intergalactic magnetic field, as well as in the magnetic field of the Milky Way. While the existence and the strength of the intergalactic magnetic field is still elusive, the existence of magnetic fields in galaxy clusters is well established through the observation of radio synchrotron emission as well as through the detection of Faraday rotation in polarized radio sources. ALP-induced effects will be searched through spectral irregularities in AGN emission that may happen from the ALP-photon conversion depending on the strength and morphology of the magnetic field. The most favorable targets are bright AGN located in galaxy clusters for which rotation measures suggest high central magnetic fields.

The search for intergalactic magnetic field will be carried out through the observation of blazars. Following the electron-positron pair production from the interaction of VHE gamma rays with EBL photons, an electromagnetic cascade may develop from the electron/positron interaction with background photons via inverse Compton scattering. In this way, the energy flux from the blazar feeds electromagnetic cascades that may lead to GeV gamma-ray emission. Provided the energy flux remains forward beamed, the level of this GeV cascade component is commensurate with the amount of the blazars intrinsic spectra absorbed by the EBL. Most favorable candidate sources exhibit both bright emission and hard spectral indices, such as extreme HBL blazars like 1ES 0229+200 and PG 1553+113.

The H.E.S.S. measurement of the CRE spectrum up to ten TeV energy range enables to probe an uncharted region so far that well complements the satellite measurements carried out with AMS02, CALET, DAMPE. Increased photon statistics in extragalactic fields will help to improve the energy spectrum determination at the highest energies provided that better electron-hadron discrimination is obtained through advanced event separation techniques.

References

[1] F. Aharonian et al. (H.E.S.S. Collaboration), Astron. Astrophys. **425**, L13 (2004)
[2] F. Aharonian et al. (H.E.S.S. Collaboration), Nature **439**, 695 (2006)
[3] A. Abramowski et al. (H.E.S.S. Collaboration), Nature **531**, 476 (2016)
[4] H. Abdallah et al. (H.E.S.S.), Phys. Rev. Lett. **117**, 111301 (2016), 1607.08142
[5] H. Abdallah et al. (H.E.S.S. Collaboration), Phys. Rev. Lett. **120**, 201101 (2018), 1805.05741
[6] L. Rinchiuso (for the H.E.S.S. Collaboration), in *These proceedings* (2018)
[7] A. Abramowski et al. (H.E.S.S. Collaboration), Phys. Rev. **D90**, 112012 (2014), 1410.2589
[8] H. Abdalla et al. (H.E.S.S. Collaboration), Submitted to JCAP (2018), 1810.00995
[9] A. Abramowski et al. (H.E.S.S. Collaboration), Astropart. Phys. **34**, 608 (2011), 1012.5602
[10] G. Amelino-Camelia, Living Rev. Rel. **16**, 5 (2013), 0806.0339
[11] G. Amelino-Camelia et al., Nature **393**, 763 (1998), astro-ph/9712103
[12] F.W. Stecker, S.L. Glashow, Astropart. Phys. **16**, 97 (2001), astro-ph/0102226
[13] U. Jacob, T. Piran, Phys. Rev. **D78**, 124010 (2008), 0810.1318
[14] H. Abdalla et al. (H.E.S.S. Collaboration), Submitted to Astrophys. J. (2018)
[15] A. Franceschini, G. Rodighiero, M. Vaccari, Astron. Astrophys. **487**, 837 (2008), 0805.1841
[16] T. Kobayashi et al., Astrophys. J. **601**, 340 (2004), astro-ph/0308470
[17] F. Aharonian et al. (H.E.S.S.), Phys. Rev. Lett. **101**, 261104 (2008), 0811.3894
[18] D. Kerszberg et al. (H.E.S.S. Collaboration), in *International Cosmic Ray Conference* (2017)
[19] A. Archer et al. (VERITAS), Phys. Rev. **D98**, 062004 (2018), 1808.10028
[20] S. Manconi, M. Di Mauro, F. Donato, JCAP **1701**, 006 (2017), 1611.06237
[21] E. Borriello, L. Maccione, A. Cuoco, Astropart. Phys. **35**, 537 (2012), 1012.0041
[22] M. Kraus et al. (H.E.S.S. Collaboration), in *International Cosmic Ray Conference* (2017)

AUTHOR INDEX

AUTHOR INDEX

AUTHOR INDEX

EDP Sciences – Web of Conferences
17, Avenue du Hoggar
Parc d'Activité de Courtabouf
BP 112
F-91944 Les Ulis Cedex A
France

ISBN 978-1-5108-8623-0

9 781510 886230

CURRAN ASSOCIATES INC.
proceedings.com